Organometallic Chemistry of Titanium, Zirconium, and Hafnium

Organometallic Chemistry of Titanium, Zirconium, and Hafnium

P. C. WAILES

R. S. P. COUTTS

H. WEIGOLD

C. S. I. R. O.
Applied Chemistry Laboratories
Melbourne, Australia

 1974

ACADEMIC PRESS   New York and London
A Subsidiary of Harcourt Brace Jovanovich, Publishers

ACADEMIC PRESS, INC.
111 Fifth Avenue, New York, New York 10003

United Kingdom Edition published by
ACADEMIC PRESS, INC. (LONDON) LTD.
24/28 Oval Road. London NW1

Library of Congress Cataloging in Publication Data

Wailes, P   C
        Organometallic chemistry of titanium, zirconium, and
hafnium.

        Bibliography: p.
        1.   Organotitanium compounds.   2.   Organozirconium
compounds.   3.   Organohafnium compounds.   4.   Organoalu-
minum compounds.   I.   Coutts, R. S. P., joint author.
II.   Weigold, H., joint author.   III.   Title.
QD412.T6W34   547'.05'51   73-9449
ISBN 0-12-730350-2

Contents

v

Preface

In this volume we have limited ourselves to those compounds having a direct metal-to-carbon linkage, be it sigma or pi. The organometallic chemistry of titanium, zirconium, and hafnium has found little mention in review articles. This is especially true of zirconium and hafnium for which there has been no comprehensive cover apart from a brief review by Larsen in 1970 (L7). Organic complexes of lower-valent titanium were reviewed in 1970 (C70) by two of the present authors, while brief mention of organotitanium compounds has been made in books by Feld and Cowe in 1966 (F10), Clark in 1968 (C30), Kepert in 1972 (K22), and in Russian reviews in 1965 (R12) and 1967 (R13). Since the last of these, a great deal of new and novel work has appeared in the literature, spurred on by the activity of these metal derivatives in Ziegler–Natta catalysis.

Interest in the organo derivatives of these metals dates back to the second half of the last century. During this period alkylation of many transition metals was attempted using zinc dialkyls and diaryls; in particular, Cahours attempted the preparation of ethyltitanium compounds in 1861 from titanium tetrachloride and diethylzinc (C1). Later, Grignard reagents and organolithium compounds were also used, but at the time the sensitivity of the expected compounds to air and to moisture was not sufficiently taken into account. In addition, the low thermal stability of the products resulted in decomposition to give, generally, reduction products of the metal halides (T24). Only when techniques were developed for working under anaerobic conditions and at low temperatures was it possible to isolate organometallic compounds of titanium, zirconium and hafnium.

The first of these was phenyltitanium triisopropoxide obtained by Herman and Nelson in 1952 (H25, H26). Two other important events occurred around this date: the discovery of ferrocene (K16) and the recognition that titanium compounds, when mixed with alkylaluminum derivatives, could accelerate the polymerization of olefins (Z6). Both of these events had a great influence on the development of organotitanium chemistry in particular. Cyclopentadienyltitanium and -zirconium halides were among the first compounds isolated during the extensive study of the action of cyclopentadienylmagnesium halides on anhydrous metal halides by Wilkinson and co-workers (W24). This unique ligand continues to dominate the titanium and zirconium scene.

Also, in the 1950's the lower-valent derivatives of these metals began to receive some attention following the preparation of Cp_2Ti^{III} derivatives (W24) and the realization that titanium(III) compounds were important in Ziegler catalysis. Titanium, zirconium, and hafnium are the first members of the transition series, with outer electronic structures of d^2, s^2. They therefore show a stable oxidation number of four. Compounds of trivalent and bivalent metals are receiving an increasing amount of attention because of the novelty of their behavior, particularly their activity in various catalytic processes.

All three metals are widely distributed in nature, occurring generally as oxides. Titanium is the ninth most abundant element on earth, occupying 0.63% of the earth's crust. It is therefore more plentiful than zinc, copper, and lead combined, and is five times more abundant than sulfur or phosphorus. Zirconium occurs to the extent of 0.022% in the lithosphere, a percentage which is roughly equivalent to that of carbon. The percentage of hafnium (0.00053%) is greater than that of tin, lead, or silver and approximately the same as that of arsenic or molybdenum. The metals are therefore more common than might be realized, their lack of availability being more a reflection of the difficulty in their refinement rather than their rarity.

This book is intended for those engaged in research. It covers the title field from its beginning, and is as factual as we could make it. The subject matter has been divided (1) by metal, zirconium, and hafnium being treated together; (2) by oxidation number; and (3) by predominant ligand.

In making each chapter complete without extensive cross-indexing, some duplication has been unavoidable, but we have endeavored to keep this to a minimum. The enormous and confusing field of olefin polymerization is covered only in sufficient depth to underline the role of organotitanium compounds in the process. Space requirements have limited a detailed treatment of some aspects, but we have endeavored to cover fully those areas of the field which have not been reviewed previously. Work received in Australia up to December 1973, including patents, has been included and, where possible, significant later results have been inserted as an Appendix.

We are indebted to C.S.I.R.O. and to Dr. S. D. Hamann in particular for encouragement to write this volume, to Mrs. Miriam Wailes for painstakingly typing the manuscript, and to Mrs. Dawn Merrill for preparing the diagrams.

P. C. WAILES
R. S. P. COUTTS
H. WEIGOLD

Chapter I

Introduction

The interest shown in the organo compounds of titanium, and to a lesser extent zirconium and hafnium, is due in no small measure to the activity of derivatives of these metals as catalysts in the polymerization of olefins and the chemical fixation of nitrogen. The presently accepted view that olefin polymerization is initiated at a titanium–carbon bond increases the importance of the chemistry of this class of compound.

By far the greater part of the work described deals with titanium because of its preeminence in Ziegler–Natta catalysis, but in the last few years organozirconium compounds have received much attention. Judging by the interesting results, they will continue to do so. Although in many cases titanium and zirconium form similar types of compounds, in some respects these compounds differ markedly from each other in their chemical behavior. The methyl derivatives of zirconium, for example, are far more sensitive to hydrolysis than the corresponding compounds of titanium, whereas the latter are more susceptible to thermal decomposition. Much of the difference in chemical behavior between zirconium and titanium can be traced to the ease of reduction of the latter.

Other differences are noted in the cyclopentadienyl derivatives. In the case of zirconium, the effect of oxygen-bonded ligands in compounds of the type, $Cp_2Zr(OR)X$ (where $Cp = \pi$-cyclopentadienyl), is to increase the lability of one of the cyclopentadienyl groups to such an extent that solvolysis or insertion of SO_2, NO, etc., can occur. No such effect is noted with the corresponding titanium compounds.

Many more differences of this type in the chemistry of titanium and zirconium are brought out in the text, but between zirconium and hafnium no such distinctions exist. Although hafnium has received little attention, it seems that the chemistry of its organo derivatives is as closely related to that of zirconium as is its inorganic chemistry. It has been pointed out (B52) that zirconium and hafnium always occur together in nature and virtually anything that has been said of the chemistry of zirconium applies almost identically to the chemistry of hafnium. The two are generally not separated in the production of commercial zirconium compounds. Because of this similarity and because so little work has been carried out on organohafnium chemistry, the two are treated together throughout.

In the case of all three metals, by far the majority of their organometallic compounds are cyclopentadienyl derivatives. As with many other transition metals, the unique stabilizing influence of this ligand has allowed the isolation of many metal–carbon-bonded compounds. In view of the importance of the cyclopentadienyl group in the chemistry of these metals, it seems desirable to make some comment on stability and bonding in compounds of this type.

The general thermal instability of alkyl transition metal compounds was originally ascribed to an inherent weakness in the metal-to-carbon bond. Thermodynamic and spectroscopic evidence suggests, however, that there is no reason to think that transition metal-to-carbon bonds are any weaker than bonds from main group metals to carbon (B78, M49). The instability of transition metal alkyl and aryl derivatives is due to facile decomposition by low activation energy pathways (M38) such as α or β elimination of metal hydride, homolysis, or coupling of ligands at the transition metal atom (B60). All these decomposition pathways are exemplified in the chemistry of titanium, zirconium, and hafnium.

It has been suggested by Chatt and Shaw (C20) that the first step in the fission of a carbon–metal bond is the promotion of a d electron from the metal into a carbon–metal nonbonding orbital, the energy required to do this being related to the stability of the compound. The stability imparted by strong π-bonding ligands such as cyclopentadienyl, phosphines, etc., was said to be due to an increase in the energy gap between these orbitals, resulting in a higher activation energy for the bond fission. These authors also pointed out that steric factors due to the associated ligands could be important, as well as the electronic features of the alkyl or aryl group (C21, W23). These last points have recently received some attention by several research schools.

Wilkinson and collaborators (M49, W23) have pointed out that the presence of π-bonded ligands is no guarantee of stability. The main function of additional ligands, whether π-bonded or otherwise, could be to block the coordination sites required for decomposition to proceed. These authors, as well as Braterman and Cross (B78) and Mingos (M38) emphasize that stability is a function of many factors, both steric and electronic, and should be considered in relation to the possible ways in which decomposition can proceed. The isolation of stable neopentyl, norbornyl, trimethylsilylmethyl, and benzyl derivatives lends support to this type of argument. All these ligands have a bulky group, but no hydrogen, on the β-carbon atom, so that possible decomposition routes have been eliminated.

A. The Cyclopentadienyl Compounds*

Cyclopentadienyl derivatives of titanium and zirconium were among the first of this type prepared by Wilkinson and co-workers in 1953 (W25). As a result bis(cyclopentadienyl)titanium and -zirconium halides have been available since then, and much of the organometallic chemistry of these metals has involved such compounds. Since the Cp ligand occurs so commonly throughout this work, its bonding modes and the possible structures of its metal derivatives are best considered here.

Many authors prefer to regard the cyclopentadienyl groups in bis-(cyclopentadienyl)metal derivatives as each occupying one vertex of a tetrahedron, a point of view which is not at variance with the topological derivation of King (K26), or with bond angles derived from X-ray crystallographic data. In bis(cyclopentadienyl) derivatives of titanium, zirconium, and hafnium the angle subtended at the metal atom by the centers of gravity, or centroids, of the cyclopentadienyl rings is normally around 130°. The configuration of these compounds is therefore essentially tetrahedral. This configuration has also been found in tetracyclopentadienylzirconium (C4, K62) and tetracyclopentadienylhafnium (K63). Bond lengths and bond angles for some bis(cyclopentadienyl)titanium compounds are shown in Table I-1; the corresponding data for zirconium compounds are shown in Table IV-5.

For a detailed treatment of the bonding between the C_5H_5 ligand and metals, the reader is referred to Rosenblum (R50) and references therein. For most purposes this bonding can be regarded as consisting of a σ bond between the A bonding orbital of the C_5H_5 and the metal, together with two π-type bonds involving two pairs of electrons from the E_1 bonding orbitals of the ligand. The bond between the so-called π-cyclopentadienyl group and a metal, therefore, involves six electrons, and the ligand can be regarded as occupying three coordination sites on the metal. Back bonding can occur but to a lesser extent than in metal carbonyls.

A structural interpretation by Ballhausen and Dahl (B8) for the hydrides, Cp_2MH_n (where $n = 1$, 2, or 3 and the angle between the ring metal axes is around 135°), is relevant also for bis(cyclopentadienyl) derivatives of titanium, zirconium, and hafnium. Of the nine hybrid metal orbitals, six

---

* Throughout this volume the π-cyclopentadienyl ligand is abbreviated as Cp.

Table I-1

Structural Data for Cp₂TiX₂ Complexes

Compound	Bond length (av. Å)				Bond angle (degrees)		
	Ti–C(Cp)	Ti–Cp(⊥)	C–C(Cp)	Ti–X	Cp–Ti–Cp(⊥)	X–Ti–X	Refs.
Cp_2TiCl_2	2.43	—	1.44	2.36	129	95.2	T38
$[(\pi\text{-}C_5H_4)_2(CH_2)_3]TiCl_2$	2.38	—	1.40	2.24	121 ± 3	100 ± 1	A5, R47
Cp_2TiS_5	2.38	2.06	—	2.36	132.6	93.7	D9
$Cp_2TiS_2C_6H_4$	2.37	2.07	1.38	2.42, 2.45	133.7	94.6	E5, E6
$Cp_2Ti(SMe)_2Mo(CO)_4$	2.37	2.06	1.38	2.42	130 ± 1	82.2	K64
	2.40	—	1.42	2.46	—	99.9	D7
Cp_2Ti (benzo-fused lactone metallacycle, with Ti–O and Ti–C bonds to the ring)	2.39	2.05	1.44	1.95 (Ti–O) 2.20 (Ti–C)	134	78	A4, K38
$Cp_2Ti(\pi\text{-}C_5H_9)$ (C_5H_9 = 1,2-dimethylallyl)	2.40	2.06	1.43	—	132	—	H6
$Cp_2Ti(h^1\text{-}C_5H_5)_2$	2.38	2.08	1.37	2.33	129.9	86.3	C4
$Cp_2Ti(C_6H_5)_2$	2.31	—	1.34	2.27	136	97.3	K34
Cp_2TiBH_4	2.35	2.03	1.40 ± 0.13, 1.37 ± 0.13	1.75	136.7	60 ± 5	M32a
$Cp_2Ti((h^2\text{-}C_5H_5)$	2.34–2.41	2.05	1.38–1.44	2.42	133	—	L23a
$(Cp_2TiSiH_2)_2$	—	2.16	1.44	2.16	126	79.2	H6a

4

are directed toward the cyclopentadienyl ligands. The remaining three can be used to house nonbonding electrons or to bond to other groups.

An alternative model was proposed by Alcock (A3) to accommodate the lone pair of electrons on rhenium in the complex $Cp(CH_3C_5H_4)Re(CH_3)_2$. Since the CH_3–Re–CH_3 angle (75°) was too small to contain a lone pair of electrons, it was proposed that they lay in an orbital along the Y axis (see Fig. I-1), symmetrically disposed on either side of the metal. More recently,

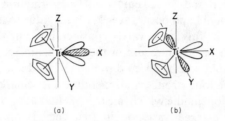

(a)   (b)

Fig. I-1. Bonding models for bis(cyclopentadienyl)titanium compounds. After (a) Ballhausen and Dahl (B8) and (b) Alcock (A3).

Green and co-workers (D29, G29) have incorporated features of both models in correlating the decrease in the X–M–X angle in Cp_2MX_2 compounds with an increase in the number of nonbonding d electrons on the metal M.

B. Spectroscopic Properties of Cyclopentadienyl Groups

The identification of cyclopentadienyl derivatives of metals is facilitated by characteristic bands in their proton magnetic resonance (PMR) and infrared spectra. All unsubstituted cyclopentadienyl ligands attached to titanium, zirconium, and hafnium with unrestricted rotation, including σ-bonded types, show a single band in the PMR spectrum between $\delta 5.5$ and 7.0. Except for tetra(cyclopentadienyl)titanium the rapid interchange in σ-C_5H_5 normally cannot be slowed down sufficiently on cooling to split the singlet. Tetra-(cyclopentadienyl)zirconium, for example, which is known to contain rings in two (K61, K62) and possibly three (C4) different hapto configurations shows a single peak in its PMR spectrum down to $-58°$ (B64). Although of great use in determining the presence of cyclopentadienyl groups, PMR spectroscopy is of little use in determining bonding modes, unless the rings are in a fixed configuration. Examples will be found in Chapter VII.

The use of PMR is restricted also, of course, to the diamagnetic tetravalent, and to a lesser extent bivalent, states. Resonance bands for trivalent titanium and zirconium compounds are rarely observed except in some polymeric compounds where coupling of the unpaired electrons has reduced the paramagnetism sufficiently.

Of greater diagnostic use is infrared spectroscopy. The bands expected for the various bonding modes of cyclopentadienyl groups have been well covered by Fritz (F27). π-Cyclopentadienyltitanium, -zirconium, and -hafnium compounds show four strong bands in their infrared spectra at ~ 3100 cm^{-1} (CH stretch), ~ 1435 cm^{-1} (CC stretch), ~ 1020 cm^{-1} (CH deformation in-plane) and ~ 820 cm^{-1} (CH deformation out-of-plane). Sometimes an additional band occurs at ~ 1120 cm^{-1} due to a ring breathing mode. The intensity of this band varies considerably, but the claim (F27) that the intensity is a reverse measure of the ionic character of the Cp ring does not appear to hold and has been disputed (B73, M26).

The presence of the four strong bands mentioned is remarkably consistent throughout the cyclopentadienyl chemistry of titanium, zirconium, and hafnium. The most marked variation in frequency is observed with the CH deformation band around 800 cm^{-1}, which moves to its lower range with decrease in oxidation number of the metal.

Monohapto-cyclopentadienyl ligands are of lower symmetry and show many more bands in their infrared spectra (F27). Davison and Rakita have pointed out (D10) that above 3000 cm^{-1}, h^1-C$_5$H$_5$ groups should show four bands whereas h^5-C$_5$H$_5$ should show only one, allowing a clear distinction to be made in this region.

Chapter II

Alkyl- and Aryltitanium(IV) Compounds

Although attempts to prepare titanium–carbon-bonded compounds date back many years (see references in T33) it was only in the early 1950's that the first example, $C_6H_5Ti(O\text{-}iso\text{-}C_3H_7)_3$, was isolated by Herman and Nelson (H26). The main synthetic difficulties encountered stem from the high reactivity and low thermal stability of the titanium–carbon linkage. Stabilization of alkyltitanium compounds by adduct formation with amines, ethers, sulfides, and phosphines is possible and has permitted the isolation and characterization of such complexes.

A. Alkyltitanium Halides

Interest in these compounds lies in their relationship to Ziegler–Natta catalysts and, indeed, their own ability to initiate the polymerization of α-olefins (B25, B26, B28, B30, B49, C11, C12, D13, E1, K10, L3) at low temperatures.

1. Monoalkyltitanium Trihalides

The preparation of the first member of this class of compound, i.e., CH_3TiCl_3, was reported in a 1958 patent (B31).

a. PREPARATION

Alkyltitanium trihalides are obtained from titanium tetrahalides in good yield (up to 75%) using alkylating agents such as dialkylzinc (B25, D18, T33), dialkylcadmium (B25), or tetraalkyllead (B18, B25, H28a). For example,

$$2TiX_4 + CdR_2 \xrightarrow[-30° \text{ to } -78°]{\text{hexane}} 2RTiX_3 + CdX_2$$

$$R = CH_3, CF_3, C_2H_5, \textit{n-}C_3H_7, CH(CH_3)_2, \textit{n-}C_4H_9, CH_2CH(CH_3)_2, C_5H_{11}$$

Other alkylating agents, especially aluminum alkyls, $[(CH_3)_nAlCl_{3-n}$, $n = 1$, 2, or 3] provide a convenient route to the alkyltitanium halides but not to the fluoride compounds (B27, B29, B31, B32, C11, D18, H3, K9, R53).

$$TiCl_4 + CH_3AlCl_2 \xrightarrow[\text{reflux 5 min}]{\text{hexane}} CH_3TiCl_3 + AlCl_3$$

Diethylaluminum bromide forms a 1:1 adduct with the titanium alkyl, $C_2H_5TiCl_3$. Gray and co-workers (G26) maintain that in the presence of excess $Al(CH_3)_3$, evidence for the formation of the transient $(CH_3)_3TiCl$ species from $TiCl_4$ can be found. However, no species containing both titanium and aluminum was formed (G26).

CH_3TiCl_3 is catalytically decomposed by aluminum alkyls (B29). Other compounds, for example, BF_3, diethyl ether (B29), tetrahydrofuran (THF) (D44), or lead alkyls (B18) accelerate decomposition. CH_3TiCl_3, free of any aluminum compounds, can be obtained by distillation (25°/2 mm Hg) or sublimation (F22) at $-10°$. In general, diphenyl ether is used to reduce the volatility of the accompanying aluminum compound. In the case of CH_3AlCl_2, reaction with sodium chloride can also be used (B27, B31), as the $Na(CH_3)AlCl_3$ so formed is nonvolatile. $Cp_2Ti(CH_3)_2$ has also been used as an alkylating agent to prepare CH_3TiCl_3 free from aluminum alkyls (B29, D12).

Jacot-Guillarmod and co-workers (B58, B59) have obtained several benzyl-titanium compounds and suggested that their most likely order of stability toward intermolecular decomposition processes via benzyl group elimination was $(C_6H_5CH_2)_4Ti > C_6H_5CH_2TiCl_3$, $(C_6H_5CH_2)_3TiCl > (C_6H_5CH_2)_2TiCl_2$. Since the rate-determining mode of decomposition of organotitanium species is dependent on the hydrocarbon moiety, this stability sequence would certainly not be expected to hold for other alkyl series. Thiele and co-workers (T33) maintain that the thermal stability of CH_3TiCl_3 exceeds that of CH_3TiBr_3 which, in turn, is greater than that of $C_6H_5TiCl_3$.

Grignard reagents have been employed to make $RTiX_3$ compounds. Interestingly, only the monoalkyl compounds CH_3TiCl_3 or CH_3TiBr_3 form in hexane even with excess Grignard reagent and independent of reaction time (T25). Use of benzene as solvent permitted the preparation of π-allyltitanium trichloride, isolated as the 2,2′-bipyridyl adduct.

Although the more reactive monoalkyl compounds such as $C_3H_7TiCl_3$ and $C_5H_{11}TiCl_3$ could only be isolated as the 2,2′-bipyridyl adducts (T33),

Bawn and Gladstone (B18) were able to prepare $C_2H_5TiCl_3$ in 40–50% yield using tetraethyllead. This titanium compound, a dark violet, crystalline solid at dry-ice temperatures and red liquid at room temperature, was purified by vacuum distillation (B25) (50°–60°/1 mm Hg). It is quite soluble in aromatic and aliphatic hydrocarbons, and in dilute solution the compound is fairly stable even at 60°–70°. In the pure state, decomposition to a trivalent titanium compound, ethane, and butane occurs readily at room temperature.

Titanium–carbon-bonded compounds have also been obtained by reaction of tetrahalides of titanium with unsaturated compounds (T10). Sharma *et al.* (S28, S29) have prepared bis(cycloheptatrienyl)- and bis(cyclooctatetraenyl)-metal dichloride complexes by refluxing $TiCl_4$, $ZrCl_4$, or $HfCl_4$ with cyclo-heptatriene or cyclooctatetraene in benzene till evolution of HCl ceased. The high thermal stability of these complexes may well mean that the ligands are held by more than one bond to the metal. Intermediates containing metal–carbon bonds can be postulated in the conversion of alkynes to alkenes with titanium tetrachloride or bromide (L16). Treatment of nitromethane with

$$TiX_4 + C_6H_5C{\equiv}CCH_3 \xrightarrow{20°} \underset{80\%}{\overset{\displaystyle C_6H_5\diagdown\qquad\diagup CH_3}{\underset{\displaystyle X\diagup\qquad\diagdown H}{C{=}C}}} + \underset{20\%}{\overset{\displaystyle C_6H_5\diagdown\qquad\diagup H}{\underset{\displaystyle X\diagup\qquad\diagdown CH_3}{C{=}C}}}$$

$TiCl_4$ under prolonged reflux (P1) led to compounds of the type $TiCl_2(CH_2NO_2)_2 \cdot 2CH_3NO_2$. $ZrCl_4$ forms the analogous compound.

b. PROPERTIES

The purple crystals of methyltitanium trichloride melt at about 29° (b.p. 37°/1 mm Hg) to a yellow liquid (D18). It is stable under an inert atmosphere at solid CO_2 temperature for months, and, depending on its purity, is stable at room temperature for many hours. The decomposition products are $TiCl_3$, methane, and an oily polymethylene (B29). In ether, in which it has a lower stability than in hydrocarbon solvents, it breaks down by an almost concentration-independent process to give $TiCl_3$, methane, and ethane, the additional protons having originated from the ether (B29). When carbon tetrachloride is also present, methyl chloride is formed in an appreciable amount. The activation of the Ti–CH_3 bond by the diethyl ether is not evident with dioxane (B29) which forms the thermally stable (at room temperature), pale violet, insoluble monoadduct $CH_3TiCl_3 \cdot C_4H_8O_2$. The mechanism for the thermal degradation of CH_3TiCl_3 is not clear (M2). Several authors (B29, D18) favor a mode of decomposition which excludes radical formation, whereas others (D44, G26, R9) believe that radicals are released from the alkyltitanium halides. Recent results by McCowan (M1a) show that the decomposition pathway is solvent dependent. Irradiation also causes decomposition (B29).

CH_3TiBr_3 (m.p. 2°–3°) decomposes only slowly at room temperature. The black-violet crystals of this compound form yellow solutions in hydrocarbon solvents and red solutions in ether or THF (T33).

The chemical behavior of CH_3TiCl_3 has been likened to that of aluminum alkyls (B29). With water, alcohols, enols, or acids the reaction is as follows.

$$ROH + CH_3TiCl_3 \longrightarrow ROTiCl_3 + CH_4$$

Exchange of the methyl group with a halide is possible with metal halides (B29, G26, R8) such as $AlCl_3$, $SnCl_4$, or $HgCl_2$. Oxygen converts CH_3TiCl_3 to the methoxides, CH_3OTiCl_3 and $CH_3OTiCl_3 \cdot (CH_3O)_2TiCl_2$. This latter product is thought to arise as a by-product by the disproportionation of CH_3OTiCl_3. The only product reported (B18) from the reaction of oxygen with $C_2H_5TiCl_3$ is the ethoxide $C_2H_5OTiCl_3$. Carbon dioxide does not react with the ethyl compound at room temperature (B18). Reaction of these compounds with iodine is rapid and quantitative (B29) and, hence, useful analytically. Holliday *et al.* (H28a) have used the reactions of vinyltitanium trichloride (a purple compound, stable at $-78°$, but decomposing rapidly above $-30°$) with chlorine and bromine to help establish its composition.

$$CH_3TiCl_3 + I_2 \longrightarrow CH_3I + ITiCl_3$$

The products of the reaction between methyltitanium trichloride and tetraethylammonium chloride or bromide in dichloromethane were found to be dependent on the stoichiometry of the reactants (C31, C31a). The anionic titanium species formed were of the type,

$\delta_{CH_3} = 2.69$, $\nu_{Ti-C} = 487$ cm^{-1} $\delta_{CH_3} = 2.43$, $\nu_{Ti-C} = 502$ cm^{-1} $\nu_{Ti-C} = 466$ cm^{-1}

Both oxygen and SO_2 can insert into the Ti–CH_3 bonds in $(C_2H_5)_4N[(CH_3)_2Ti_2Cl_7]$. All these compounds are moisture-, oxygen-, and heat-sensitive, but to a lesser extent than the parent compound.

The electronic energy levels of CH_3TiCl_3 {and $CH_3Ti[OCH(CH_3)_2]_3$} have been calculated (D24) and the lowest energy transitions assigned. The longest wavelength band at 25,000 cm^{-1} ($\varepsilon = 75$) for CH_3TiCl_3 purportedly arises from a near symmetry-forbidden $n \rightarrow \sigma^*$ transition in which the σ^* orbital is antibonding between the carbon and the titanium atoms. The adjacent band at 43,000 cm^{-1} was assigned to a transition of a nonbonding electron on the chlorine to an antibonding π level between the chlorines and metal.

There is considerable variation in the assignment of the infrared bands of CH_3TiCl_3 below 600 cm^{-1} (G25, G26, G34, K9, L1). Karapinka *et al.* (K9), Groenewege (G34), and Roshchupkina *et al.* (R52) have measured the solution spectra in noninteracting solvents, while Gray *et al.* (G26) and Gray (G25) made vapor phase measurements. In solution CH_3TiCl_3 gives rise to bands at about 530 and 460 cm^{-1}, whereas in the vapor the fundamentals are located at 550, 464, and 391 cm^{-1} and at 530 and 405 cm^{-1} for CD_3TiCl_3. Hanlan and McCowan (H3) have emphasized that no reliance could be placed on assignments based on infrared results of *in situ* preparations of CD_3TiCl_3.

The PMR spectra of several titanium alkyls have been reported (H3). The methyl protons of CH_3TiCl_3 were found at $\delta - 2.78$ in C_2Cl_4 and -2.20 in C_6H_6. In C_2Cl_4, $C_2H_5TiCl_3$ absorbed at $\delta - 2.08$ (CH$_3$) and -3.33 (CH$_2$).

2. Bis(alkyl)titanium Dihalides

The alkylating agents used in the preparation of the monoalkyl compounds, $RTiX_3$, also lend themselves to the preparation of the more unstable dialkyl compounds (B25–B27, D12, H21, H23, N8, N9, S56, S57). For example,

$$CH_3TiCl_3 + Zn(CH_3)_2 \xrightarrow[0°]{\text{hexane}} (CH_3)_2TiCl_2 + CH_3ZnCl \quad \text{(B25)}$$

$$TiCl_4 + 2CH_3MgI \xrightarrow{\text{ether}} (CH_3)_2TiCl_2 + 2MgICl \quad \text{(H21, H23, S56, S57)}$$

Dimethyltitanium dichloride is a volatile compound (b.p. 25°/0.2 mm) which forms violet to black crystals on cooling. It is sensitive to air, water, heat (decomposing to $TiCl_2$), light, and halogens (D12). The dioxane adduct is stable for several days at room temperature (B29), and is soluble in aliphatic, aromatic, and chlorinated hydrocarbons giving yellow solutions. The PMR spectrum of $(CH_3)_2TiCl_2$ has been measured ($\delta_{CH_3} = -2.00$ in C_6H_6 and -2.47 in C_2Cl_4), as has its infrared spectrum (G25, H3). An exchange, with an activation energy of 7.5 kcal/mole, occurs in C_2Cl_4 between the methyl groups of $(CH_3)_2TiCl_2$ and $Zn(CH_3)_2$. A similar exchange, which occurs slowly on the PMR time scale, also takes place between $(CH_3)_2TiCl_2$ and CH_3TiCl_3 (H3). Dimethyltitanium dichloride initiates the polymerization of α-olefins (D12).

3. Tris(alkyl)titanium Monohalides

The only tris(alkyl)titanium compound isolated is the benzyl derivative, obtained by the reaction (G5)

$$(C_6H_5CH_2)_4Ti + HCl \longrightarrow (C_6H_5CH_2)_3TiCl + C_6H_5CH_3$$

This red compound, melting with decomposition at 100°, polymerizes ethylene only slowly.

B. Adducts of Alkyltitanium Trihalides

Methyltitanium trichloride dissolves in aliphatic hydrocarbon solvents to form yellow solutions, whereas in diethyl ether or THF red solutions are formed by adduct formation. The THF adduct has been isolated (F22, R9), and is very sensitive to oxygen and moisture. Unlike the solutions of CH_3TiCl_3 in diethyl ether or THF which decompose readily (B29, D44, R9), the solid THF complex is stable at room temperature and melts with little decomposition at 107°–108°. A diminution in thermal stability in solution is a property shown by all complexes. The enhancement in the thermal stability of CH_3TiCl_3 by adduct formation is evident from the decomposition temperatures given in Table II-1. The decomposition of $CH_3TiCl_3 \cdot 2THF$ has been investigated (D44, R9); it is believed to proceed predominantly by a radical mechanism. Clark and McAlees (C32), who studied the thermal decomposition of a series of compounds and found that methyl chloride was evolved in this process, also argued in favor of a radical mode of decomposition.

The adducts show a greater resistance to oxidation and hydrolysis than do the parent compounds. This is especially the case with the 2,2'-bipyridyl complexes; hot water is required (T25) to hydrolyze $CH_3TiBr_3 \cdot 2,2'$-bipyridyl. Bulky ligands retard oxidation. In parallel with the parent compounds, the product of oxidation is the alkoxide (C32, C34). The solubility of most complexes is limited even in solvents such as diethyl ether, THF, dichloromethane, chloroform, carbon tetrachloride, and aromatic hydrocarbons.

In general, adduct formation leads to six-coordinate complexes; however, the monomeric triphenylphosphine and α-picoline adducts are exceptional in being pentacoordinate possibly owing to steric effects (F22). While dioxane coordinates through both oxygens, thioxane is thought, from the PMR spectrum of the compound, to be attached to the metal only by the sulfur. The sulfide compound, $CH_3TiCl_3 \cdot 2S(CH_3)_2$ dissociated at 0° under vacuum (F22).

Reduction of the metal by the ligand occurs in some cases, e.g., with α-picoline, and to a greater extent with trimethylamine (F22). With N,N,N',N'-tetramethyl-o-aminobenzylamine, CH_3TiCl_3 forms only the monoadduct of $TiCl_3$ (C33).

The PMR spectra of a series of adducts in methylene dichloride have been studied (C32, C33, F22), some over the temperature range 27° to −100° (C32, C33). The complexes in which the harder ligand atom lay trans to the titanium–methyl group adopted the mer configuration. The Ti–CH₃ hydrogens were found considerably further downfield than those in cyclopentadienyl compounds containing Ti–CH$_3$ groups.

The dimethyl complex $(CH_3)_2TiBr_2 \cdot$ 2,2'-bipyridyl has been made as a mixture with the monomethyl adduct (T25).

C. Phenyltitanium Trichloride

Thiele and co-workers (T33) obtained phenyltitanium trichloride as black-violet pyrophoric needles, thermally unstable above 15°, by reaction of TiCl$_4$ with either diphenylzinc or phenyllithium. Phenylation of TiCl$_4$ can also be accomplished with a Grignard reagent (H21, H23, S57). Latyaeva et al. (L9) have proposed that $C_6H_5TiCl_3$ was formed on heating TiCl$_4$ with $Hg(C_6H_5)_2$.

Coordination with aromatic nitrogen bases affords phenyltitanium trichloride a higher thermal stability. The pyridine adduct $C_6H_5TiCl_3 \cdot$ 2py is a brown powder, sparingly soluble in aliphatic hydrocarbons, but readily soluble in toluene to give yellow solutions (T33). It can be handled at room temperature, but should be stored at 0°. In the electronic spectrum (D24, T33) the band at $27,000$ cm^{-1} ($\varepsilon = 550$) was assigned to the symmetry-forbidden $n \rightarrow \sigma^*$ transition in which the antibonding orbital is associated with the Ti–C linkage. The infrared spectrum (T33) showed the Ti–Cl asymmetric stretching vibration at 433 cm^{-1}. The dark red 2,2'-bipyridyl adduct is thermally stable up to 130° and can even be handled momentarily in the air (T33). The stability of these compounds with aromatic amines contrasts sharply with that of the diethyl ether or trimethylamine adducts which are less stable than the parent compound (T32, T33). Ethereal **or**

$$C_6H_5TiCl_3 + 2(C_2H_5)_2O \xrightarrow{\text{benzene}} TiCl_3 \cdot 2(C_2H_5)_2O + C_6H_5 \cdot C_6H_5$$

trimethylamine solutions of phenyltitanium trichloride are dark red.

Pentafluorophenyltitanium trichloride has recently been made (R14) from TiCl$_4$ and $Hg(C_6F_5)_2$ or $(C_6F_5)_4Ti \cdot 2(C_2H_5)_2O$. It is monomeric, melting at 114°–118°, and volatile, subliming at $100°/10^{-4}$ mm. The slightly yellow crystals are sparingly soluble in nonpolar organic solvents, but readily soluble in water, dioxane, and THF. Unlike other alkyl- or aryltitanium compounds, the Ti–Cl bonds are less hydrolytically stable than the Ti–C$_6$F$_5$ bond. In fact, no pentafluorobenzene has been found when the compound was decomposed in HCl or alkali. It forms the white adduct $C_6F_5TiCl_3 \cdot$ 2py which is insoluble in water, dioxane, and benzene, sparingly so in cold pyridine, but soluble in dimethylformamide and concentrated HCl. Pentafluorophenyltitanium triisopropoxide has been prepared (G18).

Table II-1
Adducts of Alkyltitanium Trihalides

Complex	Color	Decomposition temperature (°C)	Ti–CH$_3$ chemical shift[a] (δ) in CH$_2$Cl$_2$	Ti–C infrared stretching frequency (cm^{-1})[b]	Other properties	Refs.
CH$_3$TiCl$_3$·2THF	Purple-red	107–108	2.19	496s	UV[c]	F22, R9, R52
CH$_3$TiCl$_3$·THF	Violet-pink	—	—	501ms	UV	F22
CH$_3$TiCl$_3$· CH$_2$OCH$_3$—CH$_2$OCH$_3$	Violet-pink	98	2.56–2.59	488ms	UV	C32, C33, F22
CH$_3$TiCl$_3$· CH$_2$OCH$_3$—CH$_2$SCH$_3$	Maroon	Room temp.	2.61	470m	—	C33
CH$_3$TiCl$_3$· CH$_2$OCH$_3$—CH$_2$N(CH$_3$)$_2$	Pink-rust	Room temp.	2.62	473vs	—	C33
CH$_3$TiCl$_3$· CH$_2$SCH$_3$—CH$_2$N(CH$_3$)$_2$	Dark pink	Room temp.	2.45	488m, 459w	—	C33
CH$_3$TiCl$_3$·[O⌬O]$_2$	Pale violet	—	—	499ms	UV	B29, F22
CH$_3$TiCl$_3$·[O⌬S]$_2$	Orange-brown	—	2.35	500m	UV	F22
CH$_3$TiCl$_3$·[S(CH$_3$)$_2$]$_2$	Red-brown	—	2.23	460m	UV	F22

14

Compound	Color	m.p.		$\tilde{\nu}$	UV	Ref.
$CH_3TiCl_3 \cdot \begin{matrix} CH_2SCH_3 \\ \| \\ CH_2SCH_3 \end{matrix}$	Maroon	110–112	2.46–2.51	475m	UV	C32, F22
$CH_3TiCl_3 \cdot \left[\text{(thiolane)} S \right]_2$ 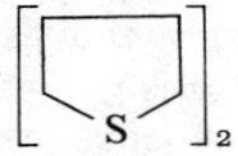	Red-brown	—	2.18	491m	UV	F22
$CH_3TiCl_3 \cdot \text{(thiane)} S \cdots S$ 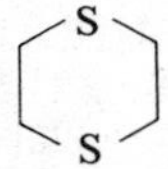	Red-violet	—	—	500m br	UV	F22
$CH_3TiCl_3 \cdot P(C_6H_5)_3$	Red	—	2.68	—	UV	B29, F22
$CH_3TiCl_3 \cdot \begin{matrix} CH_2P(C_6H_5)_2 \\ \| \\ CH_2P(C_6H_5)_2 \end{matrix} \cdot CH_2Cl_2$	Dull orange-red	180	—	446m	—	C32
$CH_3TiCl_3 \cdot \begin{matrix} CH_2P(C_6H_5)_2 \\ \| \\ CH_2P(C_6H_5)_2 \end{matrix} \cdot 1.3C_6H_5CH_3$	Orange-red	—	3.0	458m	UV	F22
$CH_3TiCl_3 \cdot 2CH_3CN$	Violet	—	2.47	491m	UV	F22, R52
$CH_3TiCl_3 \cdot \begin{matrix} CH_2N(CH_3)_2 \\ \| \\ CH_2N(CH_3)_2 \end{matrix}$	Violet	103–105	2.38	462m	—	C32
$CH_3TiCl_3 \cdot \left[\text{C}_6\text{H}_4\text{(OCH}_3\text{)(N(CH}_3\text{)}_2) \right]_2$ 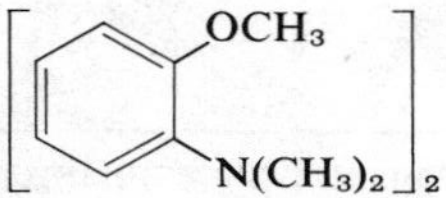	Brown-violet	d	—	—	—	C32
$CH_3TiCl_3 \cdot \left[\text{C}_6\text{H}_4\text{(N(CH}_3\text{)}_2)_2 \right]_2$ 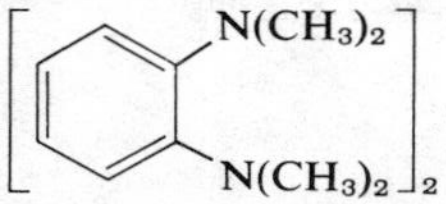	Brown-violet	d	—	—	—	C32

(continued)

15

Table II-1 (*continued*)

Adducts of Alkyltitanium Trihalides

Complex	Color	Decomposition temperature (°C)	Ti–CH$_3$ chemical shift[a] (δ) in CH$_2$Cl$_2$	Ti–C infrared stretching frequency (cm^{-1})[b]	Other properties	Refs.
CH$_3$TiCl$_3$·[py]$_2$	Violet	—	—	485m	UV	F22, R52
CH$_3$TiCl$_3$·[2-picoline]	Orange	—	—	519s	UV	F22
CH$_3$TiCl$_3$·[2,2'-bipyridyl]	Red-violet	160	—	465m	UV	F22, T25
C$_2$H$_3$TiCl$_3$·2THF	Green	Room temp.	—	—	UV	H28a
C$_2$H$_3$TiCl$_3$· $\begin{array}{c} \text{CH}_2\text{OCH}_3 \\ \| \\ \text{CH}_2\text{OCH}_3 \end{array}$	Green	Room temp.	—	—	UV	H28a
C$_3$H$_5$TiCl$_3$·[2,2'-bipyridyl][e]	Dark-blue	180	—	—	—	T25
C$_3$H$_7$TiCl$_3$·[2,2'-bipyridyl]	Blue-violet	>100	—	—	—	T33
C$_5$H$_{11}$TiCl$_3$·[2,2'-bipyridyl]	Dark-blue	>100	—	—	—	T33
CH$_3$TiBr$_3$·[2,2'-bipyridyl]	Red-brown	—	—	—	—	T25

[a] Chemical shifts are virtually temperature independent (C32).
[b] Other (Ti–Cl stretches) low frequency bands (320–400 cm^{-1}) are also given in Refs. C32, F22, and R52.
[c] UV-Visible reflectance spectra (F22).
[d] Less stable than the dimethoxyethane adduct (C32).
[e] Probably a π-allyltitanium species.

D. Alkyl- and Aryltitanium Alkoxides

Herman and Nelson (H25) in 1952 isolated the first organometallic titanium compounds incorporating a Ti–C σ bond, namely, $C_6H_5Ti[OCH(CH_3)_2]_3 \cdot$ $LiOCH(CH_3)_2 \cdot LiBr \cdot (C_2H_5)_2O$, in 76% yield. This, on treatment with $TiCl_4$ or $AlCl_3$ gave a 53% yield of the light yellow compound $C_6H_5Ti[OCH-(CH_3)_2]_3$, which although stable indefinitely below 10°, decomposed rapidly if heated above its melting point of 88°–90°. It was also synthesized (H35) from $ClTi[OCH(CH_3)_2]_3$ and C_6H_5Li in ether at $-10°$.

In the compounds $RTi(OC_4H_9)_3$, the ease of decomposition to a lower oxidation state titanium species increases in the order indenyl < α-naphthyl < phenyl < p-anisyl < ethynyl < methyl < butyl which is said to parallel the decreasing electronegativity of the R group (H27). The compounds $C_6H_5TiX_3$ show a stability order dependent on the ligand X; the stability decreasing in the order X = $OC_4H_9 > OCH_3$, Cl > F. Pyrolysis of $C_6H_5Ti(OC_4H_9)_3$ produced phenyl radicals which coupled to yield biphenyl and higher hydrocarbons. This mode of decomposition permits this compound to initiate polymerization of styrene (H21–H23, H25, N43, R7). A rate expression for this process has been evaluated by North (N43).

The potential of $C_6H_5Ti(OR)_3$ compounds to initiate polymerization has prompted the preparation of many alkyltitanium alkoxides and numerous patents claim these to be active low-temperature and low-pressure polymerization catalysts for α-olefins (C35, D15, F5, G37, K8, N8, S56–S58). $(CH_3)_2Ti(O-iso-C_3H_7)_2$ was obtained from hexane as volatile yellow crystals (m.p. 78°–88°), which sublimed at 55°/0.05 mm. The isobutoxo and ethoxo analogs were light brown and reddish oils, respectively (G37). From $(C_2H_5O)_2TiCl_2$ and allylmagnesium chloride (K8) in THF at $-70°$, $(C_2H_5O)_2Ti(\pi-C_3H_4)_2$ was obtained. Controlled alcoholysis of tetrabenzyl-titanium (G5) also led to an alkoxo compound (see Chapter II, Section F, 2).

Claims have been made (I1, S68, S69) for the syntheses of the compounds,

$$\left[(CH_3)_2\ Ti \underset{O}{\overset{O}{<}} R \right]_2$$

where R = $(CH_2)_{2, 3, or 4}$, $CH_3CH(CH_2)_2$, $(CH_3)_2CCH_2CHCH_3$, or (C_4H_9)-$CHCH(C_2H_5)CH_2$], from

$$Cl_2Ti \underset{O}{\overset{O}{<}} R$$

and methylmagnesium iodide. However, only the yellow crystalline 2,4-pentanediolato compound, obtained in 55–70% yield, has been well characterized (S70, Y5). It is quite stable below 10°, but at room temperature in the dark, blackening of the crystal surfaces is evident after several days. Decomposition is more rapid in solution and at elevated temperatures; it melts and blackens at 93°, decomposing at 98°. With water, alcohol, or hydrogen chloride it reacts according to the following equation. It is also sensitive toward oxygen.

$$[(CH_3)_2Ti(C_6H_{12}O_2)_2]_2 + 4HX \longrightarrow 2X_2Ti(C_6H_{12}O_2)_2 + 4CH_4$$

X = OH, OR, Cl

An X-ray structure determination of the compound (Y5) has confirmed that the dimer configuration found in solution (S70) exists also in the solid. Dimerization is through oxygen bridges (see Fig. II-1) so that each titanium is bonded to three oxygens and two carbons resulting in a trigonal-bipyramidal configuration about the titanium. The bridging oxygens are further from the

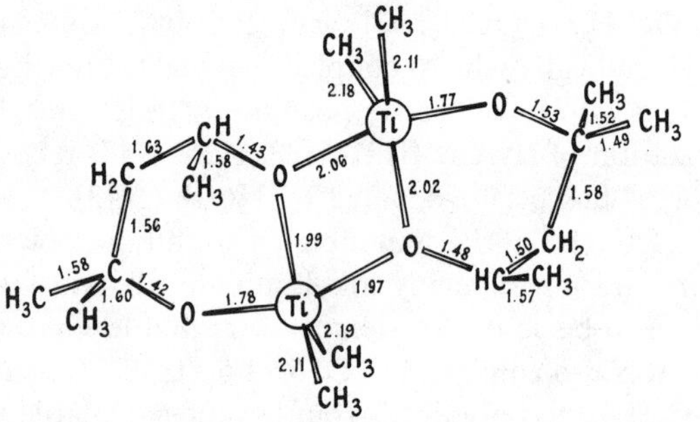

Fig. II-1.   Structure of the dimer $[(CH_3)_2Ti(C_6H_{12}O_2)]_2$ in which the metals are in a trigonal-bipyramidal coordination sphere (Y5).

metal atoms than the nonbridging oxygens. The differences in the Ti–C bond lengths (2.11 and 2.18 Å) are reflected in the two PMR lines found at δ1.20 and 1.39 for the methyl groups on the metal (S70).

Mixed alkoxo chloride compounds have been prepared (C35) in over 80% yield in methylene dichloride by equilibration reactions between $(CH_3)_nTi(O\text{-}iso\text{-}C_3H_7)_{4-n}$ ($n = 1$ or 2) and CH_3TiCl_3 or $TiCl_4$. For example,

$$2CH_3Ti[OCH(CH_3)_2]_3 + CH_3TiCl_3 \xrightarrow{\ 0° \text{ to } 20°\ } 3CH_3Ti[OCH(CH_3)_2]_2Cl$$
(volatile yellow crystals, m.p. 62°–64°)

Isopropoxomethyltitanium dichloride, a brown-violet compound (m.p. 60°–63°), decomposes very readily to a lower oxidation state species. The corresponding ethoxo compounds, $CH_3Ti(OC_2H_5)_2Cl$ and $CH_3Ti(OC_2H_5)Cl_2$, which were prepared by a parallel procedure, are both orange in color (C35).

E. Alkyl- and Aryltitanium Amides

The information accrued on the amido compounds of titanium has been due solely to the work of Bürger and Neese (B97–B99, N27). In all compounds the hydrocarbon residue is bound to the metal by a σ bond. Although they are remarkably more resistant to thermolysis than the corresponding halide species, these compounds still exhibit a high oxygen and moisture sensitivity.

Their preparation proceeds according to the reaction

$$(R_2N)_3TiBr + MR' \longrightarrow (R_2N)_3TiR' + MBr$$

$$M = Li \text{ or } MgX$$

An inconvenience is the redistribution reactions that the $(R_2N)_3TiBr$ compounds undergo. The vinyl compounds $[(C_2H_5)_2N]_3TiR$ (where R = $CH{=}CH_2$ or $CH{=}CHCH_3$) could not be isolated (N27) in the pure state. In the PMR spectra, the vinyl groups exhibit dynamic behavior. The allyl compounds $[(C_2H_5)_2N]_2Ti[CH_2C(R){=}CH_2]_2$ (where R = H or CH_3) did not possess sufficient thermal stability to permit purification by distillation.

The compounds range in color from yellow to yellow-red to olive. They are miscible with nonpolar organic solvents and exhibit no tendency to associate in solution, nor do they form adducts in THF or ether (B99). The compounds generally solidify as glasses on cooling below their melting point.

On thermolysis of the metal alkyl compounds between 70° and 160° under vacuum, the alkyl group was evolved nearly quantitatively as the alkane (B98). Dimerization of alkyl groups did not occur. Ethylene, formed in small amounts during pyrolysis of the diethylamido compounds, arose from decomposition of the amido group rather than from the Ti–R moiety. On heating $[(CH_3)_2N]_3TiCD_3$, CH_2D_2 and CD_4 together with the expected CHD_3 were found. A mechanism proposed (B98) on the basis of these findings involved formation of ions of CD_3 which reacted with $(CH_3)_2N$ groups to extract a proton and/or undergo H–D exchange. The PMR spectrum of the nongaseous reaction products from the thermolysis of $[(C_2H_5)_2N]_3TiR$ suggested that the amido-CH_2 hydrogens were the reactive ones.

F. Tetraalkyl- and Tetraaryltitanium Compounds

With the exception of tetrabenzyltitanium and to a lesser extent the tetraphenyl compound, work on this class of compounds is fraught with the problem of thermal instability. Decomposition appears to proceed via release of radicals with the consequent reduction of the titanium.

Table II-2
Alkyl- and Aryltitanium Amides

Compound	M.p. (°C)	Sublimation temp. (°C/10^{-3} mm)	Decomposition temp. (°C)	ν_{Ti-C} (cm^{-1})	PMR chemical shift (δ) of Ti–R protons[a]	J (Hz)	Refs.
[(CH$_3$)$_2$N]$_3$TiCH$_3$	8–9	20	80	499	0.47	—	B97
[(CH$_3$)$_2$N]$_3$TiCD$_3$	9–10	20	70	469	—	—	B97
[(CH$_3$)$_2$N]$_3$TiC$_2$H$_5$	≃ –5	20	70 (light-sensitive)	—	1.05(CH$_2$), 1.69(CH$_3$)	—	B97
[(C$_2$H$_5$)$_2$N]$_3$TiCH$_3$	4–5	—	120–130	≃500sh	0.43	—	B97
[(C$_2$H$_5$)$_2$N]$_3$TiCD$_3$	5	60–70	120–130	460	—	—	B97
[(C$_2$H$_5$)$_2$N]$_3$TiC$_2$H$_5$	Liquid	—	120–130	—	1.01(CH$_2$), 1.70(CH$_3$)	—	B97
[(C$_2$H$_5$)$_2$N]$_3$Ti-n-C$_3$H$_7$	13–14	—	120–130	—	—	—	B97
[(C$_2$H$_5$)$_2$N]$_3$Ti-iso-C$_3$H$_7$	< –10	—	120–130	—	—	—	B97
[(C$_2$H$_5$)$_2$N]$_3$Ti-n-C$_4$H$_9$	–4	—	120–130	—	—	—	B97
[(C$_2$H$_5$)$_2$N]$_3$Ti-$tert$-C$_4$H$_9$	< –30	—	120	—	1.39	—	B97
[(CH$_3$)$_2$N]$_3$TiCH$_2^{\beta}$CH$^{\alpha}$=CH$_2^{\beta}$	c	—	—	—	1.84(H$^{\alpha}$), 3.22(H$^{\beta}$)	—	N27
[(CH$_3$)$_2$N]$_3$TiCH$_2^{\beta}$C(CH$_3^{\alpha}$)=CH$_2^{\beta}$	c	—	—	501m	6.45(H$^{\alpha}$), 3.31(H$^{\beta}$)	—	N27
[(C$_2$H$_5$)$_2$N]$_3$TiCH$_2^{\beta}$CH$^{\alpha}$=CH$_2^{\beta}$	Liquid	—	130–135	502s	1.99(H$^{\alpha}$), 3.26(H$^{\beta}$)	—	N27
[(C$_2$H$_5$)$_2$N]$_3$TiCH$_2^{\beta}$C(CH$_3^{\alpha}$)=CH$_2^{\beta}$	Liquid	—	140–145	521s	2.26(H$^{\alpha}$), 3.46(H$^{\beta}$)	—	N27
[(C$_2$H$_5$)$_2$N]$_2$Ti[CH$_2^{\beta}$C(CH$_3^{\alpha}$)=CH$_2^{\beta}$]$_2$	c	—	—	—	—	—	N27

Compound	m.p. (°C)	b.p. (°C)		IR (cm⁻¹)	PMR (δ, ppm)	J (Hz), Ref.
$[(C_2H_5)_2N]_3TiC$ with H^β, CH_3^α, H^γ ($C=C$)	—	50–60[b]	112	533m	2.18(H^α), 5.39(H^β), 5.87(H^γ)	$^2J(H^\beta H^\gamma) = 3.4$, B99; $^4J(H^\alpha H^\gamma) = 1.6$
$[(C_2H_5)_2N]_3Ti$ with CH_3^β, CH_3^γ, H^α ($C=C$)	—	50–60[b]	110	495m	6.58(H^α), 2.00(H^β), 1.89(H^γ)	$^4J(H^\alpha H^\gamma) = 1.1$, B99; $^4J(H^\alpha H^\beta) = 0.3$
$[(C_2H_5)_2N]_3Ti$ with H^β, CH_3^α, CH_3^γ ($C=C$)	c	50–60[b]	105	—	2.22(H^α), 6.17(H^β), 1.85(H^γ)	$^3J(H^\beta H^\gamma) = 6.1$, B99; $^4J(H^\alpha H^\beta) = 0.9$, $^5J(H^\alpha H^\gamma) = 1.6$
$[(C_2H_5)_2N]_3Ti$ with CH_3^β, CH_3^α, H^γ ($C=C$)	c	50–60[b]	—	—	2.25(H^α), 2.07(H^β), 6.14(H^γ)	$^3J(H^\beta H^\gamma) = 6.7$, B99; $^3J(H^\alpha H^\gamma) = 1.5$, $^5J(H^\alpha H^\beta) = 1.3$
$[(C_2H_5)_2N]_3Ti$ with CH_3^β, CH_3^α, CH_3^γ ($C=C$)	c	80[b]	—	—	2.10(H^α), 1.81(H^β), 1.74(H^γ)	$^5J(H^\alpha H^\beta) = 1$, B99; $^5J(H^\alpha H^\gamma) = 1$
$[(C_2H_5)_2N]_3TiC{\equiv}CC_6H_5$	56–58	80	—	—	7.82	B97
$[(C_2H_5)_2N]_3TiC_6H_5$	Liquid	—	—	—	$\simeq 7.2$	B97

[a] PMR spectrum in benzene with exception of the phenyl and acetylide complexes which were measured in cyclohexane. Amido protons have also been recorded.
[b] Boiling point (at 10^{-3} mm).
[c] Not pure compounds.

21

In the solid state, tetrabenzyltitanium is stable up to 60°, and in solution to even higher temperatures. Tetraphenyltitanium can be refluxed in ether for up to 2 hr with negligible decomposition (T4); however, it should be stored below −20°. Crystals of the corresponding methyl compound (B45) must be kept below −78°. The ethyl (B58), n-butyl (B59, J1), and cyclohexyl (J1) analogs also exhibit a high thermal instability (B58, T4), and consequently have not received much attention.* Tetravinyltitanium can be kept (B58) for several hours at −80° and tetraallyltitanium is thought to be stable to −20°.

1. Tetramethyltitanium

It was first found by Clauss and Beerman (F3) that ethereal solutions of tetramethyltitanium could be prepared by reaction of titanium tetrachloride with an alkylating agent. For example,

$$TiCl_4 + 4LiCH_3 \xrightarrow[-50° \text{ to } -80°]{\text{ether}} Ti(CH_3)_4 + 4LiCl$$
$$50\text{--}75\%$$

The titanium halide was added as the bis(ether) complex and any ether which was liquid under the reaction conditions was suitable as solvent. Other alkylating agents such as the methyl derivatives of lithium, sodium, potassium, beryllium, and especially magnesium, namely, $Mg(CH_3)_2$ and CH_3MgI, have also been used (B45, C36, K23). Excess unreacted alkylating agent was destroyed by reaction with carbon dioxide at −80° to −50°.

Tetramethyltitanium is purified by distillation at temperatures preferably below −10°. The yellow ethereal solutions upon concentration do not solidify, but give an orange oil, presumably the bis(ether) adduct. However, bright yellow crystals of the pure compound were obtained by Berthold and Groh (B45) from hexane.

It is of interest that the titanium(III) halide, $TiCl_3$, does not react noticeably with lithium methyl in ether. However, in THF or dimethoxyethane it reacted readily with lithium methyl at −50° to −80° giving dark green solutions of trimethyltitanium (C37).

Tetramethyltitanium possesses a higher thermal stability in ether (owing to adduct formation) than in the solid state. Spontaneous decomposition in ether (C37) sets in near 0°, whereas the crystalline compound decomposes (B45) when warmed above −78°. All investigators agree (D43, R20, R21) that the thermal breakdown is autocatalytic, the insoluble products probably being the catalytic species. Raney nickel (R20, R21) initiates the decomposition

---

* Tetrabutyltitanium has been prepared in good yield from the bis(pyridine), quinoline, or piperidine adducts of $TiCl_4$ with a butyl Grignard reagent at −16° in ether (B59, D39, J1, T1).

of tetramethyltitanium in ether at $-20°$. In the vapor state at $80°–100°$, the compound falls apart with cleavage of all the titanium–carbon linkages and the formation of titanium metal mirrors (R20).

Six-coordinate base adducts of tetramethyltitanium with a number of ligands (M53, T28, T29) (see Table II-3) have, in most cases, been readily obtained by addition of the base to concentrated ethereal solutions of tetra-methyltitanium at $-20°$ to $-30°$. These complexes give intensely colored solutions in ether or benzene in which they have a high solubility. The monodentate adducts are also quite soluble in aliphatic hydrocarbon solvents (M53, T28)—markedly more so than the chelate complexes. Although their thermal stability shows an appreciable enhancement over the parent compound (the P,P,P',P'-tetramethylethylenediphosphine adduct, for example, can be sublimed *in vacuo* at $30°–40°$) they are acutely air-sensitive, especially when finely divided.

The resistance to thermal breakdown of the adducts varied with the ligands (M53) in the order, 1,4-dioxane $<$ $N(CH_3)_3$ $<$ $P(CH_3)_3$ $<$ pyridine $<$ $(CH_3)_2NCH_2CH_2N(CH_3)_2$ $<$ $2P(CH_3)_3$ $<$ 1,10-phenanthroline $<$ 2,2'-bipyridyl $<$ $(CH_3)_2PCH_2CH_2P(CH_3)_2$. This order was rationalized by reference to the "orbital electronegativities" of the donor atoms and the chelate effect. The bis adducts of the monodentate bases which were thought to possess a cis configuration were stable only in the solid state since in solution they dissociated to form the monoadducts.

Tetramethyltitanium reacts with water to form methane and converts

$$R_2CO + Ti(CH_3)_4 \longrightarrow R_2C(OH)CH_3$$

aldehydes or ketones to alcohols (F3). On warming with excess titanium tetrachloride (F3), a quantitative yield of titanium trichloride was obtained. The reaction of $(CH_3)_4Ti$ with iodine can be exploited analytically (F3).

$$Ti(CH_3)_4 + 4I_2 \longrightarrow TiI_4 + 4CH_3I$$

Zdunneck and Thiele (Z2) have investigated exchange reactions between tetramethyltitanium and aryl- or alkylborons. The number of groups exchanged depends primarily on the reaction temperature and the nature of the organoboron compound. With triphenylboron at $-40°$, independent of the stoichiometry of the reactants, only two methyl groups were lost from the titanium to form yellow bis(methyl)bis(phenyl)titanium. It is unstable in air

$$3Ti(CH_3)_4 + 2B(C_6H_5)_3 \xrightarrow[-40°]{\text{ether or THF}} 3(CH_3)_2Ti(C_6H_5)_2 + 2B(CH_3)_3$$

and above $0°$ pyrolyzes forming benzene and biphenyl. It should be stored at $-78°$, but can be handled momentarily at room temperature. The reactions between tetramethyltitanium and tribenzylboron which lead to

Table II-3

Adducts of Tetramethyltitanium[a]

Ligand[b]	Color	Thermal stability	Refs.
2,2'-Bipyridyl	Red crystals	Can be handled at room temp. Kept at 0°	T2, T28, T29
1,10-Phenanthroline	Red, shiny crystals	Can be handled momentarily at room temp. Decomposes slowly above 0°. Least soluble	T28, T29
N,N,N',N'-Tetramethyl-ethylenediamine[c]	Long, dark red needles	Decomposes below 0°	T2, T29
P,P,P',P'-Tetramethyl-ethylenediphosphine	Yellow platelets	Stable for some hours at room temp.	T29
1,4-Dioxane	Yellow needles	Explosively decomposes at −20°	T29
Pyridine	Large, yellow transparent crystals	Decomposes below 0°, sublimes at −20°	M53
Pyridine (bis adduct)	Red crystals	M.p. −20°. Can be handled at room temp.	M53
Trimethylamine	Yellow-orange crystals	Decomposes rapidly below 0°, sublimes at −25°	M53
Trimethylphosphine	Yellow crystals	Rapid decomposition above 0°, sublimes at −20°	M53
Trimethylphosphine (bis adduct)	Orange needles	Decomposes at room temp.	M53

[a] All are monomers in solution (benzene or o-xylene).

[b] One ligand/mole of $Ti(CH_3)_4$ unless stated otherwise.

[c] Solutions of the N,N,N',N'-tetramethylmethylenediamine adduct have also been prepared (T2).

$CH_3Ti(CH_2C_6H_5)_3$ or $Ti(CH_2C_6H_5)_4$ are described in the following section on tetrabenzyltitanium.

No reduction was observed when $Ti(CH_3)_4$ was treated with lithium methyl. From the green-yellow ethereal solution the yellow complex, $LiTi(CH_3)_5 \cdot 2C_4H_8O_2$, was precipitated on addition of dioxane (T27). It is considerably more stable than the parent organotitanium species, thermal decomposition occurring only above about 0°. It is readily soluble in THF, in which it behaves as a 1:1 electrolyte. Lithium phenyl and lithium benzyl give rise to the analogous complexes with $Ti(CH_3)_4$ on addition of dioxane (T27), as does bis(methyl)bis(phenyl)titanium and lithium phenyl.

In ether all the protons in tetramethyltitanium were observed in the PMR spectrum at δ0.68 (T2, T3). The methyl groups of $Ti(CH_3)_4$ exchange (K23), but only slowly on the PMR time scale, with those of $Al_2(CH_3)_6$ in ether at $-75°$.

The infrared (E7, T2) and electronic (M53, T29) spectra of tetramethyltitanium and a number of its adducts have been measured.

2. Tetrabenzyltitanium

Tetrabenzyltitanium is the most thermally stable of the tetraalkyl or tetraaryl compounds (T4), both in solution and in the solid state. The crystals can be kept for several months at $0°$, but decompose on heating above $61°$. In solution (G5, T4) it exhibits an enhancement in stability, such that no decomposition is evident after refluxing for several hours in ether. It can be refluxed in heptane at $100°$ for several hours before decomposition occurs. Its extraordinarily high thermal stability stems from the fact that the β-hydrogen elimination scheme is inoperative here and, further, a secondary interaction between the aromatic rings and the metal is possible.

Tetrabenzyltitanium was first prepared (B58) as an ethereal solution from titanium tetrachloride and Grignard reagent. The synthetic procedure has

$$TiCl_4 + 4C_6H_5CH_2MgX \xrightarrow[-16°]{ether} Ti(CH_2C_6H_5)_4 + 4MgClX$$

been well described by Thiele and co-workers (B96). It was first isolated by Giannini and Zucchini (G5) as red crystals (m.p. $70°–71°$) quite soluble in aliphatic hydrocarbons and other inert organic solvents. It is monomeric in benzene.

A major concern in the reaction between $TiCl_4$ and Grignard reagent is decomposition of the product with concomitant reduction of the metal. This can, to some extent, be minimized by working near solid CO_2 temperatures and in the dark. Jacot-Guillarmod and co-workers (B59, J1) noted that the presence of amine base such as pyridine, piperidine, or quinoline in the reaction mixture had a stabilizing effect. The formation of titanium(III) species in the preparation can also be minimized by replacement of the $TiCl_4$ by titanium tetrabutoxide (J1, T1, T3). Other alkoxides, in particular the tetraisopropoxide and the tetra-tert-butoxide, led to low yields of tetrabenzyltitanium with much decomposition evident (J1). The theory was advanced, in explanation of this behavior, that amines or alkoxides in the absence of steric effects, stabilize the R_4Ti species, while Lewis acids, namely, magnesium halides, produce the reverse effect (D39, J1). Dibenzylmagnesium or its bis(pyridine) adduct gives good yields of tetrabenzyltitanium with $TiCl_4$ in ether or pentane.

Zdunneck and Thiele (Z2) obtained tetrabenzyltitanium by the reaction between tetramethyltitanium and tribenzylboron at $-30°$ to $-40°$. The

$$3\text{Ti(CH}_3)_4 + 4\text{B(CH}_2\text{C}_6\text{H}_5)_3 \xrightarrow{\text{ether}} 3(\text{C}_6\text{H}_5\text{CH}_2)_4\text{Ti} \cdot 2(\text{C}_2\text{H}_5)_2\text{O} + 4\text{B(CH}_3)_3$$

ether adduct so formed is a red oil which solidifies as orange-yellow crystals at $-78°$. When the reactants were brought together in equimolar amounts at $-78°$ in ether, tribenzylmethyltitanium was formed. This compound is sensitive to light and heat [m.p. $63°$–$64°$ (dec.)] and protic reagents. It has also been prepared from tribenzyltitanium chloride (G6) and methyl-magnesium iodide. Surprisingly, with pyridine or 2,2′-bipyridyl, it failed to form isolable adducts.

The PMR spectrum of tetrabenzyltitanium in toluene-d_8 at $30°$ has a sharp singlet arising from the methylene protons at $\delta 2.57$ and two bands at $\delta 6.42$ and 6.81 assigned to the ortho and meta/para aromatic protons (G5). This spectrum is almost temperature independent down to $-70°$. The aromatic proton peaks collapse into an asymmetric singlet in the presence of

Fig. II-2.   Structure of tetrabenzyltitanium (B16).

ether or pyridine (J1, T3). It was suggested that there is an interaction, possibly of the π-allyl type, between the metal and the 1,2- and 7-benzyl carbon atoms which adduct formation was thought to preclude on steric grounds. The molecular structure determination of this compound (B16, D6) confirms an interaction between the ring of the benzyl group and the titanium (Fig. II-2), as the mean Ti–C–C bond angle is only $103°$. In the corresponding zirconium and hafnium compounds the metal–C–C bond angles are smaller still (D5, D6) (see Table IV-2). It was suggested that this interaction led to an enhancement of the thermal stabilities of these compounds.

Tetrabenzyltitanium readily forms adducts with donor molecules (B96, T2, Z2), which have a higher thermal stability than the parent compound. Thus, the 2,2′-bipyridyl adduct, made from the diethyl ether complex by a replacement reaction forms as chocolate brown crystals melting at $84°$–$85°$ with decomposition (Z2). It is not as air-sensitive as the parent compound, nor is it as soluble, especially in hexane. Other adducts with nitrogen bases

(T2), in particular, pyridine and the bidentate ligands $(CH_3)_2N(CH_2)_nN$-$(CH_3)_2$ ($n = 1$–3), have been made. These presumably octahedral adducts are all colored red to brown (T2). With triphenylphosphine, no coordination was evident (B96).

Tetrabenzyltitanium displays the expected chemical reactivity toward acidic (G6, G7) and oxidizing reagents (B59, G5). With HCl, HBr, or BF_3 in toluene, the tris(benzyl)titanium halides (m.p. 100°–103°) are formed. When it was treated with 2 moles of ethanol, the monomeric, room temperature stable bis(benzyl)bis(ethoxo)titanium (m.p. 102–103°) was obtained. Bromine converts tetrabenzyltitanium into benzyl bromide and $TiBr_4$. Reaction with oxygen gave, after hydrolysis, benzyl alcohol.

The behavior of tetrabenzyltitanium with organic carbonyl groups parallels that of benzylmagnesium chloride in giving both ortho- and normal-substituted derivatives (C13a).

Tetrabenzyltitanium does not react with hydrogen at ambient temperature and pressure. The initiation of polymerization of alkenes with this compound (G5, G42, P9) or some of its substituted analogs (P9) has been investigated. Thiele et al. (B96) have studied the infrared spectrum of tetrabenzyltitanium from 200 to 3200 cm^{-1}.

3. Tetraphenyltitanium

The methods of preparation of tetraphenyltitanium (B59, D39, R30, T1) parallel those of the corresponding benzyl compound. While the lower thermal stability of tetraphenyltitanium (m.p. 45° with decomposition) (T1) requires a more stringent control to be observed over the temperature, especially in solution (L9, R30), this condition is nowhere near as critical as in the case of tetramethyltitanium. Tetraphenyltitanium made from the metal halide and Grignard reagent precipitated from ether (B59) through complex formation with magnesium halides, possibly as $(C_6H_5)_4Ti \cdot (MgX_2)_n$. In the presence of pyridine this complex was destroyed liberating the petroleum-soluble monomeric orange-colored organometallic.

A series of adducts of tetraphenyltitanium with the amines, 2,2′-bipyridyl (orange crystals, stable at 0°), pyridine (green-brown solution), and $(CH_3)_2N$-$(CH_2)_nN(CH_3)_2$ ($n = 1$–3) (green-brown to yellow-green solutions) have been prepared (T2) and the PMR chemical shifts of the amine protons recorded. The Ti–N stretching vibration was associated with the 450 cm^{-1} band (T2). The PMR spectrum of tetraphenyltitanium itself (T3) in coordinating solvents such as ethers or amines consisted of a single peak at $\delta 7.36$. In benzene-d_6 two aromatic proton peaks were evident centered at $\delta 7.31$ and 7.00 (relative intensity 2:3).

Razuvaev *et al.* (L9, R30) have used the pyrolysis of tetraphenyltitanium to prepare the divalent titanium species $(C_6H_5)_2Ti$ (see Chapter VII). When this reaction was carried out in the presence of tolan, the main products on subsequent hydrolysis were hexaphenylbenzene and tetraphenylethylene (R15).

Tetraphenyltitanium reacts with mercuric chloride at $-70°$ in ether to give a near quantitative yield of phenylmercuric chloride. It can also be used to prepare $Cp_2Ti(C_6H_5)_2$.

$$(C_6H_5)_4Ti + 2C_5H_6 \xrightarrow[-70°]{\text{ether}} Cp_2Ti(C_6H_5)_2 + 2C_6H_6$$

Orange, moisture-sensitive crystals of the bis(ether) adduct of tetra-(pentafluorophenyl)titanium (m.p. 117°–119°), soluble in benzene and ether, were obtained in 20–30% yield (R16) by the reaction,

$$4C_6F_5Li + TiCl_4 \xrightarrow[-70°]{\text{ether/hexane}} Ti(C_6F_5)_4 + 4LiCl$$

This compound is completely stable to 100°, but decomposes explosively in the melt at 120°–130°. The pentafluorophenyl groups on the titanium are readily replaced by phenyl groups on reaction with phenyllithium or diphenylmercury (R14). Low oxidation state titanium compounds were formed when $Ti(C_6F_5)_4$ was treated with $TiCl_4$ in a 1:1 or 1:2 molar ratio. However, when the ratio was 1:3, $C_6F_5TiCl_3$ was obtained in 42% yield.

$$(C_6F_5)_4Ti + 3TiCl_4 \xrightarrow[20°]{\text{benzene}} 4C_6F_5TiCl_3$$

4. Tetraneopentyltitanium

The importance of the β-hydrogen elimination decomposition mechanism is underlined by the work of Mowat and Wilkinson (M50, M50a), who prepared a series of stable tetraneopentyl compounds. The titanium complex, $[(CH_3)_3CCH_2]_4Ti$, prepared by alkylating $TiCl_4$ with the Grignard reagent, forms yellow prisms which melt at 99°. It is soluble in benzene, petroleum, and ether and can be sublimed *in vacuo* at about 50°. However, the compound is thermally unstable at room temperature over long periods of time (> 1 week) and is also extremely moisture sensitive. It reacts with carbon monoxide to form a carbonyl compound and with nitric oxide and pyridine. The PMR absorptions are at $\delta 1.38$ (CH_3) and $\delta 2.36$ (CH_2). The metal–carbon stretching frequencies are at 540 and 505 cm^{-1}.

5. Tetrabicyclo[2.2.1]hept-1-yltitanium
[Tetra(1-norbornyl)titanium]

This yellow, diamagnetic compound was prepared in 90% yield from $TiCl_4 \cdot 2THF$ and the lithium alkyl (B60). At 100°, it has a half-life of 29 hr. In isoctane solution it is stable in the air. It does not form adducts with pyridine, presumably for steric reasons.

6. Tetra(silylmethyl)titanium Compounds

Lappert *et al.* (C41a) have recently reported the preparation of compounds of the type $(R_2R'SiCH_2)_4Ti$ (in which $R = R' = CH_3$; $R = C_6H_5$, $R' = CH_3$; $R = CH_3$, $R' = C_6H_5$ or $C_6H_5CH_2$). These pale yellow, liquid compounds are prepared by alkylating $TiCl_4$ with the Grignard reagent, lithium alkyl or, preferably, with dialkylmagnesium which minimizes reduction of the metal. The compounds are very soluble in hydrocarbon solvents, but may be crystallized from pentane solutions. Tetrakis(trimethylsilylmethyl)titanium distills at 25°–30° at 10^{-3} mm Hg. However, on heating to 80° in benzene, it forms tetramethylsilane. The higher stability in air of the titanium species compared to the corresponding zirconium and hafnium analogs was attributed to steric effects. Although the compounds are stable to some extent toward halogenated solvents, they are readily attacked by reagents containing acidic hydrogens.

The PMR chemical shifts for the methyl groups are in the range $\delta 0.3$–0.84 while the MCH_2Si protons are found in the region $\delta 2.33$–2.70. The Ti–C stretching vibrations are at $510 \pm 10 \text{ cm}^{-1}$.

Chapter III

π-Bonded Derivatives of Titanium(IV)

The large number of compounds in this category is evident from the comparative size of this chapter. All but 2% contain at least one cyclopentadienyl (or substituted cyclopentadienyl) group. The number of these ligands present, together with the type of substituent on the ring, can have a profound effect on the chemical behavior of the particular compound. An obvious example is the increase in hydrolytic stability with the number of cyclopentadienyl groups on the metal.

Whereas electron-withdrawing groups on the ring increase the sensitivity of the compound toward water (cf. π-indenyltitanium trichloride), substitution by alkyl groups has the reverse effect. Thus, $CpTiCl_3$ is destroyed by alkaline hydrolysis, but $\pi\text{-}(CH_3)_5C_5TiCl_3$ is converted only to $[\pi\text{-}(CH_3)_5C_5TiO(OH)]_n$ (R46).

As well as forming a spectrum of derivatives containing nitrogen–, phosphorus–, oxygen–, sulfur–, or selenium–metal bonds, the mono- and bis(cyclopentadienyl)titanium(IV) moieties can also bond to a range of metal and metalloid atoms. An area of growing importance is that involving σ-bonded alkyl and aryl groups and hydrides.

A. Mono(cyclopentadienyl)titanium(IV) Derivatives

1. Mono(cyclopentadienyl)titanium Trihalides

Since these halides have been the starting materials for much of the work on mono(cyclopentadienyl)titanium(IV) compounds, it is not surprising that

they constitute a large portion of the chemical literature on this type of compound. As with much of the chemistry of titanium the first reported syntheses
were in the patent literature.

a. METHODS OF PREPARATION

Method A: Halogenation. A patent was filed by Barkdoll and Lorenz in
1953, although not taken out until 1962 (B12), which included a method for
the synthesis of $CpTiBrCl_2$ and $CpTiCl_3$ by halogenation of Cp_2TiCl_2. The

$$Cp_2TiCl_2 + X_2 \longrightarrow CpTiXCl_2 + \text{"halogenated organic compounds"}$$

orange crystalline products were obtained in over 80% yield and were readily
purified by sublimation at $120°/0.2$ mm. The halogenated organic by-products
included trihalocyclopentene and pentahalopentane.

Although the halogenation reaction is useful for the preparation of $CpTiCl_3$,
the tribromide and triiodide can be obtained only in low yields by this
method (S50). They are best prepared by other means, namely, Method *D*.

Method B: Replacement of Halide by Cyclopentadienyl. The patent of
Barkdoll and Lorenz (B12) also reported the formation of $CpTiCl_3$ in low
yield from the following reaction. NaC_5H_5 (B77) and $Mg(C_5H_5)_2$ (S50, S51)

$$TiCl_4 + C_5H_5MgCl \longrightarrow CpTiCl_3 + MgCl_2$$

have also been utilized in this method of preparation.

Method C: Redistribution Reaction. This is probably the most convenient
method for the preparation of the trichloride (B77) since both starting
materials are commercially available.

$$Cp_2TiCl_2 + TiCl_4 \longrightarrow 2CpTiCl_3$$

The yield was low (13.6%) in the absence of solvent (B77), but when toluene
or xylene was employed, more than 80% crystalline $CpTiCl_3$ was obtained
(G19, G20).

Method D. Replacement of Alkoxo by Halide Using Acetyl Halide. This is
a more versatile method and was originally described by Nesmeyanov *et al.*
(N33, N34) for the preparation of $CpTiCl_3$, and later extended to cover all
halide derivatives (N37).

$$CpTi(OR)_3 + 3CH_3COX \longrightarrow CpTiX_3 + 3CH_3CO_2R$$

$$X = F, Cl, Br, I$$

$CpTiCl_3$ has also been prepared by Reid and Wailes (R37) by the reaction
of $TiCl_3$ with cyclopentadiene.

b. PHYSICAL DATA

The trihalides are all sensitive to hydrolysis (see Section A,2,d). Although the chloride and bromide are considered thermally stable, the iodide is not, and has been reported as light-sensitive (N37). The compounds, apart from the fluoride, show reasonable solubility in benzene, THF, toluene, and xylene.

Russian workers (D41, N28, N37, N41) have measured the infrared spectra in the region 400–4000 cm^{-1} for a number of mono(cyclopentadienyl)-titanium derivatives including halides and alkoxides and have used the data to study the nature of the Ti–Cp bond as a function of the ligands connected to the central atom. According to Fritz (F27), the infrared bands assigned to vibrations of the cyclopentadienyl ligands shift toward longer wavelengths with an increase in the ionic character of the metal–Cp bond.

Maslowsky and Nakamoto (M26) and Roshchupkina et al. (R51) have extended the studies of CpTiX$_3$ into the low-frequency infrared region and have coupled the investigations with Raman spectra. The Ti–Cp vibration has been placed in the region 430 ± 30 cm^{-1}, while the Ti–Cl and Ti–Br vibrations occur at about 400 and 350 cm^{-1}, respectively (R51). The near-infrared spectrum of CpTiCl$_3$ has been measured by Reid et al. (R35).

Owing to the equivalence of the five protons of π-bonded cyclopentadienyl groups, the PMR spectra of compounds containing such groups usually contain one peak (N28, N29, R42, S50), the position of which ($\delta 7.2$–6.0) is influenced by the other atoms or groups attached to the metal. Beachell and Butter (B21) have correlated ring proton chemical shifts with the composite electronegativity of the other ligands attached to titanium.

Italian workers have investigated the structures of crystalline mono-(cyclopentadienyl)titanium derivatives (A7, A8, G1, P12), including CpTiCl$_3$, which was shown to have the structure given in Fig. III-1.

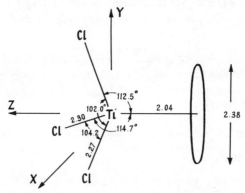

Fig. III-1.   Structure of mono(cyclopentadienyl)titanium trichloride (G1).

Electron diffraction patterns of $CpTiBr_3$ (R48) show the molecule to be a half-sandwich with equal Ti–C bond lengths and a "piano stool" configuration. It appears to be isostructural with $CpTiCl_3$ (G1) (see Table III-1).

The ^{79}Br and ^{81}Br NQR frequencies for the series Cp_nTiBr_{4-n} have been recorded (S23).

Thermochemical data for a range of organotitanium compounds (T17), including $CpTiCl_3$, indicated an average value of 74 kcal/mole for the dissociation energy of the Ti–Cp bond.

Table III-1
Structural Data of $CpTiBr_3$ and $CpTiCl_3$

Bond (or angle)	$CpTiBr_3$	$CpTiCl_3$
C–H	1.10 ± 0.03 Å	—
C–C	1.41 ± 0.02 Å	—
Ti–C	2.36 ± 0.01 Å	2.36 ± 0.05 Å
Ti–X	2.31 ± 0.0005 Å	2.27 ± 0.05 Å
X–Ti–X	101° average	103° average

The mono(cyclopentadienyl)titanium trihalides have been used mainly as starting materials for the preparation of other mono(cyclopentadienyl)-titanium derivatives. One recent communication of practical interest is the use of $CpTiCl_3$ as a catalyst in the preparation of $(C_5H_5)_2Mg$ under mild conditions (S1). The usual method for the preparation of $(C_5H_5)_2Mg$ from magnesium metal and cyclopentadiene in an inert atmosphere requires temperatures of 500°–600° (B11, R39). However, the presence of a catalytic amount of $CpTiCl_3$ in the reaction mixture of magnesium and cyclopentadiene in tetrahydrofuran facilitated the formation of $(C_5H_5)_2Mg \cdot THF$, even at 0°. Sublimation gave the unsolvated material in 60% yield.

C. SUBSTITUTED CYCLOPENTADIENYLTITANIUM TRIHALIDES

The first characterization of alkyl-substituted cyclopentadienyltitanium trihalides was by Nesmeyanov (N39), who synthesized $CH_3C_5H_4TiCl_3$ and $C_2H_5C_5H_4TiCl_3$ by the action of acetyl chloride on the corresponding trisalkoxo derivatives.

Pentamethylcyclopentadienyltitanium trichloride has been obtained from $TiCl_4$ and but-1-ene, but-2-ene, and other unsaturated hydrocarbons (R44, R46). The reactions were carried out with molar ratios of $TiCl_4$ to hydrocarbon of 1:2 to 1:5 at 300° and 30–60 atm pressure for 2 hr. The liquid reaction product was purified by fractional distillation in vacuo. The fraction boiling at 190°–220°/4–5 mm contained the organotitanium compound,

[π-$(CH_3)_5C_5$]TiCl$_3$ (20-30%). King and Bisnette (K27) prepared this compound from titanium tetrachloride and lithium pentamethylcyclopentadienide. Treatment with ethanol in the presence of triethylamine (N28) yielded a range of compounds of general formula [π-$(CH_3)_5C_5$]Ti(OC$_2$H$_5$)$_n$Cl$_{3-n}$ (n = 0-3).

The infrared data for π-$(CH_3)_5C_5$TiCl$_3$ in the range 700-4000 cm^{-1} have been reported (K27, N37) and, more recently, the low-frequency infrared spectra of both this compound and the bromine analog have been published (R51).

Samuel (S6) has prepared tetrahydroindenyltitanium trichloride by chlorination of (π-C$_9$H$_{11}$)$_2$TiCl$_2$ in CCl$_4$ (Method A). This compound can be hydrolyzed in two steps as follows.

$$(\pi\text{-C}_9\text{H}_{11})\text{TiCl}_3 \xrightarrow{\text{H}_2\text{O}} [(\pi\text{-C}_9\text{H}_{11})\text{TiCl}_2]_2\text{O} \xrightarrow{\text{H}_2\text{O}} [(\pi\text{-C}_9\text{H}_{11})\text{TiClO}]_4$$

Both products of hydrolysis are discussed in Section III,A,2,d.

Kaufman has filed a number of patents (K12, K14, N6) covering a range of compounds which fall into the category of mono(cyclopentadienyl)-titanium(IV) derivatives. They were prepared from the bis(cyclopentadienyl)-titanium dihalides by treatment with halogens in inert solvents and were formulated as (dihalocyclopentenyl)- or (tetrahalocyclopentyl)cyclopenta-dienyltitanium dihalides.

The preparation of "cyclopentadienylenetitanium dihalides," C$_5$H$_4$TiX$_2$ (X = F, Cl, or Br) by oxidation of Cp(π-C$_5$H$_3$Cl$_2$)TiX$_2$ or Cp(π-C$_5$HCl$_4$)TiX$_2$ with oxygen or peroxide-containing ether has been claimed in a patent (K13).

2. Oxygen-Bonded Compounds of Mono(cyclopentadienyl)titanium(IV)

The chemistry of oxygen-bonded derivatives of mono(cyclopentadienyl)-titanium(IV) was limited to patent literature for almost a decade prior to the mid-1960's, but since then it has received wider attention.

a. ALKOXIDES, PHENOXIDES, AND ORGANOSILOXIDES

Herman, in 1954, prepared bis(butoxo)cyclopentadienyltitanium bromide and bis(butoxo)indenyltitanium bromide (H24) from Ti(OC$_4$H$_9$)$_4$ and C$_5$H$_5$MgBr. Later, other patents reported the formation of the analogous chloride, CpTi(OC$_4$H$_9$)$_2$Cl (S55, T37), but it was not until 1960 that Gorsich published the first characterization of a compound containing an alkoxo group bonded to the CpTi^{IV} moiety (G20). This was CpTi(OCH$_3$)Cl$_2$ formed in 77% yield by dissolving CpTiCl$_3$ in hot methanol. This reaction is sur-

$$\text{CpTiCl}_3 + \text{ROH} \longrightarrow \text{CpTi(OR)Cl}_2 + \text{HCl}$$

Table III-2
Mono(cyclopentadienyl)titanium(IV) Trihalides

Compound	Method of preparation	Other data	Refs.
CpTiF$_3$	D	M.p. 130°–140°(dec) (N37); IR (N37, N41)	N37, N41
CpTiCl$_3$	A, B, C, D	M.p. 208°–211° (G20); MW, monomer (G20); X-ray powder (R37); IR (D41, F27, N41), near-IR (R35), far-IR (M25, R51, S6a); PMR (B20, N29); thermochemistry (T17); crystal structure (G1); NQR (N27a); ^{13}C NMR (N28a, N38a)	B12, B20, B21, D41, F27, G1 G20, G23, M25, N27, N27a, N28a, N29, N33, N38a, N41, R35, R37, R51, S6a, T17
CpTiCl$_2$Br	A	M.p. 102°–103° (G23)	B12, G23, G34
CpTiBr$_3$	B, D	M.p. 197°–199° (N37); IR (S50, S55, N37, N41), far-IR (R51, S6a); NQR (S23), electron diffraction (R48); ^{13}C NMR (N38a)	N37, N38a, N41, R48, R51, S6a, S23, S50, S51, S55
CpTiI$_3$	B, D	M.p. 148°–150° (N37); IR (S50, S51, N37, N41)	N37, N41, S50, S51
(π-CH$_3$C$_5$H$_4$)TiCl$_3$	C, D	M.p. 98°–99° (N39); IR (D41); PMR (N29); ^{13}C NMR (N38a)	D41, G23, N28, N29, N38a, N39
(π-CH$_3$C$_5$H$_4$)TiBr$_3$	A	—	G24
(π-C$_2$H$_5$C$_5$H$_4$)TiCl$_3$	D	B.p. 136° 1 mm (D41); IR (D41); PMR (N29)	D41, N28, N29, N39
[π-(CH$_3$)$_5$C$_5$]TiCl$_3$	B, D	M.p. 225°–227° (K27); IR (K27, N37), far-IR (R51); polarography (G35, S53); ^{13}C NMR (N38a)	G35, K27, N37, N38a, R44, R46, R51, S53
[π-(CH$_3$)$_5$C$_5$]TiBr$_3$	—	Far-IR; ^{13}C NMR (N38a)	N38a, R51
(π-C$_9$H$_{11}$)TiCl$_3$ (tetrahydroindenyl)	A	IR, PMR	S6

^{a} MW, Molecular weight.

prising in view of the ready alcoholysis of the cyclopentadienyl group in compounds of this type.

Nesmeyanov *et al.* (N34) characterized the first cyclopentadienyltitanium trisalkoxide in 1961. Following this, Russian research schools dominated this area of titanium chemistry for the next 8 years and developed many of the preparative methods listed below.

i. METHODS OF PREPARATION

Method A: Replacement of Alkoxo Groups by Cyclopentadienyl.

$$Ti(OR)_4 + C_5H_5MgBr \longrightarrow CpTi(OR)_2Br + Mg(OR)_2$$

Method B: Replacement of Halide by Cyclopentadienyl.

$$\overset{|}{\underset{|}{-Ti}}-Cl + NaC_5H_5 \longrightarrow \overset{|}{\underset{|}{-Ti}}-Cp + NaCl$$

$$\overset{|}{\underset{|}{-Ti}}-Cl + C_5H_6 + (C_2H_5)_3N \longrightarrow \overset{|}{\underset{|}{-Ti}}-Cp + (C_2H_5)_3N \cdot HCl$$

Method C: Elimination of Hydrogen Halide.

$$CpTiCl_3 + CH_3OH \longrightarrow CpTi(OCH_3)Cl_2 + HCl$$

Direct reaction of alcohols is of limited use because of the ready alcoholysis of the cyclopentadienyl group in this type of compound. Addition of triethylamine to remove the liberated HCl as amine hydrochloride has been found effective (N37, N39, R46).

A variation of this method involves the replacement of one cyclopentadienyl group, as well as one halide, in bis(cyclopentadienyl)titanium dihalides by 8-quinolinol.

$$Cp_2TiX_2 + 3C_9H_7NO \longrightarrow CpTi(C_9H_6NO)_2X + C_5H_6 + C_9H_7NO \cdot HX$$

Method D: Replacement of One Alkoxo Group by Another. This method is feasible only in those cases where alcoholysis does not occur, e.g., with some phenols (N40) or 8-quinolinol (N31).

$$CpTi(OC_2H_5)_3 + 2C_9H_7NO \longrightarrow CpTi(C_9H_6NO)_2OC_2H_5 + 2C_2H_5OH$$

The problem of alcoholysis can be overcome by using an ester instead of an alcohol. This technique has been used to prepare a series of phenoxides containing various substituents in the phenyl rings (N40).

$$CpTi(OR)_3 + 3CH_3CO_2C_6H_4R' \longrightarrow CpTi(OC_6H_4R')_3 + CH_3CO_2R$$

Method E: Replacement of Alkoxo Groups by Halide Using Acetyl Halide.
This is an excellent method for the conversion of alkoxides to halides (including fluorides), and by using stoichiometric amounts of acetyl halide, mixed derivatives can be obtained (N37). An alkyl acetate is eliminated as the only by-product.

$$CpTi(OC_3H_7)_3 + nCH_3COCl \longrightarrow CpTi(OC_3H_7)_{3-n}Cl_n + nCH_3CO_2C_3H_7$$

Method F: Redistribution Reaction. This is a very useful method and has enabled Nesmeyanov and co-workers (N37, N39) to prepare a large range of monocyclopentadienyl derivatives of titanium. The reaction is simple and

$$RTiX_3 + 2RTiZ_3 \longrightarrow 3RTiZ_2X$$
R = Cp or substituted Cp
X, Z, = halide, alkoxide, phenoxide, etc.

proceeds in most cases at ambient temperature, giving excellent yields of the product.

Method G: Displacement of Amine. Displacement of amine from an amidotitanium compound by an alcohol to give an alkoxide requires careful control of the stoichiometry to avoid alcoholysis (C17). The yield of the

$$CpTi[N(CH_3)_2]_3 + 3iso\text{-}C_3H_7OH \longrightarrow CpTi(O\text{-}iso\text{-}C_3H_7)_3 + 3(CH_3)_2NH$$

tris(isopropoxide) was 75%. Interestingly, displacement of amine by cyclopentadiene is also possible.

$$(CH_3)_2NTi(O\text{-}iso\text{-}C_3H_7)_3 + C_5H_6 \longrightarrow CpTi(O\text{-}iso\text{-}C_3H_7)_3 + (CH_3)_2NH$$

Method H: Pinacolate Formation. Mono(cyclopentadienyl)titanium(III) dihalides react with ketones to give pinacolates of $CpTi^{IV}$ (C53).

It was proposed that the yellow crystalline products were formed by an intramolecular "redox" process. Molecular weight and infrared data were in agreement with the proposed dimeric alkoxide-type structure for X = Cl or Br and R = C_6H_5; R' = H, CH_3, or R = R' = CH_3. When R = R' = C_6H_5, anomalous behavior was observed; hydrolysis or alcoholysis gave benzophenone instead of the expected dialcohol. Further, on solution in THF, $CpTi^{III}X_2 \cdot THF$ (X = Cl or Br) and free benzophenone were formed, whereas the other compounds dissolved without reaction. Decomposition to $CpTiX_2$ (X = Cl or Br) and benzophenone occurred in benzene or toluene at 40°, whereas all the remaining compounds dissolved in boiling toluene without decomposition. The authors (C53) contend that this anomalous behavior may be a result of the reversible coupling of the benzophenone as the intermediate ketyl can be stabilized by electron delocalization, which offers an opportunity to reverse the reaction under suitable conditions.

ii. PHYSICAL DATA.   The alkoxides and phenoxides of $CpTi^{IV}$ are generally colorless to yellow. Those which have been tested appear to be sensitive to water and alcohols. With HCl they form $CpTiCl_3$. Hydrolysis generally results in rupture of the Ti–Cp bond as well as the alkoxide bonds. From the rate of alcoholysis of a group of cyclopentadienyltitanium trisphenoxides, it was found that the nature of the substituents on the phenoxide exerts a noticeable influence on the character of the Ti–Cp bond (N40).

Nesmeyanov and co-workers made a comprehensive study of the nature of the Ti–Cp bond in these compounds using PMR (N28, N29, N40), infrared (D41), and mass spectrometric (N27) techniques. The high-resolution PMR spectra of a number of compounds of type $CpTi(OR)_3$ showed that the electron density at the C–H bonds of the cyclopentadienyl ring was very similar in all cases, the sensitivity of PMR spectroscopy being too low to enable any distinctions to be made (N38). A noticeable difference was observed in the ^{19}F NMR spectra of the compounds $RTi(O-4-FC_6H_4)_3$ [R = Cp, $\pi-CH_3C_5H_4$, or $\pi-(CH_3)_5C_5$] measured in chloroform. As the number of methyl groups on the cyclopentadienyl group increased, the ^{19}F NMR signal moved upfield (N38).

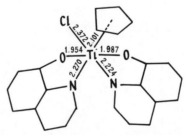

Fig. III-2.   Structure of mono(cyclopentadienyl)bis(8-quinolinolato)titanium chloride (M30).

The 8-quinolinolato ligand in $CpTi(C_9H_6NO)_2Cl$ has been reported (N32, S24) as being bidentate, giving the complex octahedral stereochemistry. This has now been confirmed by a three-dimensional X-ray diffraction study (M30, M31) (see Fig. III-2).

The only siloxide in this class of compound was isolated as a by-product from the reaction of sodium cyclopentadienide with $(CH_3)_3SiOTiCl_3$ in THF. The main product of the reaction was $Cp_2Ti(Cl)OSi(CH_3)_3$, and from the mother liquor remaining after its separation $CpTi[OSi(CH_3)_3]_3$ was obtained as a colorless mobile oil distilling at $92.5°–93.5°/1$ mm (N35).

b. CARBOXYLATES

i. METHODS OF PREPARATION. Tris(carboxylato) derivatives of mono-(cyclopentadienyl)titanium(IV) are normally prepared by the action of a metal salt of a carboxylic acid on $CpTiCl_3$, e.g.,

$$CpTiCl_3 + 3AgO_2CC_6H_5 \xrightarrow[\Delta]{benzene} CpTi(O_2CC_6H_5)_3 + 3AgCl$$

An unusual synthesis of the tris(benzoate) (in 72% yield) was reported by a Russian group (R18). The reaction was depicted as shown below, but the source of the proton on benzoic acid was not indicated.

$$CpTi(O_2CC_6H_5)_2 + (C_6H_5CO_2)_2 \xrightarrow{ether} CpTi(O_2CC_6H_5)_3 + C_6H_5CO_2H$$

ii. PHYSICAL DATA. The tris(carboxylates) are yellow to orange solids which are sparingly to moderately soluble in polar organic solvents. They are considerably less stable thermally than the bis(cyclopentadienyl) derivatives, and they hydrolyze readily with elimination of the carboxylic acid. With anhydrous HCl the carboxylato groups are removed with formation of the trichloride, $CpTiCl_3$.

c. β-DIKETONATES

Although there are many zirconium and hafnium β-diketonato derivatives of the type $CpM(diketonato)_2X$ (where X = halide) this is not the case for titanium where such compounds are rare and have only recently been prepared.

Sodium ferrocenoyltrifluoroacetonate was found to cause extensive decomposition of Cp_2TiCl_2, which was accompanied by the generation of an intense red-violet color. Chromatographic separation gave a dark violet material melting at 140°, with elemental analyses corresponding to CpTiCl-(ferrocenoyltrifluoroacetonate)_2 (R11). The UV-visible spectra data from this unstable compound have been discussed in relation to a range of similar metal complexes.

Table III-3

Mono(cyclopentadienyl)titanium(IV) Alkoxides, Phenoxides, and Siloxides

Compound	Method of preparation	Other data[a]	Refs.
CpTi(OCH₃)₃	C, D	M.p. 50°–52°; b.p. 88°/1.5 mm; MW, monomer(benzene)	N39
CpTi(OC₂H₅)₃	B	B.p. 54°–80° 0.03 mm (B77), 103°/2 mm (N34); MW, monomer(benzene) (N34); n_D^{20}1.5466 (B77); 1.5500 (N34); IR (D41); PMR (neat) δ6.22(Cp), 4.20(CH₂), 1.10(CH₃) (N29); ¹³C NMR (N28a, N38a)	B77, D41, N27–N29, N34, N38a
CpTi(O-n-C₃H₇)₃	B, D	B.p. 106°–107°/0.5–1 mm; n_D^{20} 1.5310;; MW, monomer(benzene)	N34, N39
CpTi(O-n-C₄H₉)₃	B	B.p. 124.5°–125.5°(dec)/0.5–1 mm; n_D^{20} 1.5224; d_4^{20} 1.0250; MW, monomer(benzene)	N35
CpTi(O-tert-C₄H₉)₃	A, C, D	B.p. 102°/1 mm; n_D^{20} 1.5065	N39
CpTi(OC₆H₁₃)₃	D	B.p. 177°–181°/1 mm; n_D^{20} 1.5082	N39
CpTiF(OC₂H₅)₂	F	B.p. 110°/1 mm; n_D^{20} 1.5550; IR	N37
CpTiCl(OC₂H₅)₂	F	B.p. 113°–114° 0.5 mm (N36), 109°–111°/1 mm (N39); n_D^{20} 1.5812 (N36), 1.5818 (N39); IR (D41); MS (N27); d_4^{20} 1.2253 (N36); PMR δ6.40(Cp), 4.43(CH₂), 1.15(CH₃) (N29); ¹³C NMR (N38a)	D41, N27, N29, N36, N38a, N39
CpTiBr(OC₂H₅)₂	F	B.p. 118°–119°/1 mm (N37); n_D^{20} 1.5986 (N37); NQR (S23); IR (N37)	N37, S23
CpTiI(OC₂H₅)₂	F	IR	N37
CpTiCl(O-iso-C₃H₇)₂	F	B.p. 132°–145°(dec)/1 mm (N34); 82°–84°/0.5 mm (N37); MW, monomer(benzene) (N34); d_4^{23} 1.1510 (N37); n_D^{20} 1.5562 (N37)	N34, N37
CpTiCl(OC₄H₉)₂	B	B.p. 145°–150°/2 mm	N5, S55
CpTiBr(OC₄H₉)₂	A	B.p. 136°–145°/0.8 mm	H24, N3
CpTiCl₂(OCH₃)	C	M.p. 93°–96°; IR	G20
CpTiF₂(OC₂H₅)	F	B.p. 120°–123°/1 mm; IR	N37
CpTiCl₂(OC₂H₅)	A, F	M.p. 48°–49° (N36); IR (D41); MS (N27); PMR δ6.79(Cp); 1.28(CH₃) (N29); ¹³C NMR (N28a)	D41, N27, N28a, N29, N36
CpTiBr₂(OC₂H₅)	F	M.p. 51°–55°(dec) (N37); NQR (S23); IR (N37)	N37, S23

Compound	Method	Properties	Ref.
$CpTiI_2(OC_2H_5)$	F	IR	N37
$CpTiCl_2(O\text{-}n\text{-}C_3H_7)$	B	B.p. 159°–161°/2 mm; MW, monomer(benzene)	N34
$CpTiCl_2(O\text{-}iso\text{-}C_3H_7)$	F	M.p. 113°–114.5°	N37
$\pi\text{-}CH_3C_5H_4Ti(OC_2H_5)_3$	A	B.p. 80°–81°/1 mm; n_D^{20} 1.5401 (N39); d_4^{20} 1.0780 (N39); IR (D41); PMR (N28)	D41, N28, N39
$\pi\text{-}CH_3C_5H_4TiCl(OC_2H_5)_2$	F	B.p. 143°–145°/2 mm (N39); n_D^{20} 1.5730 (N39); IR (D41); PMR (N28)	D41, N28, N39
$\pi\text{-}CH_3C_5H_4TiCl_2(OC_2H_5)$	F	IR (D41); PMR (N28)	D41, N28, N39
$\pi\text{-}C_2H_5C_5H_4Ti(OC_2H_5)_3$	A	B.p. 101°–102°/2 mm (N28); n_D^{20} 1.5359 (N28); d_4^{20} 1.0717 (N28); IR (D41)	D41, N28
$\pi\text{-}(CH_3)_5C_5Ti(OC_2H_5)_3$	F, D	B.p. 115°–117°/1 mm; n_D^{23} 1.5308; d_4^{23} 1.0479	N37
$\pi\text{-}(CH_3)_5C_5TiCl(OC_2H_5)_2$	F	B.p. 125°/2 mm; n_D^{20} 1.5986; IR	N37
$\pi\text{-}(CH_3)_5C_5TiCl_2(OC_2H_5)$	F	M.p. 46°–47°; b.p. 135°–139°/1 mm; IR	N37
$\pi\text{-}(Indenyl)Ti(OC_4H_9)_2$	A	—	N2
$\pi\text{-}(Indenyl)TiBr(OC_4H_9)_2$	A	—	H24, N3
$(CpTiCl_2OCH_2{}^-)_2C(CH_3)_2$	C	M.p. 157°–159°; IR	G20

Cp–Ti(X)(X)–O–C(R')(R)–O–Ti(X)(X)–Cp

	Method	Properties	Ref.
X = Cl, R = CH₃, R' = CH₃	C, H	IR	C53, C72
X = Br, R = CH₃, R' = CH₃	H	IR; MW, dimer(toluene)	C53
X = Cl, R = CH₃, R' = C₆H₅	H	IR; MW, dimer(toluene)	C53
X = Br, R = CH₃, R' = C₆H₅	H	IR; MW, dimer(toluene)	C53
X = Cl, R = C₆H₅, R' = H	H	IR; MW, dimer(toluene)	C53
X = Br, R = C₆H₅, R' = H	H	IR; MW, dimer(toluene)	C53
X = Cl, R = C₆H₅, R' = C₆H₅	H	IR	C53
X = Br, R = C₆H₅, R' = C₆H₅	H	IR	C53
$CpTi(OC_6H_5)_3$	D	M.p. 102°–104°; IR; PMR(THF) δ6.54(Cp), 6.79–7.53(C₆H₅); ¹³C NMR (N38a)	N38, N38a, N40
$CpTi(O\text{-}4\text{-}FC_6H_4)_3$	D	M.p. 86°–87°; IR; PMR(THF) δ6.49(Cp) 6.89, 7.02(C₆H₄); ¹⁹F NMR(THF) δ13.30 (CHCl₃) δ8.80	N38, N40

41

(continued)

Compound	Method of preparation	Other data[a]	Refs.
CpTi(O-4-ClC$_6$H$_4$)$_3$	D	M.p. 102°–103°; IR; PMR(CDCl$_3$) δ6.39(Cp), 677, 7.04(C$_6$H$_4$); ^{13}C NMR (N38a)	N38, N38a, N40
CpTi(O-4-NO$_2$C$_6$H$_4$)$_3$	D	M.p. 256°–257°; MW, monomer(nitrobenzene); IR; PMR(CDCl$_3$) δ6.54(Cp), 6.90, 8.17(C$_6$H$_4$)	N38, N40
CpTi(O-3-NO$_2$C$_6$H$_4$)$_3$	D	M.p. 140°–141°; IR	N38, N40
CpTi[O-4-(CH$_3$)$_2$NC$_6$H$_4$]$_3$	D	PMR(THF) δ6.34(Cp), 2.80, 2.91(CH$_3$), 6.74, 6.94(C$_6$H$_4$)	N38, N40
CpTi[O-3,5-(CH$_3$)$_2$C$_6$H$_3$]$_3$	D	M.p. 80°–81°; IR; PMR(THF) δ6.47(Cp), 2.28(CH$_3$), 6.64(C$_6$H$_3$)	N38, N40
CpTi[O-2,4-(CH$_3$)$_2$C$_6$H$_3$]$_3$	D	M.p. 64°–65°; ^{13}C NMR	N38a
CpTi(O-4-CH$_3$C$_6$H$_4$)$_3$	D	M.p. 145°–147°; IR; PMR(THF) δ6.48(Cp), 2.28(CH$_3$), 6.89, 7.10(C$_6$H$_4$); ^{13}C NMR (N38a)	N38, N38a, N40
CpTi(O-3-CH$_3$C$_6$H$_4$)$_3$	D	M.p. 65°–66°; IR; PMR(THF) δ6.50(Cp), 2.33(CH$_3$), 6.62–7.42(C$_6$H$_4$)	N38, N40
π-(CH$_3$)$_5$C$_5$Ti(OC$_6$H$_5$)$_3$	D	B.p. 214°–216°/1 mm; IR; PMR(THF) δ2.05(CH$_3$), 6.43–7.25(C$_6$H$_5$)	N38, N40
CpTiCl(OC$_6$H$_5$)$_2$	F	M.p. 39°–42°; IR; PMR(THF) δ6.92(Cp)	N38, N40
CpTiCl$_2$(OC$_6$H$_5$)	F	IR; PMR(THF) δ6.67(Cp)	N38, N39, N40
CpTiCl(O-4-NO$_2$C$_6$H$_4$)$_2$	F	M.p. 212°–214°(dec) IR; PMR(CDCl$_3$) δ6.63(Cp) 6.84, 8.08(C$_6$H$_4$)	N38, N40
π-CH$_3$C$_5$H$_4$Ti(O-4-FC$_6$H$_4$)$_3$	D	B.p. 226°–237°/1 mm; IR; PMR(THF) δ2.06(CH$_3$), 6.60–7.05(C$_6$H$_4$); ^{19}F NMR(THF) δ13.30 (CHCl$_3$), δ9.15	N38, N40
π-(CH$_3$)$_5$C$_5$Ti(O-4-FC$_6$H$_4$)$_3$	D	M.p. 65°–67°; b.p. 258°–262°/2 mm; IR; ^{19}F NMR(THF) δ13.30 (CHCl$_3$), δ10.10	N38, N40
CpTi(OC$_2$H$_5$)(C$_9$H$_6$NO)$_2$	C	M.p. 158°–159°; MW, monomer(benzene)	C17, N32, N36
CpTiCl(C$_9$H$_6$NO)$_2$	C	M.p. 231°–232° (N32); IR (N32, S24); crystal structure (M30, M31); conductance (C17); MW, monomer (C17); MS (C17); PMR (C17)	C17, M30, M31, N32, S24
CpTiBr(C$_9$H$_6$NO)$_2$	C	IR; PMR; MS; conductance	C17
CpTi[OSi(CH$_3$)$_3$]$_3$	B	B.p. 138°–139°/4 mm; n_D^{20} 1.4582; d_4^{20} 0.9436; MW, monomer	N35

[a] MS, Mass spectra.

Table III-4

Mono(cyclopentadienyl)titanium(IV) Carboxylates

Compound	Data	Refs.
$CpTi(O_2CCH_3)_3$	M.p. 115°–117° (N39); IR (D41); PMR (N29)	D41, N29, N38
$CpTi(O_2CCF_3)_3$	M.p. 179°–181° (dec) (N40); ^{19}F NMR (N40)	N39, N40
$CpTi(O_2CC_6H_5)_3$	IR	R18
$CpTi(O_2C-4-NO_2C_6H_4)_3$	150° (dec); IR	R19
$\pi\text{-}(CH_3)_5\overset{\cdot}{C}_5Ti(O_2CCH_3)_3$	M.p. 136°–138°; IR	N37
$\pi\text{-}CH_3C_5H_4Ti(O_2CCF_3)_3$	^{19}F NMR (N38)	N38, N40
$\pi\text{-}(CH_3)_5C_5Ti(O_2CCF_3)_3$	M.p. 243°–246° (dec) (N40); ^{19}F NMR (N38)	N38, N40
$CpTi(OC_2H_5)_2(O_2CCH_3)$	PMR (N39)	N39, N40

Treatment of Cp_2TiCl_2 with β-diketones in the molar ratio 1:2 in acetonitrile containing triethylamine gave the bis(β-diketonates), $CpTiCl(acac)_2$ and $CpTiCl(bzac)_2$, as orange solids in about 65% yield (F23). The compounds were nonconducting and were considered to be monomeric. Infrared and variable temperature PMR measurements indicated that the benzoylacetonate existed as the cis form in solution, while two isomeric forms of the acetylacetonate were evident.

d. POLYTITANOXANES

This section deals with compounds containing the mono(cyclopentadienyl)-titanium(IV) moiety connected to another such group, or to different groups, via oxygen bridges between the metal atoms.

Low molecular weight polytitanoxanes, composed exclusively of recurring units with a single π-cyclopentadienyl ligand, are accessible by several routes. The first polymer of this type was reported as $(CpTiClO)_4$ in a patent in 1958 (B77). In 1959 a yellow, poorly soluble oligomer was obtained (G20) by controlled hydrolysis of $CpTiCl_3$ in aqueous acetone at 56°. The reaction

$$2nCpTiCl_3 \xrightarrow{nH_2O} n(CpTiCl_2)_2O \xrightarrow{nH_2O} 2(CpTiClO)_n$$

apparently proceeded via $(CpTiCl_2)_2O$, as this compound could be identified under milder conditions. Recently, Coutts et al. (C55) have shown that polytitanoxanes, $(CpTiXO)_n$, can be formed from $CpTi^{III}X_2$ (X = Cl, Br, or I) by a redox reaction in acetone.

Earlier patent applications had claimed $(CpTiClO)_n$ to be either a trimer (B12) or a tetramer (B77, G22). Saunders and Spirer (S11), using chemical, light scattering, and X-ray single crystal methods, suggested that it was a tetramer, a finding confirmed by a complete crystal structure determination (S47, S48) which showed that $(CpTiClO)_4$ is indeed tetrameric with titanium tetrahedrally coordinated. Figure III-3 shows that the eight-membered Ti–O–Ti ring is far from planar. The arrangement of the four titanium atoms is such that the distances Ti-1–Ti-2 and Ti-3–Ti-4 are nearly equal, the angle between the planes containing Ti-1, Ti-2, Ti-3 and Ti-1, Ti-2, Ti-4 is 146°, as is the angle between Ti-3, Ti-4, Ti-1 and Ti-3, Ti-4, Ti-2. The Ti–O–Ti angle is 162° ± 3°. The chlorine and Cp groups are alternatively above and below the ring of metal and oxygen atoms.

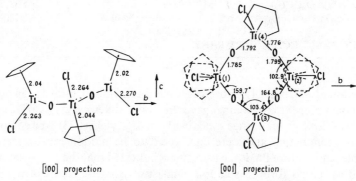

[100] projection   [00ī] projection

Fig. III-3. Structure of $(CpTiClO)_4$ (S47).

Samuel showed that tetrahydroindenyltitanium trichloride, like $CpTiCl_3$, hydrolyzed in two steps giving oxygen-bridged derivatives (S6). The inter-

$$\pi\text{-}C_9H_{11}TiCl_3 \xrightarrow{H_2O} (\pi\text{-}C_9H_{11}TiCl_2)_2O \xrightarrow{H_2O} (\pi\text{-}C_9H_{11}TiClO)_4$$

mediate compound was not isolated, but the tetrameric end product crystallized as bright yellow needles.

When $\pi\text{-}(CH_3)_5C_5TiCl_3$ was hydrolyzed in aqueous alkaline solutions, only the halogen atoms were replaced giving yellow solid polymeric compounds, $[\pi\text{-}(CH_3)_5C_5TiO(OH)]_n$ ($n = 4$–6) (R45, R46). These are soluble in organic solvents, particularly THF. Partial hydrolysis leads to the chlorine-containing polymer, $[\pi\text{-}(CH_3)_5C_5TiClO]_n$.

Hydrolysis of $CpTiCl_3$ in methanol with dilute sulfuric acid gave a polytitanoxane of average composition, $[CpTiO(SO_4)_{1/2}\cdot H_2O]_n$. This insoluble, cross-linked material was examined by chemical analysis, ionic decomposition in sulfuric acid, and electron micrography. It was considered to be built up of $(CpTiO)_n^{n+}$ chains joined by sulfate groups (S11). The electron microscope photographs show that the complex forms laminae resembling asbestos and a similar type of structure has been suggested (see Fig. III-4).

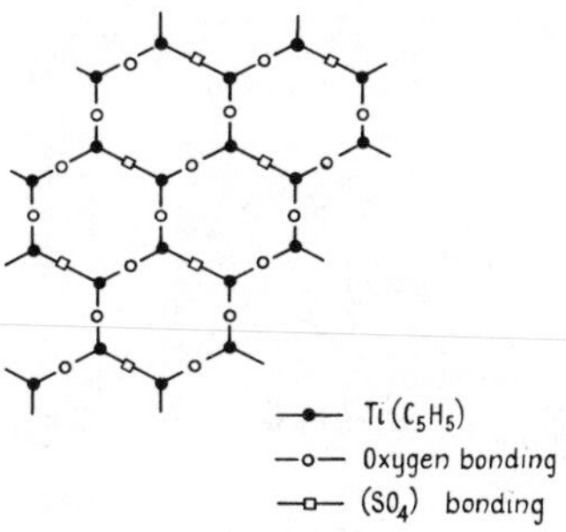

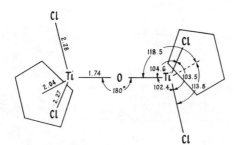

— Ti (C$_5$H$_5$)
—o— Oxygen bonding
—□— (SO$_4$) bonding

Fig. III-4.   Structure of [CpTiO(SO$_4$)$_{1/2}$·H$_2$O]$_4$ (S11).

The intermediate (CpTiCl$_2$)$_2$O in the hydrolysis of CpTiCl$_3$ to (CpTiClO)$_4$ has aroused considerable interest from a structural point of view. The complete X-ray study (A7, A8) confirms the presence of a colinear Ti–O–Ti bond, the short Ti–O bond length (1.74 Å) suggesting some double bond character. As seen in Fig. III-5, the ligands are tetrahedrally coordinated to the titanium atom.

Fig. III-5.   Structure of (CpTiCl$_2$)$_2$O (S11).

Another bridged species reported (N35) to contain the mono(cyclopentadienyl)titanium(IV) entity was formed by the hydrolysis reaction.

$$2Cp_2Ti(O_2CCH_3)_2 + H_2O \longrightarrow [CpTi(O_2CCH_3)_2]_2O + 2C_5H_6$$

3. Sulfur-Bonded Compounds of Mono(cyclopentadienyl)titanium(IV)

Compounds of this type were first mentioned by Giddings (G8) in a 1961 patent application in which CpTi(SCH$_3$)$_2$Cl, CpTi(SCH$_3$)$_3$, and CpTi-(SC$_{10}$H$_7$)Cl$_2$ were said to form by the reaction of the appropriate sodium thioalkoxide with CpTiCl$_3$ in toluene. Köpf and Block (K43) have used

Table III-5
Monocyclopentadienyl Polytitanoxanes

Compound	Highest yield (%)	Other data	Refs.
(CpTiClO)$_4$	Quantitative	M.p. 260° (S11); IR (C55, S11); X-ray (S11, S48, S49); MW, tetramer (B77, G22, S11), trimer (B12); PMR (C55)	A8, B12, B77, C43, C55, G20, G22, M47, M48 S11, S48, S49
(CpTiBrO)$_n$	30	IR; PMR	C55
(CpTiIO)$_n$	30	IR; PMR	C55
[π-(CH$_3$)$_5$C$_5$TiO(OH)]$_n$ (n = 4–6)	82–97	—	R45, R46
[CpTiO(SO$_4$)$_{1/2}$ · H$_2$O]$_n$	—	IR; electron micrography	S11
(CpTiCl$_2$)$_2$O	91	M.p. 149°–151° (G20, G22); MW, monomer; X-ray (A8, C43)	A7, A8, C43, G20, G22, P12
(CpTiBr$_2$)$_2$O	—	MW, monomer	G22
[CpTi(O$_2$CCH$_3$)$_2$]$_2$O	97	PMR	N35
(π-C$_9$H$_{11}$TiCl$_2$)$_2$O	Not isolated	IR; MW, tetramer; PMR	S6
(π-C$_9$H$_{11}$TiClO)$_4$	—		S6

the reactions to form phenylthiolato products. When $n = 3$ and the reaction

$$CpTiCl_3 + nHSC_6H_5 + n(C_2H_5)_3N \longrightarrow CpTi(SC_6H_5)_nCl_{3-n} + n(C_2H_5)_3N \cdot HCl$$

is carried out in ether at $-25°$, a deep red solution is obtained from which an unstable orange crystalline solid, believed to be $CpTi(SC_6H_5)_3$, precipitates at $-78°$. The addition of the reactants in an equimolar ratio yields 35% stable $CpTi(SC_6H_5)Cl_2$.

Coutts and Wailes (C72) have reacted $CpTiCl_2$ with dialkyl (diaryl) disulfides and a range of thiuram disulfides to form organotitanium(IV) derivatives by oxidative-addition reactions

$$2CpTi^{III}Cl_2 + RS-SR \longrightarrow 2CpTi^{IV}Cl_2(SR)$$

$$2CpTi^{III}Cl_2 + R_2NCS-SCNR_2 \longrightarrow 2CpTi^{IV}Cl_2(S_2CNR_2)$$
$$\qquad\qquad \overset{\|}{S} \quad\ \overset{\|}{S}$$

The yellow-orange thiolato derivatives are four-coordinate monomeric compounds and extremely sensitive to hydrolysis. The orange-red dithiocarbamato derivatives on the other hand are comparatively stable to hydrolysis, unlike most monocyclopentadienyltitanium(IV) dihalo derivatives. Molecular weight determinations coupled with infrared and conductivity data showed the compounds to be nonconducting monomers containing bidentate dithiocarbamato groups, suggesting that the titanium(IV) atom is in the rare five-coordinate environment. PMR studies indicate isomers are present in the reaction products, but by heating and cooling solutions of the compounds only the more stable isomer is isolated. The suggested stereochemistry is trigonal bipyramidal and not square pyramidal.

Anionic species containing the monocyclopentadienyltitanium(IV) entity bonded through sulfur were formed from Cp_2TiCl_2 with 2 moles of the maleonitrile dithiolate, $Na_2S_2C_2(CN)_2$ (L18). The complex $[(C_2H_5)_4N]^+-\{CpTi[S_2C_2(CN)_2]_2\}^-$ was considered to have five-coordinate titanium atoms. McCleverty et al. (M1) examined some of the species reported (L18)

using voltammetry in dichloromethane solution. The existence of $\{CpTi-[S_2C_2(CN)_2]_2\}^-$ may be in some doubt (M1), since the diffusion current of the reduction wave in the voltammogram of the monoanion is more than twice that of the corresponding oxidation wave.

The displacement of Cp rings from Cp_2TiCl_2 by the dithiolato ligands, $[S_2C_6Cl_4]^{2-}$ or $[S_2CN(CH_3)_2]^-$, was studied by James and McCleverty (J6). Treatment of Cp_2TiCl_2 with 1 mole of $[S_2C_6Cl_4]^{2-}$ gave the green Cp_2TiS_2-C_6Cl_4, which was converted to the magneta ionic complex, $[CpTi(S_2C_6Cl_4)_2]^-$ with a further mole of reagent. Both Cp ligands can be replaced if the reaction is carried out under reflux in acetone–ethanol.

4. Nitrogen-Bonded Compounds of Mono(cyclopentadienyl)titanium(IV)

The only monocyclopentadienyltitanium(IV) pseudohalide derivative reported is $CpTi(NCS)_3$. This red compound was considered to be bonded via the nitrogen atom of the ambidentate ligand and so is reported in this section (K43).

Tetrakis(dimethylamido)titanium reacted exothermally when treated with an excess of cyclopentadiene, but only one amido group was replaced, the product being cyclopentadienyltris(dimethylamido)titanium (C15). In a

$$Ti(NR_2)_4 + C_5H_6 \text{ (excess)} \longrightarrow CpTi(NR_2)_3 + R_2NH$$

similar way the ethyl homolog was obtained. The methyl derivative is a low-melting red solid, whereas the ethyl compound is a reddish brown, viscous liquid. With methylcyclopentadiene the low-melting red solid, π-$CH_3C_5H_4Ti[N(CH_3)_2]_3$, was formed.

With monosubstituted acetylenes, ethynyltitanum compounds result (J7), e.g.,

$$\pi\text{-}CH_3C_5H_4Ti[N(CH_3)_2]_3 + HC{\equiv}CC_6H_5 \longrightarrow$$
$$\pi\text{-}CH_3C_5H_4Ti(C{\equiv}CC_6H_5)[N(CH_3)_2]_2 + (CH_3)_2NH$$

Although the product from this reaction, a dark brown viscous liquid, was not analytically pure, its presence was substantiated by infrared, PMR, and mass spectroscopy.

A series of tris(dialkylamido) organotitanium compounds of general formula, $(R_2N)_3TiR'$, have been prepared by Bürger and Neese (B97) (see Chapter II). One of the compounds (R = CH_3, R' = Cp) falls into this category of nitrogen-bonded derivatives of mono(cyclopentadienyl)titanium-(IV).

Bradley and Kasenally (B62) treated $CpTi[N(CH_3)_2]_3$ with metal carbonyls to form derivatives of mono(cyclopentadienyl)titanium(IV) containing dissimilar metals. When the reagents were mixed in cyclohexane under UV

$$CpTi[N(CH_3)_2]_3 + M(CO)_6 \longrightarrow CpTi[N(CH_3)_2]_3M(CO)_3 + 3CO$$
$$M = Cr, Mo, \text{ and } W$$

Table III-6
Mono(cyclopentadienyl)titanium(IV) Compounds with Sulfur-Bonded Ligands

Compound	Color	Other data	Refs.
CpTi(SCH$_3$)$_2$Cl	—	—	G8
CpTi(SCH$_3$)$_3$	—	—	G8
CpTi(SCH$_3$)Cl$_2$	Yellow-orange	M.p. 143°–146°; PMR (CDCl$_3$) δ6.80 (Cp), 3.71(s) (CH$_3$)	C72
CpTi(SC$_{10}$H$_7$)Cl$_2$	—	—	G8
CpTi(SC$_6$H$_5$)$_3$	Orange	Unstable	K43
CpTi(SC$_6$H$_5$)Cl$_2$	Orange	M.p. 104°–105° (K43); 120°–125° (C72); sublimes 50°–110° 1/mm; MW, monomer (benzene) (K43); PMR (CDCl$_3$) δ6.85(s) (Cp), 7.35(m) (C$_6$H$_5$) (C72)	C72, K43
CpTi[S$_2$CN(CH$_3$)$_2$]Cl$_2$	Orange-red	M.p. 252°–256°; IR; MW, monomer (CHCl$_3$); nonconducting (CH$_3$CN); PMR (CDCl$_3$) δ6.82(s) (Cp), 3.40(s) (CH$_3$)	C72
CpTi[S$_2$CN(C$_2$H$_5$)$_2$]Cl$_2$	Red	M.p. 181°–184°; IR; MW, monomer (benzene); nonconducting (CH$_3$NO$_2$); PMR (CDCl$_3$) δ6.85(s) (Cp), 3.82(q) (CH$_2$), 1.31(t) (CH$_3$)	C72
CpTi[S$_2$CN(n-C$_4$H$_9$)$_2$]Cl$_2$	Orange-red	M.p. 76°–78°; IR; MW, monomer (benzene); PMR (CDCl$_3$) δ6.83(s) (Cp), 3.75(t), 1.55(m), 0.95(m) (n-C$_4$H$_9$)	C72
CpTi[S$_2$CN(CH$_2$)$_5$]Cl$_2$	Orange	M.p. ~170°(dec); IR; nonconducting (CH$_3$NO$_2$); PMR (CDCl$_3$) δ6.85(s) (Cp), 3.9(m), 1.74(m) (CH$_2$)	C72
[(C$_2$H$_5$)$_4$N]$^+$\{CpTi[S$_2$C$_2$(CN)$_2$]$_2$\}$^-$	Black	M.p. 204°–205°; cond. 1:1 (acetone); IR; PMR (acetone) δ6.31 (Cp), 1.40 (CH$_3$), voltammetry (M1)	L18, M1
[(C$_2$H$_5$)$_4$N]$^+$[CpTi(S$_2$C$_6$Cl$_4$)$_2$]$^-$	Dark red	M.p. 208°–212°; UV-visible; conductance 1:1 (acetone)	J6
[(C$_6$H$_5$)$_4$P]$^+$[CpTi(S$_2$C$_6$Cl$_4$)$_2$]$^-$	Red-black	PMR(CDCl$_3$) δ7.6 (C$_6$H$_5$), 6.20 (Cp)	J6

s = Singlet, t = triplet, m = multiplet, q = quartet.

49

irradiation, black microcrystalline products resulted. It was thought that the molecular structure involved three bridging –N(CH$_3$)$_2$ groups between titanium and the other metal which also contained three terminal carbonyl groups. The infrared data agreed with this structure, but the PMR spectra

$$\text{Cp—Ti} \overset{\displaystyle N(CH_3)_2}{\underset{\displaystyle N(CH_3)_2}{-N(CH_3)_2-}} M \overset{\displaystyle CO}{\underset{\displaystyle CO}{-CO}}$$

showed that the amido groups were in different magnetic environments, a result which could not be reconciled with the proposed structure.

Cyclopentadienyltitanium trichloride is soluble in liquid ammonia and methylamine at $-36°$ (A13). Tensimetric studies in these systems revealed the presence of 1 mole of NH$_4$Cl (or CH$_3$NH$_2 \cdot$HCl) per mole of CpTiCl$_3$, respectively, so that an average of one Ti–Cl bond is broken under these conditions.

5. Alkyl and Aryl Compounds of Mono(cyclopentadienyl)-titanium(IV)

The first account of a σ-carbon-bonded derivative was from Giannini and Cesca (G4) who, in 1960, isolated CpTi(CH$_3$)$_3$ in 40% yield, from the reaction between CpTiCl$_3$ and methyllithium. Clauss and Bestian (C38) obtained CpTi(CH$_3$)$_3$ by the same route in high yield (77%) as a lemon-colored crystalline solid, monomeric in benzene (G4). In air at room temperature the compound is pyrophoric (C38). Unlike Cp$_2$Ti(C$_6$H$_5$)$_2$ and Cp$_2$Ti(CH$_3$)$_2$, CpTi(CH$_3$)$_3$ decomposes in water or alcohols liberating methane.

Razuvaev and co-workers (R32) synthesized CpTi(C$_6$H$_5$)$_3$ from CpTiCl$_3$ and phenyllithium at $-78°$. The orange compound is stable in solution at $-70°$, but not at room temperature when decomposition occurs to give benzene, biphenyl, and titanium(II) compounds (R30).

B. Bis(cyclopentadienyl)titanium(IV) Derivatives

1. Bis(cyclopentadienyl)titanium(IV) Halides

The importance of this group of compounds and in particular Cp$_2$TiCl$_2$ is evident from the work on titanium described in this volume. Bis(cyclopentadienyl)titanium compounds constitute by far the majority of organotitanium derivatives so far prepared, and these have generally been prepared from the halides described in this section. It must be emphasized that mention

Table III-7
Nitrogen-Bonded Derivatives of Mono(cyclopentadienyl)titanium(IV)

Compound	Physical data	Refs.
$CpTi(NCS)_3$	M.p. 162°–165°; IR; MW, monomer(benzene); sublimes 150°/10^{-3} mm	K43
$CpTi[N(CH_3)_2]_3$	B.p. 95°/0.01 mm; 140°(dec), (B97); PMR, δ6.14(Cp) 3.27(CH$_3$) (B97)	B97, C15
$CpTi[N(C_2H_5)_2]_3$	—	C15
$\pi\text{-}CH_3C_5H_4Ti[N(CH_3)_2]_3$		C15, J7
$CpTi[N(CH_3)_2]_3Cr(CO)_3$	IR; MW, monomer(benzene); PMR(C$_6$D$_6$) δ5.74(Cp), 2.36, 2.71, 2.76, 3.0(CH$_3$)	B62
$CpTi[N(CH_3)_2]_3Mo(CO)_3$	IR; MW, monomer(benzene); PMR(C$_6$D$_6$) δ5.77(Cp), 2.32, 2.76, 2.82, 3.03(CH$_3$)	B62
$CpTi[N(CH_3)_2]_3W(CO)_3$	IR; MW, monomer(benzene); PMR(C$_6$D$_6$) δ5.73(Cp), 2.20, 2.80, 2.86, 2.89(CH$_3$)	B62
$\pi\text{-}CH_3C_5H_4Ti(C\equiv CC_6H_5)[N(CH_3)_2]_2$	IR; MS; PMR(C$_6$D$_6$) δ7.25(C$_6$H$_5$), 5.82(Cp), 3.1(N–CH$_3$), 2.27(C–CH$_3$)	J7

of bis(cyclopentadienyl)titanium dihalides in the literature is so extensive that only those contributions which are constructive to this specific class of compounds will be discussed.

a. METHODS OF PREPARATION

Bis(cyclopentadienyl)titanium dihalides, Cp_2TiX_2, were among the first organotitanium compounds to be isolated. Wilkinson *et al.* (W25) prepared the first member of this series (where X = Br) in 1953.

Method A: Salt Elimination.

$$TiCl_4 + C_5H_5MgBr \text{ (excess)} \xrightarrow{\text{toluene}} Cp_2TiBr_2 + MgBrCl$$

LiC_5H_5 (S72), TlC_5H_5 (H41), and $Mg(C_5H_5)_2$ (S50) have also been used in this reaction.

Method B: Ligand Exchange. Cp_2TiF_2 has been prepared by a ligand exchange reaction employing Cp_2TiBr_2 in 12 N hydrofluoric acid.

$$Cp_2TiBr_2 + 2HF \longrightarrow Cp_2TiF_2 + 2HBr$$

Samuel (S5) observed that in an aquous medium Cp_2TiCl_2 would undergo ligand exchange reactions with alkali metal salts. This is a more convenient method for preparing Cp_2TiF_2 as it avoids the use of the HF solutions (W24).

$$Cp_2TiCl_2 + 2K(Na)X \longrightarrow Cp_2TiX_2 + 2K(Na)Cl$$
$$X = OCN, NCS, F$$

Method C: BCl_3 Elimination. The best synthesis of bis(cyclopentadienyl)-titanium dihalides (where X = Br or I) is that involving boron trichloride elimination (D35). The reaction, normally carried out in methylene dichloride, offers a simple route to the dibromide and diiodide derivatives. Moreover, it is essentially quantitative after a few minutes at ambient temperature and pressure, and the products are easily isolated since the only by-product, BCl_3, is readily removed by pumping (BCl_3, b.p. 12°/760 mm).

$$3Cp_2TiCl_2 + 2BX_3 \xrightarrow{CH_2Cl_2} 3Cp_2TiX_2 + 2BCl_3$$

b. PHYSICAL DATA

Possibly the most useful physical technique for the characterization of these compounds has been infrared spectroscopy. The normal vibrations in the region 650–4000 cm^{-1} of a π-bonded cylopentadienyl ligand when attached to a metal atom are found at characteristic frequencies which have been considered in Chapter I and will not be discussed here. The range of infrared spectral data has been extended above 4000 cm^{-1} by Reid *et al.* (R35), who measured the near-infrared combination reflection spectra of a number of cyclopentadienylmetal compounds including that of Cp_2TiCl_2. More recently, the low-frequency infrared region and Raman spectra of Cp_2TiX_2 compounds have been studied (D36, F27, M25, R51) (see Table IV-6).

Like infrared spectroscopy, interest in polarographic reductions has persisted from the earliest work on organotitanium derivatives up to the present time. This continual interest stems presumably from the use of derivatives of cyclopentadienyltitanium in homogeneous catalysis where the oxidation–reduction reactions are thought to be important.

Although the polarographic reduction of the Cp_2TiX_2 derivatives has been the subject of many papers (D17, G39–G41, H38, H39, K54, V1, W24), only two compounds (X = Cl or Br) had been studied until recent times. Gubin and Smirnova have carried out an electrochemical investigation on the Cp_2TiX_2 compounds (X = F, Cl, Br, and I) (G35, G36, S53) in both protic and aprotic solvents and have shown that the polarographic reduction proceeds with successive additions of one electron to the Ti–X bond.

Some of the first reported electronic spectral data was due to kineticists who utilized the variation of peak positions to monitor their reaction progress (J11, J12, L2). The electronic spectra of some bis(cyclopentadienyl)metal compounds, including Cp_2TiX_2 (X = Cl, Br, and I), have been studied by Chien (C25).

Dillard and Kiser (D26) measured the mass spectra of Cp_2TiCl_2 and Cp_2ZrCl_2 and examined the energetic processes involved in ion formation. The unimolecular decomposition processes leading to the formation of the most abundant ions in the mass spectrum were

$$Cp_2TiCl_2^+ \longrightarrow CpTiCl_2^+ \longrightarrow CpTiCl^+ \longrightarrow C_3H_3TiCl^+$$
$$\searrow Cp_2TiCl^+ \qquad\qquad C_5H_5^+ \longrightarrow C_3H_3^+$$

For other fragment ions, removal and fragmentation of the Cp group was favored, and ions retaining the chlorine substituent were the most abundant in a given series. Dillard (D25) attempted to establish the identity of the negative ions observed in the mass spectrum and to measure the energy associated with electron capture processes.

Hunt *et al.* (H42) have measured the methane chemical ionization mass spectra of a range of organometallic compounds including Cp_2MCl_2 (M = Ti, Zr, Hf).

Proton magnetic resonance is of little diagnostic use with Cp_2TiX_2 compounds, apart from monitoring the purity of the samples. The Cp resonance is always a sharp singlet at $\delta 6.5$–7.0 in THF with the exception of the fluoride, Cp_2TiF_2, which shows (D35) a triplet due to ^{19}F–^1H interaction ($J = 1.7$ Hz).

Giddings and Best (G14) were the first to offer evidence in support of the expected "angular" sandwich structure of the bis(cyclopentadienyl)titanium dihalides. They reported a high value (6.3 D) for the dipole moment of Cp_2TiCl_2 in benzene. This was taken to indicate that the structure is approaching the tetrahedral configuration depicted in Fig. III-6.

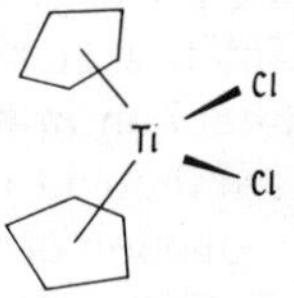

Fig. III-6. Proposed structure of Cp_2TiCl_2.

Electron diffraction studies (A5, R47) on the structure of the Cp_2TiCl_2 molecule showed that the compound was indeed of the structural type proposed (G14), and indicated that the Cp–Ti–Cp angle is $121° \pm 3°$, the Cl–Ti–Cl angle is $100° \pm 1°$, and the bond lengths are Ti–C(Cp), 2.38 Å and Ti–Cl, 2.24 Å.

A recent X-ray study of Cp_2TiCl_2 by Tkachev and Atovmyan (T38) has given the values: Ti–Cl, 2.36 Å; Ti–C, 2.43 Å; C–C, 1.44 Å; angle Cl–Ti–Cl, 95.2°; angle Cp–Ti–Cp, 129°. These values are at variance with those from electron diffraction above, but compare favorably with the corresponding values for the 1,1′-trimethylene compound shown in Fig. III-7 (D9).

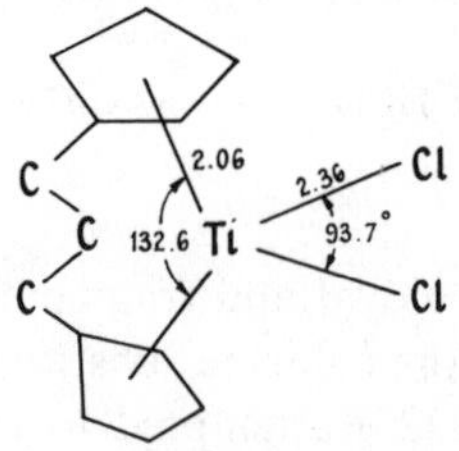

Fig. III-7. Structure of $(CH_2)_3(C_5H_4)_2TiCl_2$ (D9).

Needless to say, such a common reagent as Cp_2TiCl_2 is bound to be examined at various times by a range of techniques. Thus, we find investigations concerning the paper chromatographic separation of Cp_2TiCl_2 from $CpTiCl_3$ or $Cp_2Ti(C_6H_5)_2$ (K55, S67), the photocolorimetric determination of Cp_2TiCl_2 (S22), para-hydrogen conversion in solution (S18, S19) by Cp_2TiCl_2, conductivity measurements (S64), and atomic absorption spectros-

Table III-8
Bis(cyclopentadienyl)titanium(IV) Halides

Compound	Color	Method of preparation	Other data	Refs.[a]
Cp$_2$TiF$_2$	Yellow	B	M.p.(dec); IR (D35), far-IR (D36, M26, R51); polarographic reduction (G35, and refs. therein); MS (D35); PMR(THF) δ6.51 (D35); ^{13}C NMR (N38a); ^{19}F NMR (N38a)	D35, D36, G35, M26, N38a, R51, W24
Cp$_2$TiCl$_2$	Red	A, B	M.p. 289°–291°; IR (D35), near IR (R35), far-IR (D36, M26, R51); polarographic reduction (G35, and refs. therein); kinetics (J12, and refs. therein); UV-visible (C25); MS (D26, D35); dipole moment (G16); electron diffraction (R47); conductance (S64); PMR (CCl$_4$) δ6.60 (B21), (THF) δ6.70 (D35), (CDCl$_3$) δ6.55, (acetone) δ6.61 (C72); hydrolysis (I2); ^{13}C NMR (N38a); NQR (N27a)	B21, C25, C72, D26, D35, D36, G14, G35, I2, J12, M26, N27a, N38a, R35, R47, R51, S64, W24
Cp$_2$TiBr$_2$	Dark red	A, B, C	M.p. 309°–313°; IR (D35), far-IR (D36, M26, R51); polarographic reduction (G35, and refs. therein); kinetics (J12, and refs. therein); UV-visible (C25); BrNQR (S23); PMR (CCl$_4$) δ6.67 (B21), (THF) δ6.83 (D35)	B21, C25, D35, D36, G35, J12, M26, R51, S23, S50, W24, W25
Cp$_2$TiI$_2$	Black	A, B, C	M.p. 315°–317°; IR (D35), far-IR (D36, M26, R51); polarographic reduction (G35, and refs. therein); UV-visible (C25); MS (D35); PMR (CCl$_4$) δ6.76 (B21), (THF) δ7.0 (D35)	B21, C25, D35, D36, F21, G35, K44, M26, R51, S50
Cp$_2$Ti(Br)Cl	—	—	PMR	D35

[a] Only the reference which is considered the most informative or to contain most relevant references has been given in this Table.

55

copy (S10). Semin *et al.* (S23) have recorded the ^{79}Br and ^{81}Br nuclear quadrupole resonance spectra of some organotitanium bromides, including Cp_2TiBr_2. Tel'noi *et al.* (T17) studied the thermochemistry of several organotitanium compounds including Cp_2TiCl_2.

C. RING-SUBSTITUTED CYCLOPENTADIENYLTITANIUM(IV) DIHALIDES

The simplest ring-substituted compound known is the red deuterated species, $(\pi\text{-}C_5D_5)_2TiCl_2$ (D30), which was prepared from NaC_5D_5 and $TiCl_4$ in THF. An alternative preparation is from hydrogenated (methylallyl)-dicyclopentadienyltitanium(III) which exchanges its ring hydrogen atoms with deuterium gas. The deuterated complex so obtained is then oxidized with hydrogen chloride and air (M17).

The first alkyl-substituted cyclopentadienyltitanium(IV) dihalide reported was the methylcyclopentadienyl derivative, $(\pi\text{-}CH_3C_5H_4)_2TiCl_2$, prepared by Reynolds and Wilkinson (R41) in 1958 from $TiCl_4$ and $NaC_5H_4CH_3$ in THF. To obtain mono- and bis(substituted-cyclopentadienyl)titanium dihalides, Sullivan and Little (S71) utilized the reactions of fulvenes with $LiAlH_4$, LiC_6H_5, and $Li\text{-}n\text{-}C_4H_9$ to produce alkylcyclopentadienide salts which were then treated with $TiCl_4$ to produce bis(substituted-cyclopentadienyl)titanium dichlorides, or with $CpTiCl_3$ to produce (substituted-cyclopentadienyl)-cyclopentadienyltitanium dichloride derivatives.

Katz and Acton (K11) have obtained compounds which contain two cyclopentadienyl rings linked by methylene bridges by reacting the bis-(cyclopentadienyl)methane anion with $TiCl_4$ to give a 30% yield of the monomeric bridged dichloride. The product with lithium naphthalene in THF

fixes nitrogen, and with diethylaluminum chloride polymerizes ethylene. Hillman and Weiss (H31) prepared the trimethylene-bridged analog of the above compound by a similar reaction. Its molecular structure is shown in Fig. III-7. More recently the neutron diffraction structure of this same compound has been studied to obtain accurate information for those structural parameters involving hydrogen atoms (E5a).

Bercaw *et al.* (B43) found that from $TiCl_4$ with a dimethoxyethane solution of $(CH_3)_5C_5Na$ in a 1:4 molar ratio, a mixture of the red-orange $[\pi\text{-}(CH_3)_5C_5]TiCl_3$ and purple-brown $[\pi\text{-}(CH_3)_5C_5]_2TiCl_2$ (in 14% yield) was obtained.

Higher yields (up to 40%) of $[\pi\text{-}(CH_3)_5C_5]_2TiCl_2$ were obtained when $TiCl_3$ was first reacted with saturated THF solutions of $(CH_3)_5C_5Na$ to form $[\pi\text{-}(CH_3)_5C_5]_2TiCl$, which was then oxidized with HCl.

Bis(1-methyl-3-phenylcyclopentadienyl)- and bis(1-methyl-3-p-bromo-phenylcyclopentadienyl)titanium dichlorides have been prepared (H4) from the lithium salts of the corresponding substituted cyclopentadienes and $TiCl_4$. Attempts to prepare the related p-methoxy and p-methyl derivatives were unsuccessful.

The first fused ring derivatives characterized were the bis(π-indenyl)-titanium dihalides, $(\pi\text{-}C_9H_7)_2TiX_2$, X = Cl, Br, and I. They were prepared by reaction of the appropriate titanium tetrahalide with lithium indenide in diethyl ether in mole proportions 1:2 (M8).

Later preparations of $(\pi\text{-}C_9H_7)_2TiCl_2$ (J15, S9) showed the compound to be brown (m.p. 210°) whereas the initial preparation (M8) yielded a deep green product (312°–313°). The dark brown crystalline product on hydrogenation with hydrogen in the presence of Adams catalyst in dimethoxyethane (S5) gave orange bis(π-tetrahydroindenyl)titanium(IV) dichloride. Although bis(indenyl)titanium dichloride has a high thermal stability (m.p. 210°), it decomposes readily in organic solvents in the air. Bis(tetrahydroindenyl)titanium dichloride, on the other hand, is a sublimable solid which can be recrystallized from organic solvents. It shows a remarkable resistance to hydrolysis, and is completely insoluble in boiling water. The mixed cyclopentadienyl(indenyl)titanium dichloride, which is a dark brown crystalline solid, was obtained from sodium indenide and $CpTiCl_3$ (S6).

The reaction of Cp_2TiCl_2 with anhydrous hydrogen chloride in CCl_4 has been claimed to yield cyclopentenylcyclopentadienyl(cyclopentadienyl)-titanium dihalides (N4, T36).

Some substituted cyclopentadienyl derivatives which do not contain the normal alkyl, aryl, or fused ring systems described above have also been reported in the literature. These have for the most part ill-defined structures and properties and, in some instances, doubt arises as to their exact composition.

d. METAL-SUBSTITUTED CYCLOPENTADIENYL DERIVATIVES

1,1'-Bis(triphenylgermylcyclopentadienyl)titanium dichloride was prepared in less than 1% yield from lithium triphenylgermylcyclopentadienide and $TiCl_4$. The red-violet crystals decomposed at 244° (S27).

Drozdov et al. (D32) prepared a silyl derivative by reacting sodium dimethylethoxysilylcyclopentadienide with $TiCl_4$ in THF. The expected bis-substituted metal derivative was obtained as a red, crystalline, air-stable compound in ~60% yield.

Table III-9

Bis(substituted-cyclopentadienyl)titanium Dihalides

Compound	Other data[a]	Refs.
$(\pi\text{-}C_5D_5)_2TiCl_2$	MS (D30); IR (D30, M17)	D30, M17
$(\pi\text{-}CH_3C_5H_4)TiF_2$	M.p. 189°–191°; ^{19}F NMR; ^{13}C NMR	N38a
$(\pi\text{-}CH_3C_5H_4)_2TiCl_2$	M.p. 217°–218°(dec); IR	R41
$\left[\begin{array}{c}R'-C\underset{R}{\overset{R}{-}}\text{(cyclopentadienyl)}\end{array}\right]_2 TiCl_2$		
$R = CH_3, R' = H$	M.p. 170°–172°; PMR(CDCl$_3$) δ6.36(m)(C$_5$H$_4$), 3.23(C–H), 1.21(d)(CH$_3$)	S71
$R = C_6H_5, R' = H$	M.p. 199°–201.5°; PMR(CDCl$_3$) δ6.18(t)(C$_5$H$_4$), 5.84(t)(C$_5$H$_4$), 7.16(m)(C$_6$H$_5$), 5.7(s)(C–H)	S71
$R = CH_3, R' = C_6H_5$	M.p. 217°–218°; PMR(CDCl$_3$) δ6.48(t)(C$_5$H$_4$), 6.10(t)(C$_5$H$_4$), 7.24(m)(C$_6$H$_5$), 1.78(s)(CH$_3$)	S71
$R = CH_3, R' = n\text{-}C_4H_9$	M.p. 142°–145°; PMR(CDCl$_3$) δ6.47(m)(C$_5$H$_4$), 1.32(s)(CH$_3$), 1.8–0.7(m)(C$_4$H$_9$)	S71
$R'-C\underset{R}{\overset{R}{-}}\text{(cyclopentadienyl)} CpTiCl_2$		
$R = CH_3, R' = H$	M.p. 145°–152°; PMR(CDCl$_3$) δ6.55(s)(Cp), 6.46(m)(C$_5$H$_4$), 3.23(C–H), 1.21(d)(CH$_3$)	S71
$R = C_6H_5, R' = H$	M.p. 210°–212°; PMR(CDCl$_3$) δ6.17(s)(Cp), 6.55(t)(C$_5$H$_4$), 6.35(t)(C$_5$H$_4$), 7.21(m)(C$_6$H$_5$), 5.68(s)(C–H)	S71
$R = CH_3, R' = C_6H_5$	M.p. 194°–196°; PMR(CDCl$_3$) δ6.31(s)(Cp), 6.62(t)(C$_5$H$_4$), 6.45(t)(C$_5$H$_4$), 7.25(s)(C$_6$H$_5$), 1.77(s)(CH$_3$)	S71
$R = CH_3, R' = n\text{-}C_4H_9$	M.p. 150°–152°; PMR(CDCl$_3$) δ6.53(s)(Cp), 6.50(m)(C$_5$H$_4$), 1.31(s)(CH$_3$), 1.6–0.7(m)(C$_4$H$_9$)	S71

[π-(CH$_3$)$_5$C$_5$]CpTiF$_2$	M.p. 163°–165°; ^{19}F NMR; ^{13}C NMR	N38a
[π-(CH$_3$)$_5$C$_5$]$_2$TiCl$_2$	M.p. 273°(dec); IR; PMR(CDCl$_3$) δ2.00(s)(CH$_3$); ^{13}C NMR (N38a)	B43, N38a
(1-CH$_3$-3-C$_6$H$_5$C$_5$H$_3$)$_2$TiCl$_2$	M.p. 224°–225°; UV	H4
[1-CH$_3$-3-(4-BrC$_6$H$_4$)C$_5$H$_3$]$_2$TiCl$_2$	M.p. 234°–237°	H4
(π-Indenyl)$_2$TiCl$_2$	M.p. 210° (J15, S9); IR (S5, S9); Visible (S5)	J15, S5, S9
(π-tetrahydroindenyl)$_2$TiCl$_2$	M.p. 220°; IR; sublimable; PMR(CDCl$_3$) δ6.30(t), 5.78(d), (C$_5$ ring signals)	S5, S6

	M.p. 262°–264°(dec)	J15

Cp(π-indenyl)TiCl$_2$	IR; PMR(CDCl$_3$) δ6.20(Cp), 6.86(t), 6.66(d) (C$_5$ ring protons in C$_9$H$_7$), 7.2–7.9(m)(C$_6$ ring protons in C$_9$H$_7$)	S6
[(C$_6$H$_5$)$_3$GeC$_5$H$_4$]$_2$TiCl$_2$	M.p. 244°(dec.); IR	S27
[(CH$_3$)$_2$(C$_2$H$_5$O)SiC$_5$H$_4$]$_2$TiCl$_2$	M.p. 165°–166.5°; MW monomer(benzene); UV; IR	D32
(C$_6$H$_5$N=N—C$_5$H$_4$)$_2$TiCl$_2$	M.p. 198°–204°(dec)	K33, P2
(CH$_3$CHN=N—C$_5$H$_4$)$_2$TiCl$_2$	—	P2

$n = 1$	M.p. >360°; MS; PMR δ6.84(t, $J = 2.7$ Hz), 5.92(t, $J = 2.7$ Hz), 4.16(s)	K11
$n = 3$	IR (H30); UV (H30); PMR (H30), crystal structure (D9); neutron diffraction (E5a)	D9, E5a, H30

^{a} s, Singlet; d, doublet; t, triplet; m, multiplet; br, broad.

59

2. Pseudohalide Derivatives of
Bis(cyclopentadienyl)titanium(IV)

The two mixed halido–pseudohalido derivatives of bis(cyclopentadienyl)-titanium(IV), brown $Cp_2Ti(Cl)NCS$ and orange-red $Cp_2Ti(Cl)CN$, were prepared in 1963 (M4) by a ligand exchange reaction, e.g.,

$$Cp_2TiCl_2 + AgX \xrightarrow[\Delta]{toluene} Cp_2Ti(Cl)X + AgCl$$

Kinetic studies by Jensen (J9) on the substitution reactions of Cp_2TiX_2 compounds led to the preparation of the first bis(pseudohalido) derivatives of bis(cyclopentadienyl)titanium(IV), namely, $Cp_2Ti(SCN)_2$. Langford and Aplington (L2) made this and the corresponding azido compound in low yield from Cp_2TiCl_2 and excess sodium salt in boiling acetone. Molecular weights determined in benzene and tetramethylene sulfone solutions gave values less than monomeric (H28).

Coutts and Wailes (C62) have established that the compounds Cp_2MX_2 (M = Ti or Zr, X = NCO, NCS) are monomeric in acetone, although Jensen (J9) had earlier believed that $Cp_2Ti(NCS)_2$ is dimeric in this solvent. The products are reasonably stable in air and sublime at low pressure with some decomposition. From the infrared data the ligands were considered to be bound to the metal through nitrogen as isothiocyanato and isocyanato derivatives (C62). Samuel (S5), who also prepared the same two compounds by ligand exchange in aqueous medium, made the reverse bonding assignments, and was supported by Giddings (G12) in the case of $Cp_2Ti(NCS)_2$.

More recently, Burmeister et al. (B100, B101, J10) have reinvestigated these derivatives in an attempt to confirm the bonding modes of the ambidentate ligands. They concluded that the thiocyanato and selenocyanato complexes contain N-bonded groups. The situation regarding the mode of attachment of the cyanato group is less straightforward as evidence for a particular mode of attachment of this group based solely on infrared data cannot be regarded as conclusive. However, recent dipole moment studies suggest that $Cp_2Ti(NCO)_2$ is N-bonded, whereas $Cp_2Zr(OCN)_2$ is O-bonded (J10).

The bis(azido) derivatives, $Cp_2Ti(N_3)_2$ and $Cp_2Zr(N_3)_2$, were formed by Thayer (T22, T23) by the action of HN_3 on the corresponding dichlorides in a chloroform–water solution. On reinvestigation of this reaction in water (C71) the mixture of products obtained included $Cp_2Ti(N_3)_2$ (70%), $(Cp_2TiN_3)_2O$, and a polymeric product which could be separated by fractional crystallization.

Table III-10

Bis(cyclopentadienyl)titanium(IV) Pseudohalides

Compound	Color	Highest yield (%)	Other data	Refs.
$Cp_2Ti(Cl)NCS$	Brown	—	M.p. 186°–190°	M5
$Cp_2Ti(Cl)CN$	Orange	—	M.p. 208°–222°	M5
$Cp_2Ti(OMe)CN$	Yellow	23	M.p. 165°–166°; IR	N37
$Cp_2Ti(N_3)_2$	Orange	Quantitative (L2)	M.p. 145°–146° (C71); sublimes 200°/10^{-3} mm (C71); IR (C71, T22); MS (C71); UV-visible (L2); PMR(CDCl$_3$) $\delta6.31$,(THF) $\delta6.51$ (C71)	C71, L2, T22, T23
$Cp_2Ti(NCO)_2$	Red-orange	69 (B100)	M.p. 275°–277° (C62); sublimes 160°/10^{-3} mm (C62); MW, monomer(acetone) (C62); IR (B100, B101, C62, S5); conductance, nonelectrolyte (B100, B101); MS (B100, B101); PMR(acetone) $\delta6.62$ (C62)	B100, B101, C62, S5
$Cp_2Ti(NCS)_2$	Maroon	79 (G12)	M.p. 303°–307° (C62, G12); sublimes 200°/10^{-3} mm (C62); MW, monomer(acetone) (C62); IR (B100, B101, C62, G12, S5); UV-visible (B100, B101, L2); MS (B100); Conductance, nonelectrolyte (B101); PMR(acetone) $\delta6.77$ (C62)	B100, B101, C62, G12, L2, S5, W27
$Cp_2Ti(NCSe)_2$	Green	66 (K46)	M.p. 140° (K46); IR (B100, B101, K46); UV-visible (B100, B101, K46); Conductance, nonelectrolyte (B101)	B100, B101, K46

61

3. Oxygen-Bonded Compounds of
Bis(cyclopentadienyl)titanium(IV)

There is little doubt that the titanium–oxygen-bonded compounds have received more attention than any other group of titanium compounds (F10). Tetraalkyl- and tetraaryltitanates (alkoxides and phenoxides) of general formula $Ti(OR)_4$ have been known since 1875 (D16). However, the real interest in alkoxides of titanium was stimulated by papers published by Kraitzer, McTaggart, and Winter (K56, K57) describing the use of alkoxides in heat-resistant paints.

Uses claimed for organotitanium compounds lie to a great extent in this field of paint and varnish manufacture and in the treatment of various types of fibers, such as textiles, wood, and paper. Most of these applications are centered around organic titanium compounds formed from alcohols, polyols, carboxylic acids, and siloxo derivatives of titanium, all of which contain the titanium–oxygen bond (F10).

a. ALKOXIDES, PHENOXIDES, ORGANOSILOXIDES, AND
ORGANOPHOSPHINOXIDES

i. ALKOXIDES AND PHENOXIDES.   All the preparative methods for these compounds involve replacement of labile halides on titanium compounds. The first example reported was the monoethoxo derivative, $Cp_2Ti(OC_2H_5)Cl$, described by Brantley *et al.* in a 1958 patent application (B77) and later by Nesmeyanov *et al.* (N34).

Method A: Replacement of halide by Cyclopentadienyl. The reaction of sodium cyclopentadienide with bis(ethoxo)titanium dichloride resulted in the formation of two organotitanium compounds (N34). The monoethoxo

$$3NaC_5H_5 + 2(CH_3CH_2O)_2TiCl_2 \longrightarrow$$
$$CpTi(OCH_2CH_3)_3 + Cp_2Ti(Cl)OCH_2CH_3 + 3NaCl$$

derivative, obtained in only 17% yield from this reaction, became the major product (61%) when $(CH_3CH_2O)TiCl_3$ was used as starting material.

$$2NaC_5H_5 + (CH_3CH_2O)TiCl_3 \longrightarrow Cp_2Ti(Cl)OCH_2CH_3 + 2NaCl$$

Method B: Replacement of Halide by Alkoxo. Treatment of Cp_2TiCl_2 in a solvent with an alcohol or phenol in the presence of a base (tertiary amine or sodamide) replaced one or both chlorides. Although unsuccessful in the

$$Cp_2TiCl_2 + nROH + nbase \longrightarrow Cp_2Ti(OR)_nCl_{2-n} + nbase \cdot HCl$$

presence of pyridine and excess butanol (W24), the method has led to a series of bis(phenoxo) derivatives obtained by Andrä in yields of 65–93%

(A14, A15) and to the bis(fluorophenoxo) compounds (N40), $R_2Ti(OC_6H_4F)_2$ [$R = Cp$, $\pi\text{-}CH_3C_5H_4$, $\pi\text{-}(CH_3)_5C_5$].

Pyrocatechol and 2,2'-dihydroxybiphenyl react similarly to form cyclic compounds (A15),

Bis(alkoxides) have also been prepared using 2 moles of the sodium alkoxide, e.g., trifluoroethoxide (D34) or hexafluoroisopropoxide (B51). A similar reaction using sodium pentafluorophenoxide yielded only the monosubstitution product, $Cp_2Ti(OC_6F_5)Cl$ (B51).

The alkoxides and phenoxides are yellow to orange monomeric solids, moderately soluble in organic solvents. A study of the ^{19}F NMR spectra of the fluorophenoxo derivatives prepared by Nesmeyanov *et al.* (N40) revealed that the ^{19}F signal moved upfield with increase in substitution of the cyclopentadienyl ring (N38).

From the reaction of Cp_2TiCl_2 with lithium acetylpentacarbonylchromate in dichloromethane at room temperature (F15), two red monomeric carbene complexes, $Cp_2Ti(Cl)OC(CH_3)Cr(CO)_5$ and $Cp_2Ti[OC(CH_3)Cr(CO)_5]_2$, were obtained.

ii. ORGANOSILOXIDES. Preparative methods for trisorganosiloxides are essentially the same as those for alkoxides and phenoxides described above, namely, from Cp_2TiCl_2 and a silanol, in the presence of base, or a sodium siloxide (S49). Both mono and bis(siloxo) derivatives are formed depending on the reaction conditions. Noltes and van der Kerk (N42) found that from $(C_6H_5)_3SiONa$ and Cp_2TiCl_2 in toluene at 70° for 10 min, $Cp_2Ti[OSi(C_6H_5)_3]$-Cl was obtained in 76% yield, whereas at 110° for 30 min the bis(triphenylsiloxo) compound was formed in 63% yield.

Many of these reactions were accompanied by extensive decomposition with formation of $[(C_6H_5)_3SiO]_4Ti$ or titanium dioxide (G7, G38).

The preparation of $Cp_2Ti[OSi(CH_3)_3]Cl$ from $[(CH_3)_3SiO]TiCl_3$ and NaC_5H_5 has been described by Nesmeyanov *et al.* (N35).

The stabilities of these yellow to orange siloxo compounds have been related to the basicities of the siloxo groups (S49). The triphenyl derivatives appear to be stable in air (N42).

iii. ORGANOPHOSPHINOXIDE. The only organophosphinoxide of a transition metal reported is the bis derivative shown below which was made in near-quantitative yield from Cp_2TiCl_2 and the diphosphane (R34).

$$Cp_2TiCl_2 + 2(CF_3)_2POP(CF_3)_2 \longrightarrow 2(CF_3)_2PCl + Cp_2Ti[OP(CF_3)_2]_2$$

Table III-11

Bis(cyclopentadienyl)titanium(IV) Alkoxides, Phenoxides, and Siloxides

Compound	Method of preparation	Other data	Refs.
$Cp_2Ti(Cl)OC_2H_5$	A, B	M.p. 88°–93° (all refs.); MW, monomer(benzene) (N35)	B77, N35
$Cp_2Ti(Cl)OCH_2CH=CH_2$	B	M.p. 57°–58°; MW, monomer(benzene) (N35); IR (K53)	K53, N35
$Cp_2Ti(Cl)OC_6H_5$	B	M.p. 71°–73°; MW, monomer(benzene) (N35)	N35
$Cp_2Ti(OC_2H_5)O_2CC(CH_3)=CH_2$	A	M.p. 165°–166°; MW, monomer(benzene) (N35)	N35
$Cp_2Ti(C_6F_5)OC_2H_5$	A	M.p. 117° (C23); IR (C23)	C22, C23
$Cp_2Ti(C_6F_5)OCH_3$	—	M.p. 144°–145°; IR	C22
$Cp_2Ti(OCH_2CF_3)_2$	B	M.p. 47°–48°; UV	D34
$Cp_2Ti[OCH(CF_3)_2]_2$	B	MW, monomer(mass spectra); IR	B51
$Cp_2Ti(Cl)OC_6F_5$	B	M.p. 240°–245°(dec); ^{19}F NMR; IR	B51
$Cp_2Ti(OR)_2$			
R = C_6H_5	A	M.p. 142°; MW, monomer(benzene); reduction potential, –1174 mV; IR	A14, A15
R = 2-ClC_6H_4	A	M.p. 145°–147°; MW, monomer(benzene); reduction potential, –1180 mV; IR	A15
R = 2-$CH_3C_6H_4$	A	M.p. 162°; MW, monomer(benzene); IR	A15
R = 2-$NO_2C_6H_4$	A	M.p. 122°–124°; MW, monomer(benzene); IR	A15
R = 4-ClC_6H_4	A	M.p. 125°–127°; MW, monomer(benzene); reduction potential –1150 mV; IR	A15
R = 4-$CH_3C_6H_4$	A	M.p. 164°–167°; MW, monomer(benzene); reduction potential –1155 mV; IR	A15
R = 4-$C_6H_5C_6H_4$	A	M.p. 136°–137°; MW, monomer(benzene); IR	A15
R = 3,5-$(CH_3)_2C_6H_3$	A	M.p. 162°–163°; MW, monomer(benzene); IR	A15
$Cp_2Ti(O-4-FC_6H_4)_2$	A	M.p. 174°–176° (N40); ^{19}F NMR(CHCl$_3$) 13.10, (THF) 13.53 (N38); ^{13}C NMR (N38a)	N38, N38a, N40
$Cp_2Ti(O-3-FC_6H_4)_2$	A	M.p. 100°–111.5° (N40); ^{19}F NMR(CHCl$_3$)	N38, N40
$(\pi$-$CH_3C_5H_4)_2Ti(O-4-FC_6H_4)_2$	A	M.p. 178°–179° (N40); ^{19}F NMR(CHCl$_3$) 13.30, (THF) 13.50 (N38)	N38, N40

Compound	Method	Data	Ref.
$[\pi\text{-}(CH_3)_5C_5]CpTi(O\text{-}4\text{-}FC_6H_4)_2$	A	M.p. 126°–128° (N40); ^{19}F NMR(CHCl$_3$) 14.40, (THF) 14.44 (N38)	N38, N40
$Cp_2Ti(O\text{-}4\text{-}NO_2C_6H_4)_2$	A	M.p. 232°(dec) (N40); PMR(CDCl$_3$) δ6.58(Cp) 6.78, 8.25(C$_6$H$_4$) (N38); IR (N40)	N38, N40
(structure) Cp_2Ti	A	M.p. 100°–103°; IR	A15
(structure) Cp_2Ti	A	M.p. 208°–211°; IR	A15
(structure) Cp_2Ti	—	M.p. 155°; PMR[(CD$_3$)$_2$CO] δ6.1(Cp), 7.5, 8.0, 8.7 (all phenanthrylene); MW, monomer(benzene); MS	F21
$Cp_2Ti(Cl)OSi(CH_3)_3$	A	M.p. 137°–137.5° (N35); MW, monomer (N35); UV-visible (S49); IR (S49)	N35, S49
$Cp_2Ti(Cl)OSi(C_6H_5)(CH_3)_2$	B	M.p. 56°; MW, monomer(benzene); IR	T9
$Cp_2Ti(Cl)OSi(CH_3)(C_6H_5)_2$	A, B	M.p. 160°–161.5° (T9) MW, monomer(benzene) (T9); visible (S49); IR (S53, T9); thermal analysis (S49); X-ray powder pattern (T9)	S49, T9
$Cp_2Ti(Cl)OSi(C_6H_5)(C_4H_9)_2$	B	M.p. 97°–99°; MW, monomer(benzene); IR; X-ray powder pattern	T9
$Cp_2Ti(Cl)OSi(C_6H_5)_3$	A, B	M.p. 210°–212° (N42); MW, monomer(benzene) (N42); visible (S49); IR (N42, S49); Thermal analysis (S49)	N42, S49
$Cp_2Ti[OSi(CH_3)_2C_6H_5]_2$	B	MW, monomer(benzene); IR	T9
$Cp_2Ti[OSi(CH_3)(C_6H_5)_2]_2$	B	M.p. 113.5°–114.5°; MW, monomer(benzene); IR	T9
$Cp_2Ti[OSi(C_6H_5)_3]_2$	A, B	M.p. 199°–200° (G7, S49, 208°–210° (N42); MW, monomer(benzene) (N42, T9); visible (S49); IR (N42, S49); thermal analysis (S49); X-ray powder pattern (T9)	G7, N42, S49, T9
$Cp_2Ti[OP(CF_3)_2]_2$	—	MW, monomer; MS; PMR (toluene) δ6.37(Cp); IR	R34

b. CARBOXYLATES

Many carboxylato derivatives have been prepared since the first examples were described in a 1961 patent by the National Lead Company (N10).

Method A: Replacement of Halide by Carboxylato. This is the most useful reaction. Bis(cyclopentadienyl)titanium dihalide is refluxed with a silver or alkali metal acylate in THF under anaerobic conditions (N35, R23). Per-

$$Cp_2TiX_2 + 2NaO_2CR \longrightarrow Cp_2Ti(O_2CR)_2 + 2NaX$$

fluorocarboxylato derivatives also can be made in this way in benzene or acetone (D33, K29, N40).

Method B: Oxidative Addition. A number of arylcarboxylato derivatives have been prepared by oxidation of "titanocene" with benzoyl (or substituted benzoyl) peroxide (R23). No carbon dioxide was liberated.

$$\text{"Titanocene"} + (RCO_2)_2O \longrightarrow Cp_2Ti(O_2CR)_2$$

Method C: Displacement of Phenyl by Carboxylato. Arylcarboxylato derivatives have also been obtained by reaction of $Cp_2Ti(C_6H_5)_2$ with benzoyl peroxide or a nitrobenzoyl peroxide in isopropanol, THF, or benzene at low temperatures ($20° \pm 5°$) over a period of 1–2 days (R18–R20). Surprisingly,

$$Cp_2Ti(C_6H_5)_2 + (RCO_2)_2O \longrightarrow Cp_2Ti(O_2CR)_2 + 2C_6H_5\cdot$$

when this reaction is carried out with benzoyl or acetyl peroxide in benzene, ether, or chlorobenzene, a titanium(III) carboxylate is formed (R17, R18). The reaction of carbon dioxide with $Cp_2Ti(C_6H_5)_2$ at 80°–90° in xylene yielded the red crystalline metallocyclic carboxylate shown in Fig. III-12 (p. 100).

The carboxylates are thermally unstable compounds, colored orange or red, and are generally readily hydrolyzed (N39). The perfluoro analogs, on the other hand, are air-stable, and in solution are thermally more robust than their monocyclopentadienyltitanium counterparts.

A study by Tel'noi *et al.* (T16) of the thermochemical properties of $Cp_2Ti(O_2CC_6H_5)_2$ suggested that the Ti–O bond strength was of the order of 110 kcal/mole.

c. β-DIKETONATES

In 1967 Doyle and Tobias (D30) reported the synthesis of a series of compounds of the type, $[Cp_2TiL]^+X^-$ [where L is the conjugate base of acetylacetone (acac), benzoylacetone, dibenzoylmethane, dipivalomethane, and tropolone and X^- is ClO_4^-, BF_4^-, PF_6^-, AsF_6^-, SbF_6^-, or $CF_3SO_3^-$]. In all cases the β-diketonate ligand appears to be chelating, the titanium atom being essentially tetrahedral.

Since only one β-diketone ligand coordinates to the Cp_2Ti^{IV} moiety, it was suggested that the ease of preparation of the mono-β-diketonate complexes resulted from the very low solvation energy of the complex cation. In the PMR spectra of these complexes the $=C–H$ resonance of the acetylacetonate appears to be the lowest field value observed for acetylacetonate complexes (D30).

The complex anions, $FeCl_4^-$, $CoCl_4^{2-}$, $ZnCl_3(H_2O)^-$, $SnCl_3^-$, $CdCl_4^{2-}$, and $SnCl_6^{2-}$ could be precipitated from aqueous solutions of the metal chlorides in the presence of excess chloride ion by the cation $[Cp_2Ti(acac)]^+$ (D31).

In the ethyl acetoacetate complex, $[Cp_2Ti(C_6H_9O_3)]ClO_4$, it was suggested (D31) that bonding to the metal was via the keto oxygen and the "ether"-type oxygen of the OC_2H_5 group. Reinvestigation by White (W20) showed only the infrared bands characteristic of the normal coordination form, namely,

The reaction between Cp_2TiCl_2 and acetoacetanilide or dibenzoylmethane in refluxing benzene gave, respectively, acetoacetanilidobis(cyclopentadienyl)-titanium chloride or bis(cyclopentadienyl)-1,3-diphenyl-1,3-propanedionato-titanium chloride (S24). These reaction products are probably best formulated as ionic complexes analogous to those of Doyle and Tobias.

d. POLYTITANOXANES

Compounds containing the bis(cyclopentadienyl)titanium(IV) moiety connected to another such group, or to different groups, via oxygen bridges between the metal atoms are described in this section.

Table III-12
Bis(cyclopentadienyl)titanium(IV) Carboxylates

Compound	Method of preparation	Other data	Refs.
$Cp_2Ti(O_2CCH_3)_2$	A	M.p. 126°–130° (N35, R23); MW, monomer (N35); PMR, $\delta 6.54$(Cp), 1.97(CH$_3$) (N29)	N10, N29, N35, N36, R23
$Cp_2Ti(O_2CCH_2Cl)_2$	A	M.p. 98°–99° (N39)	N10, N39
$Cp_2Ti(O_2CCH_2OH)_2$	A	—	N10
$Cp_2Ti(O_2CCH_2SH)_2$	A	—	N10
$Cp_2Ti(O_2CCHCl_2)_2$	A	—	N10
$Cp_2Ti(O_2CCCl_3)_2$	A	M.p. 192°–194° (N39); ^{13}C NMR (N38a)	M4, N10, N38a, N39
$Cp_2Ti(O_2CCF_3)_2$	A	M.p. 175°–179° (D33, S11), 130° (K29); MW (monomer) (D33, S11); UV-visible (D33); IR (K29, S11); sublimes 110°–120°/0.1 mm (K29); ^{19}F NMR (CHCl$_3$), $\delta -3.90$, (THF) -2.94 (N38)	D33, K29, N38, N40, S11
$(\pi\text{-}CH_3C_5H_4)_2Ti(O_2CCF_3)_2$	A	M.p. 110°–111° (N40); ^{19}F NMR(CHCl$_3$) $\delta -3.84$, (THF) -2.94 (N38)	N38, N40
$[\pi\text{-}(CH_3)_5C_5]_2Ti(O_2CCF_3)_2$	A	M.p. 191°–195° (N40); ^{19}F NMR(CHCl$_3$) $\delta -3.72$, (THF) -2.54 (N38)	N38, N40
$Cp_2Ti(O_2CC_2H_5)_2$	A	M.p. 114°–116° (N39); PMR $\delta 6.49$(Cp), 1.03(CH$_3$) 2.16(CH$_2$) (N29)	N29, N39
$Cp_2Ti(O_2CC_2F_5)_2$	A	M.p. 112°–113°; sublimes 90°–100°/0.1 mm; IR, ^{19}F NMR(CDCl$_3$); PMR(CDCl$_3$) $\delta 6.66$(Cp)	K29
$Cp_2Ti(O_2CC_3H_7)_2$	A	M.p. 82°–84°	N39
$Cp_2Ti(O_2CC_3F_7)_2$	A	M.p. 110°–111° (D33), 141°–142° (K29); MW, monomer (D33, K29); UV-visible (D33); IR (K29); sublimes 90°–100°/0.1 mm (K29); ^{19}F NMR(CDCl$_3$) (K29)	D33, K29

Compound	Method	Properties	Ref.
$Cp_2Ti[O_2C(CH_2)_{16}CH_3]_2$	A	—	N10
$Cp_2Ti[O_2C(CH_2)_7CH=CH(CH_2)_7CH_3]_2$	A	—	N10
$Cp_2Ti(OCH_2CH_3)O_2CC=CH_2$ with CH_3	A	M.p. 165°–166°; MW; monomer(benzene)	N35
$Cp_2Ti(Cl)O_2CC=CH_2$ with CH_3	A	IR	R2
$Cp_2Ti(O_2CC=CH_2)_2$ with CH_3	A	M.p. 135°–150° (N35); MW (benzene) (N35); IR (R2); thermogravimetric analysis (R2)	N35, R2
$Cp_2Ti(O_2CC_6H_5)_2$	A	M.p. 188° (R23); MW, monomer (R23); PMR $\delta 6.77$(Cp), 7.67, 8.17(C_6H_5) (N29); IR (R23, V32)	N29, R23, V32
$Cp_2Ti(O_2C-4-NO_2C_6H_4)_2$	B, C	M.p. 180°(dec); IR	N40, R19
$Cp_2Ti(O_2C-3-NO_2C_6H_4)_2$	C	M.p. 150°(dec); IR	R19
$Cp_2Ti(O_2CC_6H_5)O_2C-4-NO_2C_6H_4$	C	—	R19
	—	M.p. 192°–195°(dec) (A4); diamagnetic (A4); IR (A4); crystal structure (A4, K38); PMR, $\delta 6.5-7.55$(C_6H_5), 6.45(Cp)	A4, K38
	A	See Table III-15	S25, S26
$(\pi\text{-Indenyl})_2Ti(O_2CCCl_3)_2$	A	—	N10

Table III-13

Bis(cyclopentadienyl)titanium β-Diketonates

Compound[a]	Color	Yield (%)	Other data	Refs.
[(π-C$_5$D$_5$)$_2$Ti(acac)]ClO$_4$	—	—	IR	D30
[Cp$_2$Ti(acac)]ClO$_4$	Violet	70	IR; conductance (1:1 nitrobenzene); PMR (DMF) δ6.90(Cp), 6.33(CH=), 2.27(CH$_3$)	D30
[Cp$_2$Ti(acac)]F$_3$CSO$_3$	—	—	IR	D30
[Cp$_2$Ti(acac)]BF$_4$	—	—	—	D30
[Cp$_2$Ti(acac)]Cl	Red	20	M.p. 239°–240°(dec); IR	S24
[Cp$_2$Ti(acac)]X X = FeCl$_4^-$, SnCl$_3^-$, HgCl$_3^-$, ZnCl$_3^-$, $\frac{1}{2}$CdCl$_4^{2-}$, $\frac{1}{2}$SnCl$_6^{2-}$	—	—	IR; conductance (1:1 nitrobenzene for monoanions, 2:1 for dianions)	D31
[Cp$_2$Ti(bzac)]ClO$_4$	Purple	51	IR; conductance (1:1 nitrobenzene); PMR(DMF) δ6.97(Cp), 2.47(CH$_3$)	D30
[Cp$_2$Ti(dbzm)]ClO$_4$	Violet	21	IR; conductance (1:1 nitrobenzene); PMR(DMF) δ7.05(Cp)	D30
[Cp$_2$Ti(dbzm)]Cl	Orange	22	IR	S24
[Cp$_2$Ti(dpm)]ClO$_4$	Purple	27	IR; conductance (1:1 nitrobenzene); PMR(DMF) δ6.89(Cp), 1.30[–C(CH$_3$)$_3$]	D30
[Cp$_2$Ti(trop)]ClO$_4$	Green	55	IR; conductance (1:1 nitrobenzene); PMR(DMF) δ6.83(Cp)	D30
[Cp$_2$Ti(C$_6$H$_9$O$_3$)]BF$_4$	Purple	—	IR; PMR(CDCl$_3$) δ6.75(Cp), 5.46(C–H) 4.35, 1.33(CH$_2$CH$_3$), 2.13(CCH$_3$)	W20
[Cp$_2$Ti(C$_6$H$_9$O$_3$)]PF$_6$	Purple	—	IR; PMR(SO$_2$) δ6.72(Cp), 5.51(C–H), 4.33, 1.35(CH$_2$CH$_3$), 2.15(CCH$_3$)	W20
[Cp$_2$Ti(C$_6$H$_9$O$_3$)]B(C$_6$H$_5$)$_4$	Brown	33	IR; PMR(SO$_2$) δ6.53(Cp), 5.50(C–H) 4.27, 1.33(CH$_2$CH$_3$), 2.12(CCH$_3$)(CDCl$_3$) δ5.97(Cp), 5.35(C–H) 4.00, 1.23(CH$_2$CH$_3$), 1.98(CCH$_3$)	W20
[Cp$_2$Ti(C$_6$H$_9$O$_3$)]ClO$_4$	Purple	84	IR (D31, W20); PMR(SO$_2$) δ6.73(Cp), 5.37(C–H), 4.34, 1.37(CH$_2$CH$_3$), 2.15(CCH$_3$)(CDCl$_3$), δ6.78(Cp), 5.57(C–H), 4.37, 1.37(CH$_2$CH$_3$), 2.16(CCH$_3$)	D31, W20

[a] acacH, Acetylacetone; bzacH, benzoylacetone; dbzmH, dibenzoylmethane; dpmH, dipivalomethane; tropH, tropolone; C$_6$H$_{10}$O$_3$, ethyl acetoacetate.

Allyloxobis(cyclopentadienyl)titanium chloride (K53) can be homopolymerized by heating the monomer in toluene at 100° with benzoyl peroxide as initiator. The dark orange product was found to be trimeric.

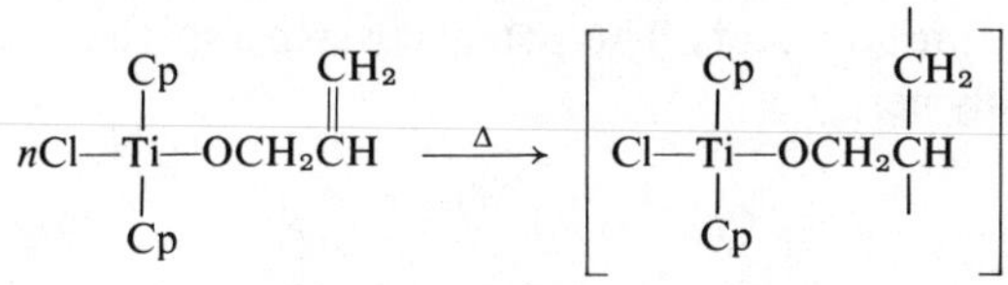

$$n\text{Cl}\overset{\displaystyle \text{Cp}}{\underset{\displaystyle \text{Cp}}{\vert\ \text{Ti}\ \vert}}\text{—OCH}_2\overset{\displaystyle \text{CH}_2}{\text{CH}} \xrightarrow{\ \Delta\ } \left[\text{Cl}\overset{\displaystyle \text{Cp}}{\underset{\displaystyle \text{Cp}}{\vert\ \text{Ti}\ \vert}}\text{—OCH}_2\overset{\displaystyle \text{CH}_2}{\text{CH}}\right]_n$$

Similarly, copolymers of allyloxobis(cyclopentadienyl)titanium chloride were prepared (K53) using styrene and methyl methacrylate as comonomers in the ratio 1:10.

In dimethylformamide (DMF) at 80°, Cp$_2$TiCl(methylacrylate) and bis-(cyclopentadienyl)titanium bis(methylacrylate) could be homopolymerized when initiated by benzoyl peroxide (R2). The proposed structure of the polymer is shown below.

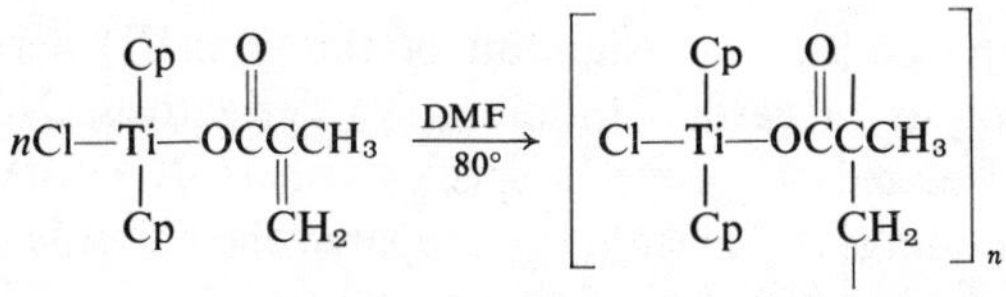

$$n\text{Cl}\overset{\displaystyle \text{Cp}}{\underset{\displaystyle \text{Cp}}{\vert\ \text{Ti}\ \vert}}\text{—OC}\overset{\displaystyle \text{O}}{\underset{\displaystyle \text{CH}_2}{\text{C}}}\text{CH}_3 \xrightarrow[80°]{\text{DMF}} \left[\text{Cl}\overset{\displaystyle \text{Cp}}{\underset{\displaystyle \text{Cp}}{\vert\ \text{Ti}\ \vert}}\text{—OC}\overset{\displaystyle \text{O}}{\underset{\displaystyle \text{CH}_2}{\text{C}}}\text{CH}_3\right]_n$$

Neither homopolymer showed impressive thermostability; decomposition began below 220°. Copolymerization of the two monomers with methylacrylate gave copolymers with mechanical properties resembling those of commercial "Plexiglas."

Giddings (G9) in 1962 claimed that linear or cross-linked polymers, useful as pigments, could be prepared from R$_2$TiX$_2$, in which R is a Cp or lower alkyl-substituted Cp group and X is Cl, Br, or I.

Schramm and Frühauf (S17) made several polymers from bis(cyclopentadienyl)-, or bis(substituted-cyclopentadienyl)titanium dichloride and the alkali metal salts of 4,4′-dihydroxybiphenylene sulfone or of dibasic acids such as maleic and terephthalic acid, in boiling amide-type solvents. The products were insoluble oxygen-bridged polytitanoxanes of the type shown below.

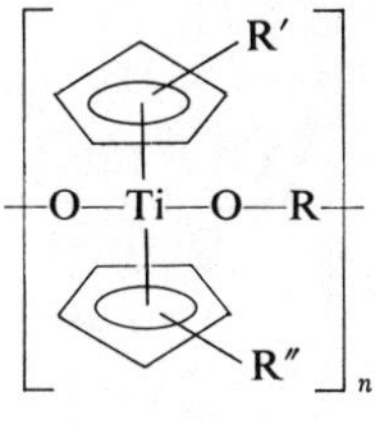

More recently Carraher and Bajah (C9a) have extended the range of polymers of the type shown above, using interfacial and aqueous solution techniques.

McHugh and Smith (M4) took out a patent for extreme-pressure additives for high-temperature lubricants. This patent covered a spectrum of derivatives, one being the polymer,

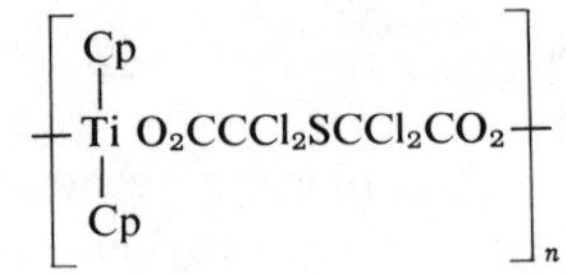

$$\left[\begin{array}{c} Cp \\ | \\ Ti\ O_2CCCl_2SCCl_2CO_2 \\ | \\ Cp \end{array}\right]_n$$

Dimeric polytitanoxanes of the type,

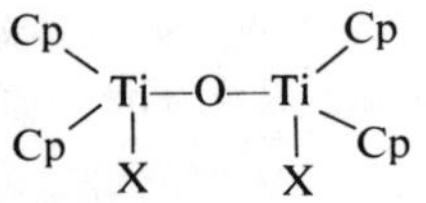

$$\begin{array}{cc} Cp \diagdown \quad \quad \diagup Cp \\ \quad Ti-O-Ti \\ Cp \diagup\ | \quad \quad |\ \diagdown Cp \\ X \quad \quad X \end{array}$$

can be formed by controlled oxidation of titanium(III) derivatives or by controlled hydrolysis of certain titanium(IV) derivatives. Nöth and Hart-wimmer (N45) isolated the dimeric species (where X = Br) by oxidation of $Cp_2Ti^{III}Br$. Giddings (G11) similarly prepared the chloride analog. Coutts and Wailes observed that the pseudohalide derivatives, $Cp_2Ti^{III}X$ (X = CN, NCS, OCN), gave yellow polymeric products when oxidized with oxygen (C63).

Giddings (G10–G13) attempted the synthesis of compounds of type

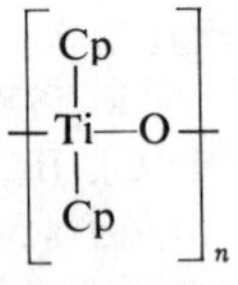

$$\left[\begin{array}{c} Cp \\ | \\ Ti-O \\ | \\ Cp \end{array}\right]_n$$

Yellow titanoxane polymers in the form of relatively soft infusible powders were obtained (G10, G13) from reaction of zinc or silver oxide with Cp_2TiX_2, or bis(substituted-cyclopentadienyl)titanium dihalides, in the presence of liquid ethers or ketones under anhydrous conditions, in the presence of oxygen. Elemental analysis obtained by three different methods (G11) suggested the formula

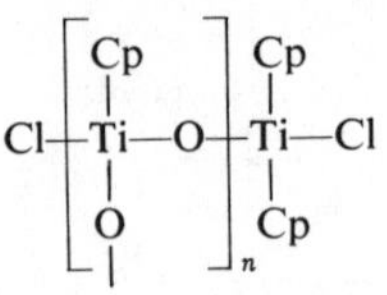

$$Cl\left[\begin{array}{c} Cp \\ | \\ Ti-O \\ | \\ O \\ | \end{array}\right]_n \begin{array}{c} Cp \\ | \\ Ti-Cl \\ | \\ Cp \end{array}$$

Coutts *et al.* (C61) found that hydrolysis of $Cp_2Ti(Cl)SR$ compounds in moist air led to the formation of the orange oxo-bridged compound, $(Cp_2$-$TiCl)_2O$, and the free mercaptan. The bis(thiolato) derivatives were far more stable to hydrolysis.

Treatment of $Cp_2Ti(NCS)_2$ with butylamine in acetone in air gave an orange crystalline product (G12), which was identified as the dinuclear compound, $Cp_2Ti(NCS)OTi(NCS)Cp_2$.

Samuel (S5) showed that reaction of Cp_2TiCl_2 in water with NaBr, NaI, or $NaNO_3$ gave orange crystalline products which were found to be hydrated bridged oxides, $(Cp_2TiX)_2O \cdot nH_2O$ (X = Br, I, or NO_3). On heating *in vacuo* or by recrystallizing from acetone, the anhydrous compounds were obtained.

When aqueous solutions of sodium azide and Cp_2TiCl_2 were mixed (C71), a 70% yield of the diazide was obtained together with a yellow residue consisting chiefly of the oxygen-bridged dinuclear derivative (1) and a small amount of a more insoluble polymeric material (2) analogous to the polymer obtained from the corresponding chloride (G11). In air the oxo-bridged

(1)   (2)

compound slowly converted into the polymer. The IR, PMR, and mass spectra of these compounds were recorded.

Polymeric $(C_{10}H_{10}TiNO)_n$ was prepared by Salzmann (S3) from the reaction between titanocene and NO in toluene at room temperature, while reaction with molecular oxygen gave $(C_{10}H_{10}TiO_2)_n$.

e. MISCELLANEOUS DERIVATIVES

Sulfur dioxide reacts in the expected manner with alkyl- and arylbis(cyclopentadienyl)titanium compounds to yield red monomeric *O*-sulfinates (M50b, W7); e.g., with $Cp_2Ti(CH_3)_2$,

$$Cp_2Ti(CH_3)_2 + 2SO_2 \xrightarrow[-78°]{solvent} Cp_2Ti(O_2SCH_3)_2$$

Wailes *et al.* (W8) found that $Cp_2Ti(CH_3)_2$ reacts with NO in benzene at room temperature to give the insertion product, $\left[Cp_2Ti(CH_3)ONCH_3 \atop \qquad\qquad NO \right]_n$.

4. Sulfur-Bonded Compounds of
Bis(cyclopentadienyl)titanium(IV)

It is interesting to note that whereas early attempts to prepare bis(cyclopentadienyl)titanium(IV) alkoxides were unsuccessful, the corresponding thioalkoxides, $Cp_2Ti(SR)_2$, were readily prepared and characterized. The ease of preparation of sulfur-bonded titanium derivatives led to numerous publications in this area of organotitanium chemistry.

a. THIOALKOXIDES AND SELENOALKOXIDES

In a 1961 patent Giddings claimed an extensive range of compounds, only two of which, $Cp_2Ti(SR)_2$ (where $R = CH_3$ or C_6H_5) were characterized (G8). Methods of preparation which have been used are

Method A: Salt Elimination.

$$Cp_2TiCl_2 + 2NaSR \longrightarrow Cp_2Ti(SR)_2 + 2NaCl$$

Although the sodium mercaptide is normally used (C61, G8) a convenient variation is the use of an amine with the mercaptan. Köpf and Schmidt found that the reaction of Cp_2TiCl_2 with mercaptans and thiophenols in benzene or ether solution in the presence of triethylamine gave good yields of the thio complex (K52; see also G12).

The parent monomeric thioalkoxo derivative, $Cp_2Ti(SH)_2$, was prepared in almost quantitative yields by this method from hydrogen sulfide (K51). It is surprisingly stable toward oxidation and hydrolysis. $Cp_2Ti(SD)_2$ was similarly prepared using D_2S.

The bis(selenoalkoxides) had been prepared earlier by the same workers using the amine method (K49).

Method B: Oxidative Addition. Coutts et al. (C61) showed that $Cp_2Ti^{III}Cl$ cleaves dialkyl and diaryl disulfides to give the monothio compounds, $Cp_2Ti(Cl)SR$, in high yields.

$$2Cp_2Ti^{III}Cl + RSSR \longrightarrow 2Cp_2Ti(Cl)SR$$

Oxidative addition reactions can also be employed to form bis(thioalkoxides) and bis(selenoalkoxides) of Cp_2Ti^{IV} in low yields (K49) from titanocene and dialkyl disulfides or dialkyl diselenides.

Method C: Trimethylsilylchloride Elimination. The ability of the (alkyl/arylthio)trimethylsilanes to effect the replacement of halide by alkyl/arylthio groups was demonstrated (A1) by the formation of $Cp_2Ti(SR)_2$ ($R = CH_3$ or C_6H_5) compounds in high yields by the reaction,

$$Cp_2TiCl_2 + 2(CH_3)_3SiSR \longrightarrow Cp_2Ti(SR)_2 + 2(CH_3)_3SiCl$$

Table III-14

Bis(cyclopentadienyl)titanium Bis(thioalkoxides) and Bis(selenoalkoxides)

Compound	Method of preparation	Other data	Refs.
$Cp_2Ti(Cl)SCH_3$	B	M.p. 179°(dec); PMR(CDCl$_3$) δ6.34(Cp) 3.12(CH$_3$)	C61
$Cp_2Ti(Cl)SC_2H_5$	B	M.p. 108°(dec); PMR(CDCl$_3$) δ6.34(Cp), 3.62(CH$_2$), 1.31(CH$_3$)	C61
$Cp_2Ti(Cl)SCH_2C_6H_5$	B	M.p. 134°–135.5°(dec); PMR(CDCl$_3$) δ7.16(C$_6$H$_5$), 6.29(Cp), 4.63(CH$_2$)	C61
$Cp_2Ti(Cl)SC_6H_5$	B	M.p. 140°(dec); PMR(CDCl$_3$) δ7.32(C$_6$H$_5$), 6.21(Cp)	C61
$Cp_2Ti(SH)_2$	A	M.p. 150°–160°(dec); MW, monomer(CHCl$_3$); PMR(CHCl$_3$) δ6.34(Cp), 3.44(SH)	K51
$Cp_2Ti(SCH_3)_2$	A, B, C	M.p. 184°–185° (A1, C61), 193°–197° (G8, G12), 202°–209° (C22, K44); PMR(CDCl$_3$) δ6.12(Cp), 2.62(CH$_3$)(C61); (CS$_2$) δ6.08(Cp), 2.62(CH$_3$) (K47); IR (C22, K44)	A1, C22, C61, G8, G12, K44
$Cp_2Ti(SC_2H_5)_2$	A	M.p. 107°–117°; IR (K49); MW, monomer (K49); PMR(CDCl$_3$) δ6.10(Cp), 3.07(CH$_2$), 1.18(CH$_3$) (C61); CS$_2$ δ6.09(Cp), 3.12(CH$_2$), 1.24(CH$_3$) (K49); IR (K49)	C61, K49
$Cp_2Ti(SC_3H_7)_2$	A	M.p. 88°–93°; IR; MW, monomer	K49
$Cp_2Ti(SCH_2C_6H_5)_2$	A	M.p. 172°–178°; IR; MW, monomer (K49); PMR(CDCl$_3$) δ7.15(C$_6$H$_5$), 6.12(Cp), 4.19(CH$_2$) (C61); (CH$_2$Cl$_2$) δ7.18(C$_6$H$_5$), 6.18(Cp), 4.22(CH$_2$) (K49)	C61, K49
$Cp_2Ti(SCH_2CH_2C_6H_5)_2$	A	M.p. 92°–94°; IR; MW, monomer; PMR(CS$_2$) δ7.18(C$_6$H$_5$), 6.05(Cp), 3.02(CH$_2$)	K49
$Cp_2Ti(SC_6H_5)_2$	A, B, C	M.p. 199°–206° (all refs.); IR (K49); MW, monomer (G8, K49); polarography (D17); PMR(CDCl$_3$) δ7.40(C$_6$H$_5$), 6.03(Cp) (C61); (CS$_2$) δ7.31(C$_6$H$_5$), 6.00(Cp) (K44); (DMSO) δ7.38(C$_6$H$_5$), 6.13(Cp) (K49)	A1, C61, D17, G8, G13, K44, K49, S26
$Cp_2Ti(SC_6H_4CH_3)_2$	A	M.p. 198.5°–199.5°; IR; MW, monomer; PMR(CS$_2$) δ7.16(C$_6$H$_5$), 5.92(Cp), 2.43(CH$_3$)	K49
$Cp_2Ti(SC_6H_4Cl)_2$	A	M.p. 178°–182°; IR; MW, monomer	K49
$Cp_2Ti(SeC_2H_5)_2$	A, B	M.p. 99°–101°; IR; MW, monomer; UV	K46
$Cp_2Ti(SeC_6H_5)_2$	A, B	M.p. 120°–122°; IR; MW, monomer; UV; PMR(CS$_2$) δ7.45(C$_6$H$_5$) 5.99(Cp)	K44, K46

b. METALLOCYCLES CONTAINING Ti–S BONDS

McHugh and Smith in a 1963 patent application (M4) claimed the preparation of the dark red solid, Cp_2TiS_5, and $(\pi\text{-}CH_3C_5H_4)_2TiS_x$, as dark purple crystals melting at $165°$–$170°$. No characterization was offered for these derivatives.

In 1966 Samuel obtained the former compound, Cp_2TiS_5, from Cp_2TiCl_2 and ammonium sulfide, and proposed the structure

$$
\begin{array}{c}
Cp \\
\quad \diagdown \\
\qquad Ti \\
\quad \diagup \\
Cp
\end{array}
\begin{array}{c}
S\!-\!S \\
\diagup \qquad \diagdown \\
\qquad\qquad S \\
\diagdown \qquad \diagup \\
S\!-\!S
\end{array}
$$

The preparation of like compounds, $(Cp_2TiS_x)_n$ (where $x = 3$ or 4), was later claimed by Ralea and co-workers (R1) from Cp_2TiCl_2 and various chlorosulfides and sodium sulfides, e.g.,

$$nCp_2TiCl_2 + nNa_2S_x \longrightarrow [Cp_2TiS_x]_n + 2nHCl$$

However, on reinvestigation of this reaction using sodium and ammonium polysulfides, Köpf and Block (K45) established that in all cases only Cp_2TiS_5 could be isolated. This compound has been a useful synthetic reagent in the controlled synthesis of the thermodynamically unstable new homoatomic compounds, S_7, S_9, S_{10}, and S_{20} (D10a, S15, S16).

The crystal and molecular structure of Cp_2TiS_5 has been determined from diffraction data (E5, E6) (see Fig. III-8).

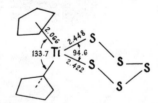

Fig. III-8. Structure of Cp_2TiS_5 (E6).

The asymmetry introduced into the molecule by the chair configuration of the TiS_5 ring suffices to make the PMR of the two cyclopentadienyl rings nonequivalent at $30°$. At elevated temperatures ($120°$) only one PMR peak is found owing to the rapid conformational changes (K47). The corresponding selenium analog, Cp_2TiSe_5, showed a similar behavior.

Another type of metallocyclic ring system containing sulfur atoms attached to the titanium atom has been made from $Cp_2Ti(SH)_2$ with dichlorotetramethyldisiloxane, dichlorohexamethyltrisiloxane, and dichlorooctamethyl-tetrasiloxane in acetone with trimethylamine (U3). Cyclic structures were proposed.

From Cp_2TiCl_2 and 1,2-dithiols in the presence of a base (or as the disodium dithiolates), a number of five-membered heterocycles have been synthesized including phenyl and substituted phenyl derivatives, e.g.,

$R = H$ or CH_3 (C22, K50), as well as the aliphatic derivatives,

where $R = R' = CN$ (C22, K50, L18); $R = C_6H_5$, $R' = H$(K40), and $R = R' = H$ (K28, K40).

The possible conformations and inversion processes occurring in the five-membered TiS_2C_2 rings in solution have been studied by Köpf (K42) by PMR spectroscopy. At $-50°$ two sharp singlets of equal intensity due to the Cp protons indicated the presence of the conformers,

When the temperature was raised the signals broadened and finally coalesced indicating mutual interconversion.

The molecular and crystal structure of $Cp_2TiS_2C_6H_4$ has been determined by Kutoglu (K64). The $S_2C_6H_4$ plane of the molecule is folded out of the TiS_2 plane by an angle of $46°$ (see Fig. III-9).

An electrochemical study of $Cp_2TiS_2C_6H_3CH_3$ showed that its reduction is electrochemically reversible (D17).

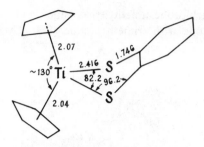

Fig. III-9.   Structure of $Cp_2TiS_2C_6H_4$ (K64).

Köpf and Schmidt found (K50) that Cp_2TiCl_2 reacts with alkane α,ω-dithiols in the presence of triethylamine to form unstable oligomers which still contain chlorine (see Section B,4,c). However, ethane-1,2-dithiol has been shown to form a stable chloride-free heterocycle, $Cp_2TiS_2(CH_2)_2$, in chloroform in the presence of gaseous ammonia (S25, S26). This compound differed in color, thermal stability, and solubility from the compound of identical formulation prepared earlier (C22) from the disodium dithiolate. The monomeric parent ions were present in the mass spectra of this compound and the thiosalicylato derivative (S26).

The PMR spectrum of the latter compound in deuterochloroform showed a singlet centered at δ6.30 (Cp protons) and three multiplets centered at δ7.32[H(c)], δ7.52[H(b)], and δ8.37[H(a)], with relative intensities appropriate to the formulation,

The α-aminobenzenethio compound contained three molecules of ligand per titanium atom and could be assigned either structure **3** or **4**.

(3)   (4)

Formulation **4** was supported by thermogravimetric analysis, which showed loss of two molecules of the ligand in a single step between 104° and 385°.

A patent (J16) in 1968 described the compounds $Cp_2Ti(SC_6H_5)_2Mo(CO)_4$ and $Cp_2Ti(SC_6H_5)_2PdCl_2$ which were synthesized by displacement of a neutral ligand from its metal complex, e.g.,

$$Cp_2Ti(SCH_3)_2 + (nor\text{-}C_7H_8)Mo(CO)_4 \longrightarrow Cp_2Ti(SCH_3)_2Mo(CO)_4$$

nor-C_7H_8 = norbornadiene

$$Cp_2Ti(SCH_3)_2 + (cyclooctadiene \cdot CuCl)_2 \longrightarrow [Cp_2Ti(SCH_3)_2CuCl]_2$$

The intense blue titanium–molybdenum complex was diamagnetic. Infrared data indicated that the molybdenum atom had dispersed some of the charge and of the two alternative structures (**5** or **6**), **6** was favored.

$$(5) \qquad (6)$$

Other workers (K48) prepared these bimetallic complexes with chalcogen-bridging ligands by displacing CO from metal carbonyls or acetonitrile from $(CH_3CN)_2Mo(CO)_4$.

The crystal and molecular structures of $Cp_2Ti(SMe)_2Mo(CO)_4$ have been determined (D7). The complex contains a planar four-membered ring system,

with a Ti–Mo distance of 3.32 Å, (Ti–S = 2.46 Å and Mo–S = 2.56 Å). The complex is the cis isomer, as both SCH_3 groups lie on the same side of the heterocyclic ring (see Fig. III-10). The interaction of these thiol groups with the mutually staggered cyclopentadienyl rings probably accounts for the stability of the cis relative to the trans isomer. This structure confirmed the earlier PMR investigation (B80). To account for the acute Ti–S–Mo angle (83°) and the large S–Ti–S angle (99.9°), it was necessary to postulate a metal–metal interaction. The similarity between solution spectra and diffuse reflectance spectra for this and related complexes (B80, B81), and the general similarities in the physical properties indicate that there is no major change

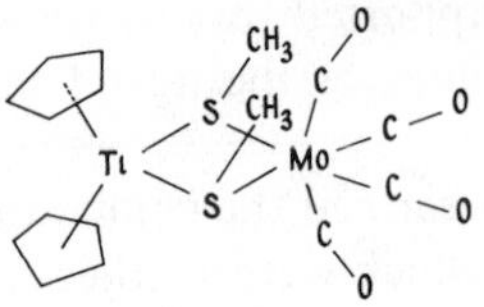

Fig. III-10.   Structure of $Cp_2Ti(SCH_3)_2Mo(CO)_4$ (D7).

on dissolution, so that all these complexes may be presumed to have a basic-ally similar structure.

Evidence for Cu–Ti bonds in the complexes, $[Cp_2Ti(SR)_2CuX]_n$, came from their electronic spectra (B79). The molecular structures and degree of poly-merization of diolefin copper complexes have been found to depend on the diolefin used with the halogens acting as bridging groups (B4, V2). Similarly in the complexes $[Cp_2Ti(SR)_2CuX]_n$, n may well vary from complex to complex.

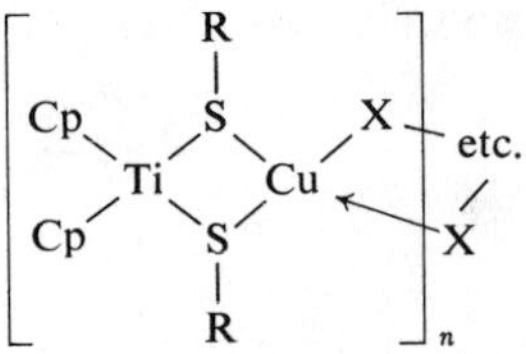

The diffuse reflectance spectra of these complexes show a low energy band not present in the spectra of $Cp_2Ti(SR)_2$. This new, intense low energy band was assigned to a transition (presumably $Cu^I \rightarrow Ti^{IV}$ charge transfer) involving the two metal atoms.

c. POLYMERIC DERIVATIVES

Polymeric compounds containing oxygen linked to the titanium in the main chain are numerous and have received considerable attention in recent years. The occurrence of polymers containing Cp_2Ti units linked by sulfur

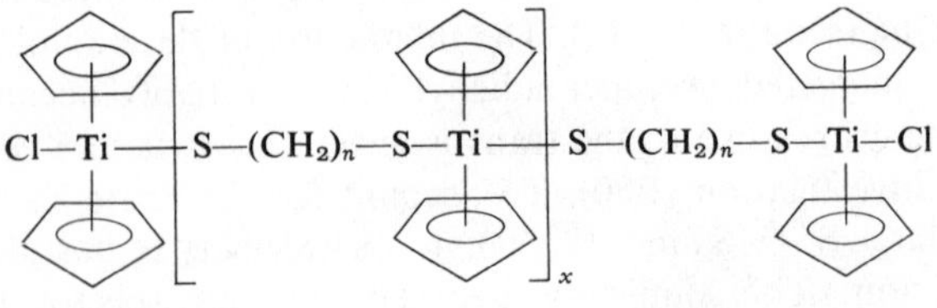

$n = 2, 3,$ or 4

are relatively rare. When Cp_2TiCl_2 and alkane α,ω-dithiols were reacted in boiling toluene (K50) in the presence of triethylamine to neutralize liberated hydrogen chloride, scarlet-red condensation products were obtained. These light- and air-sensitive compounds of only moderate thermal stability, have been formulated as straight-chain oligomers with some chlorine still present. The parent compound ($n = 2$) is monomeric (i.e., $x = 0$) (S25, S26), but the limited solubility of the other polymers prevented the determination of their molecular weights.

Carraher and Nordin (C9b) have also studied polymers of this type using aqueous solution synthesis.

Polytitanosiloxanes containing sulfur bridges were formed from $Cp_2Ti(SH)_2$ and dichlorosiloxanes under mild conditions in the presence of triethylamine (U3). Reaction of Cp_2TiCl_2 at room temperature with dichlorohexamethyl-trisiloxane in acetone solution afforded a linear hydrolytically sensitive oligomer containing about six recurring units. A similar compound was

obtained from the corresponding tetrasiloxane derivative. From the atomic ratios, C:Ti and S:Ti, it appeared that the elimination of cyclopentadiene was not extensive.

5. Alkyl and Aryl Compounds of Bis(cyclopentadienyl)titanium(IV)

Efforts to prepare σ-bonded organotitanium compounds have continued for over a century. In 1861 Cahours (C1) reported the attempted isolation of compounds containing titanium–carbon bonds, by reacting diethylzinc with titanium tetrachloride. However, it was not until 1952 that the first organotitanium compound was prepared (H25).

It has generally been accepted that the carbon–titanium sigma bond is weak. Recently, the question of bond strength has been the center of much discussion (B78, M49), and the results suggest that too often researchers have correlated the chemical or thermal stability of this type of compound with the strength of the bond.

Table III-15
Metallocycles of Bis(cyclopentadienyl)titanium^{IV} with Ti–S Bonds

Compound	Color	Highest yield (%)	Other data	Refs.
$(CH_3C_5H_4)_2TiS_x$	Dark purple	—	—	M4
Cp_2TiS_3	Red	70	M.p. 205°–210°; MW, monomer(dioxane)	R1
Cp_2TiS_4	Dark red	80	M.p. 185°–190°; MW, monomer(dioxane)	R1
Cp_2TiS_5	Maroon-red	97	M.p. 201°–202° (K41, K45, K47, 248° (S4); IR; UV; MW, monomer(benzene) (K47); PMR at 30°(CS$_2$) δ6.42, 6.10(Cp); [(CH$_3$)$_2$S$_2$] δ6.11(Cp) (K47); crystal structure (E5, E6)	E5, E6, K41, K45, K47, M4, S5, S15, S16
Cp_2TiSe_5	Violet	87	M.p. 211°; MW, monomer(CS$_2$); IR; UV; PMR at 30°(CS$_2$) δ6.44, 5.98(Cp); [(CH$_3$)$_2$S$_2$] δ6.40, 5.98(Cp)	K47
Cp_2Ti with S–Si(CH$_3$)$_2$–O–Si(CH$_3$)$_2$–S ring	Orange	85	M.p. 190°–192°; MW, monomer(dioxane); IR	U3
Cp_2Ti with S–Si(CH$_3$)$_2$–O–Si(CH$_3$)$_2$–O–Si(CH$_3$)$_2$–S ring	Orange	50	M.p. 217°–218°; MW, monomer(DMSO)	U3
Cp_2Ti with bis-siloxane–S bridged ring	Orange	—	M.p. 236°–238°; MW, monomer(DMSO); IR	U3

(continued)

Structure	Color	No.	Properties	References
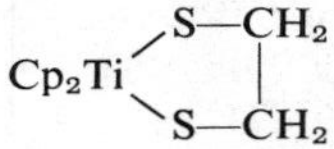 Cp_2Ti (S–CH$_2$ / S–CH$_2$)	Violet	20	M.p. 50°(dec) (C22); 140°(dec) (S26); MS (S26)	C22, S25, S26
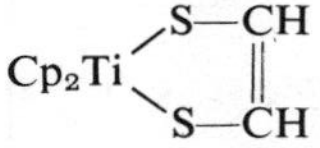 Cp_2Ti (S–CH= / S–CH=)	Green-black	63	M.p. 177°–179° (K28), 202°–204° (K40); IR (K28, K40); UV-visible (K28); sublimes 150° (K40); 110°–120° at 0.1 mm (K28); MW, monomer(CS$_2$) (K40), (benzene) (K28); PMR (CDCl$_3$) δ7.43(S$_2$C$_2$H$_2$), 5.96(Cp) (K28); (CS$_2$) δ7.57(S$_2$C$_2$H$_2$), 6.06(Cp) (K40)	K28, K40
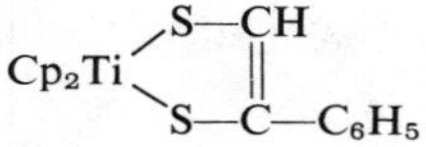 Cp_2Ti (S–CH= / S–C–C$_6$H$_5$)	Deep green	52	MW, monomer(benzene); PMR(CS$_2$) δ7.90(S$_2$C$_2$H), 7.45(C$_6$H$_5$), 6.00(Cp)	K40
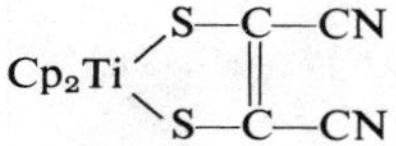 Cp_2Ti (S–C=CN / S–C–CN)	Deep green	86	M.p. >260° (C22, K50, L18); IR (C22, K50, L18; conductance, nonelectrolyte (CH$_3$NO$_2$) (L21); voltammetry (M1); MW, monomer(benzene) (K50); PMR δ6.43(Cp) (L18)	C22, K50, L18, M1
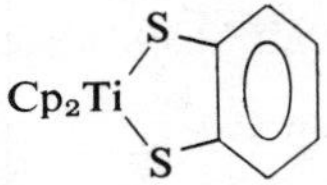 Cp_2Ti (benzene, S / S)	Green-black	92	M.p. 222°–224°; MW, monomer(benzene)	K50
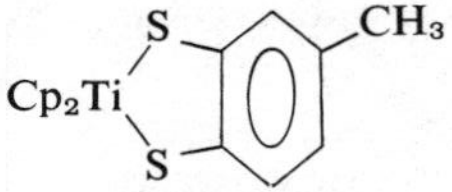 Cp_2Ti (toluene, CH$_3$, S / S)	Green	75	M.p. 158°–160° (all refs.); MW, monomer (C22, K50); IR (C22, K50); electrochemical study and ESR (D17); PMR(CS$_2$) δ7.08(C$_6$H$_3$), 5.90(Cp), 2.43(CH$_3$) (K50)	C22, D17, K50, S26
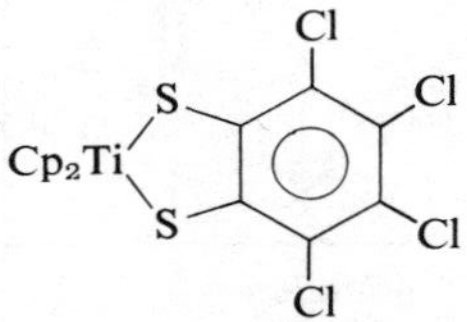 Cp_2Ti (tetrachlorobenzene, Cl, S / S)	Green	73	M.p. 259°–261°; MS; UV-visible; conductance, nonelectrolyte (acetone)	J6
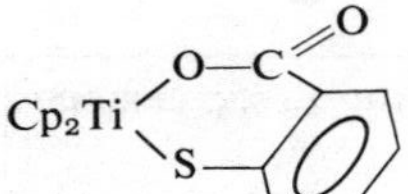 Cp_2Ti (O–C(=O) benzene / S)	Green	37	M.p. 145°(dec); MS; PMR	S25, S26

83

Table III-15 (continued)

Metallocycles of Bis(cyclopentadienyl)titanium^{IV} with Ti–S Bonds

Compound	Color	Highest yield (%)	Other data	Refs.
Cp_2Ti (S / N-H ring) $\cdot 2$ (benzene ring with SH, NH_2)	Magenta	—	M.p. 93°–95°; MW, monomer; thermogravimetric data; PMR	S25, S26
$Cp_2Ti(SCH_3)_2Cr(CO)_4$	Green	30	M.p. >300°; IR; MW, monomer(acetone); UV-visible; PMR(C_6H_5Cl, 20°) δ5.21, 5.14, 5.01(Cp), 2.45(SCH_3)	B80
$Cp_2Ti(SCH_3)_2Mo(CO)_4$	Blue	60	M.p. >300°; IR (B80, C6, J17); diamagnetism (J17); MW, monomer(acetone) (B80); UV-visible (B80, C6); PMR(C_6H_5Cl, 20°) δ5.26, 5.06(Cp), 2.50(SCH_3) (B80, C6); crystal structure (C6, D7)	B80, C6, D7, J17
$Cp_2Ti(SCH_3)_2W(CO)_4$	Blue	72	M.p. >300°; IR, MW, monomer(acetone); UV-visible; PMR(C_6H_5Cl, 20°) δ5.22, 5.12,5.00(Cp), 2.57(SCH_3)	B80
$Cp_2Ti(SC_6H_5)_2Mo(CO)_4$	Bluish black	51	M.p. >300°; IR (B80, K48); MW, monomer-(benzene) (K48), monomer(acetone) (B80); UV-visible (B80, K48); PMR(C_6H_5Cl) δ5.18(Cp) (B80); (CS_2) δ5.48(Cp) (K48)	B80, J16, K48
$Cp_2Ti(SC_6H_5)_2W(CO)_4$	Blue	25	M.p. >300°; IR; UV-visible; MW, monomer-(acetone); PMR(C_6H_5Cl, 20°) δ5.15(Cp)	B80
$Cp_2Ti(SeC_6H_5)_2Mo(CO)_4$	Bluish black	66	IR; MW, monomer(benzene); UV-visible; PMR(CS_2) δ5.37(Cp)	K48
$[Cp_2Ti(SCH_3)_2CuCl]_n$	Black	—	M.p. 175°(dec) (J17); 170° (B79); IR (B79, J17); UV-visible (B79)	B79, J17
$[Cp_2Ti(SCH_3)_2CuBr]_n$	Black	—	M.p. >186°(dec); UV-visible	B79
$[Cp_2Ti(SC_6H_5)_2CuCl]_n$	Brown	—	M.p. >135°(dec); UV-visible	B79
$[Cp_2Ti(SC_6H_5)_2CuBr]_n$	Brown	—	M.p. >174°(dec); UV-visible	B79
$Cp_2Ti(SC_6H_5)_2PdCl_2$	—	—	—	J16

a. BIS(ARYL)BIS(CYCLOPENTADIENYL)TITANIUM(IV)
DERIVATIVES

Summers and Uloth (S72) isolated the first bis(aryl) derivatives of bis-(cyclopentadienyl)titanium(IV) by the reaction,

$$Cp_2TiCl_2 + 2LiR \longrightarrow Cp_2TiR_2 + 2LiCl$$

The products were obtained in up to 81% yield when R = phenyl, m-tolyl, p-tolyl, and p-dimethylaminophenyl (S72, S73). Attempts to isolate the o-tolyl and α-naphthyl derivatives were unsuccessful, presumably owing to steric hindrance. Beachell and Butter (B21) have extended the range of compounds to include $Cp_2Ti(m\text{-}CF_3C_6H_4)_2$ and $Cp_2Ti(p\text{-}XC_6H_4)_2$ (where X = OCH_3, F, Cl, Br, CF_3) and have examined their PMR spectra. It is interesting to note here that the alkylating agent, phenyllithium, can be prepared conveniently and in high yield by metallation of benzene with n-butyllithium-N,N,N',N'-tetramethylethylenediamine (R5). Reagent prepared by this method has enabled Rausch and Ciappenelli to obtain Cp_2Ti-$(C_6H_5)_2$ in 96% yield.

The parent compound bis(cyclopentadienyl)bis(phenyl)titanium has been the subject of many physical and chemical investigations. Tel'noi et al. (T17) established from thermochemical experiments that the average dissociation energy of the Cp–Ti bond was 74 kcal/mole and that of the Ti–C_6H_5 bond in $Cp_2Ti(C_6H_5)_2$ was 84 ± 5 kcal/mole. These data illustrate that Ti–C σ bonds are not as weak as was first believed. The polarographic reduction of $Cp_2Ti(C_6H_5)_2$ was found to be concentration dependent and the reduction was assumed to be a one-electron process involving cleavage of the Ti–C_6H_5 bonds (G36).

In the crystal, the molecular structure of $Cp_2Ti(C_6H_5)_2$ has two staggered cyclopentadienyl rings π-bonded to titanium with an average Ti–C distance of 2.31 Å (K34). Their planes are tilted at an angle of 44° (see Fig. III-11).

The interest in $Cp_2Ti(C_6H_5)_2$ as a catalyst component with either $TiCl_4$ or an alkylaluminum compound for polymerization of ethylene is evident in numerous publications (F19, G3, N25, N26, P7, P8, R54).

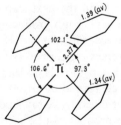

Fig. III-11.   Structure of bis(cyclopentadienyl)bis(phenyl)titanium (K34).

A much studied reaction of $Cp_2Ti(C_6H_5)_2$ has been its thermal decomposition under various conditions. When first prepared by Summers and Uloth (S73), pyrolysis experiments showed that at 105° under N_2 benzene was formed. The formation of benzene during its thermolysis in hydrocarbon solvents has been confirmed by Dvorak *et al.* (D40), who also established that a phenylene–titanium complex was produced for each molecule of benzene evolved.

$$Cp_2Ti(C_6H_5)_2 \xrightarrow{\text{benzene}} Cp_2Ti(C_6H_4) + C_6H_6$$

The nature of the $Cp_2TiC_6H_4$ intermediate is not apparent since either a phenylene Ti^{IV} compound (7) or a benzyne complex of Ti^{II} (8) is possible.

(7)   (8)

The reaction of $Cp_2Ti(C_6H_5)_2$ with diphenylacetylene in refluxing benzene gave 1,1-bis(cyclopentadienyl)-2,3-bis(phenyl)titanaindene (M23). Dvorak (D40) showed that this reaction probably proceeds via the *o*-phenylene intermediate, e.g.,

The reaction can be simply visualized as the insertion product of C≡C into a Ti–C bond.

On reaction of $Cp_2Ti(C_6H_5)_2$ with carbon monoxide in benzene at room temperature and 200 atm, benzophenone was obtained in good yields (M23). It was considered that the carbon monoxide initially coordinated to the metal through the vacant orbital of titanium. Insertion of carbon monoxide into the $Ti–C_6H_5$ bond would occur in the next step, followed by migration of another phenyl ligand to the acyl group producing benzophenone. An identical reaction has been observed with $Cp_2Ti(CH_2C_6H_5)_2$ and CO in which $Cp_2Ti(CO)_2$ and dibenzylketone were formed (F1).

The reaction of CO_2 with $Cp_2Ti(C_6H_5)_2$ leads to carboxylation of the phenyl ring and to metallocycle formation (K38). The molecular structure of the compound has been determined by X-ray studies (A4, K38) (see Section B,5,g). The formation of the product can be seen as occurring via insertion of CO_2 into a Ti–C bond of the phenylenetitanium complex.

$$Cp_2Ti(C_6H_5)_2 \xrightarrow{80°-90°} Cp_2Ti \underset{\text{benzyne}}{\bigcirc} \xrightarrow{CO_2} Cp_2Ti\underset{O}{\overset{}{\diagup}}C{=}O$$

In addition to CO_2 and N_2 (see Chapter VIII), other molecules with multiple bonds undergo this type of insertion reaction. For example, Kolomnikov et al. (V15) found that $Cp_2Ti(C_6H_5)_2$ reacted with phenylisocyanate and benzophenone to yield the respective products,

$$Cp_2Ti\underset{O}{\overset{}{\diagup}}C{=}NC_6H_5 \qquad Cp_2Ti\underset{O}{\overset{}{\diagup}}C\underset{C_6H_5}{\overset{C_6H_5}{\diagdown}}$$

The above results negate the radical mechanism for the thermal decomposition of $Cp_2Ti(C_6H_5)_2$, which was favored during earlier work by Razuvaev and co-workers (L12, R17–R19, R22, R28), who assumed that the $Ti–C_6H_5$ bond initially split homolytically to yield free phenyl radicals.

It would appear that in no instance where $Cp_2Ti(C_6H_5)_2$ has been used as a source of so-called "titanocene" has the actual compound been isolated and characterized (L12, R22–R30). These investigators have claimed that reaction of the "titanocene" with halogenated compounds yields approximately quantitative amounts of the normal red Cp_2TiCl_2 and that the "titanocene" does not react with alcohols. In all reports of authenticated, stable "titanocene" species (see Chapter VII), the reactions with halogenating compounds result in low yields of the expected red Cp_2TiCl_2, and all the various titanocenes react vigorously with alcohols.

Summers and Uloth (S72) observed that $Cp_2Ti(C_6H_5)_2$ dissolved in an excess of an ethereal solution of C_6H_5Li, a behavior which was attributed

$$Cp_2Ti(C_6H_5)_2 + C_6H_5Li \rightleftharpoons [Cp_2Ti(C_6H_5)_3]^-Li^+$$

to the formation of an anionic complex. In a more recent and detailed investigation (L10, L11, R31) the reaction was found to take place with reduction of Ti^{IV} to Ti^{III} and then to Ti^{II}. The reaction products in the presence of excess C_6H_5Li were $(C_6H_5)_2Ti$, C_5H_5Li, C_6H_6, metallic lithium, and small amounts of biphenyl and triphenylene. To explain the presence of triphenylene in the reaction mixture it was suggested that at some stage in the reduction dehydrobenzene was formed as an intermediate (L11) which gave triphenylene on trimerization. Titanium compounds were found to catalyze the decomposition of C_6H_5Li to C_6H_6 and Li.

Table III-16

Bis(aryl)bis(cyclopentadienyl)titanium Derivatives

Compound	Highest yield (%)	Other data	Refs.
$Cp_2Ti(C_6H_5)_2$	80 (S73)	M.p. 142°–148° (R5, R18, R24, S73); MW, monomer(benzene) (S72); IR (S41); crystal structure (K34); thermochemistry (D40, T17); polarography (G36); PMR($CDCl_3$) $\delta6.93(C_6H_5)$, $6.22(Cp)$ (R5); (CCl_4) $\delta6.11(Cp)$, $6.79(C_6H_5)$ (B21); MS (G18)	B21, D40, F19, G3, G18, G36, K34, K38, K55, L10–L12, M23, N23, N25, P7, P8, R5, R6, R17–R19, R22–R31, R54, S35, S41, S72, S73, T17, V26, V28, W7
$Cp_2Ti[C_6H_4\text{-}p\text{-}N(CH_3)_2]_2$	35 (S73)	M.p. 137°–139° (B19); PMR(CCl_4) $\delta6.08(Cp)$, $2.81(CH_3)$	B21, S72, S73
$Cp_2Ti(C_6H_4\text{-}p\text{-}CH_3)_2$	77 (S73)	M.p. 133°–134° (B19); PMR(CCl_4) $\delta6.09(Cp)$, $2.18(CH_3)$	B21, S72, S73
$Cp_2Ti(C_6H_4\text{-}p\text{-}CF_3)_2$	—	M.p. 142°–143°(dec); PMR(CCl_4) $\delta6.18$	B21
$Cp_2Ti(C_6H_4\text{-}p\text{-}F)_2$	—	M.p. 120°(dec); PMR(CCl_4) $\delta6.12$	B21
$Cp_2Ti(C_6H_4\text{-}p\text{-}Cl)_2$	—	M.p. 130°(dec); PMR(CCl_4) $\delta6.13$	B21
$Cp_2Ti(C_6H_4\text{-}p\text{-}Br)_2$	—	M.p. 130°(dec); PMR(CCl_4) $\delta6.16$	B21
$Cp_2Ti(C_6H_4\text{-}m\text{-}CH_3)_2$	87 (S73)	M.p. 135°–140°(dec); PMR(CCl_4) $\delta6.10(Cp)$, $2.16(CH_3)$	B21, S72, S73
$Cp_2Ti(C_6H_4\text{-}m\text{-}CF_3)_2$	—	M.p. 145°–146°(dec); PMR(CCl_4) $\delta6.19$	B21
$Cp_2Ti(C_6H_4\text{-}p\text{-}OCH_3)_2$	—	PMR(CCl_4) $\delta6.07(Cp)$, $3.65(OCH_3)$	B21
$(\pi\text{-Indenyl})_2Ti(C_6H_5)_2$	>50	PMR	R4

b. PENTAFLUOROPHENYL DERIVATIVES OF
BIS(CYCLOPENTADIENYL)TITANIUM(IV)

A series of pentafluorophenyl derivatives of transition metals was prepared by Stone and co-workers (T40) in 1963 by the reaction,

$$Cp_2TiCl_2 + 2LiC_6F_5 \longrightarrow Cp_2Ti(C_6F_5)_2 + 2LiCl$$

Rausch (R4), at about the same time, reported that the Grignard C_6F_5MgBr was an alternative alkylating agent for the conversion of Cp_2TiCl_2 into $Cp_2Ti(C_6F_5)_2$. The monohalides $Cp_2Ti(C_6F_5)Cl$ and $Cp_2Ti(C_6F_5)Br$ were found as by-products.

These fluorinated aryl derivatives are stable in air for months and do not decompose appreciably below their melting points, which are over 200°. Prolonged pyrolysis of $Cp_2Ti(C_6F_5)_2$ (C23) *in vacuo* at 150° results in partial decomposition to give the fluoride, $Cp_2Ti(C_6F_5)F$. Tamborski *et al.* (T11) also prepared $Cp_2Ti(C_6F_5)_2$ by a salt elimination reaction and their findings were in total agreement with the earlier work (C23, T40). It is evident from the foregoing properties that the fluorinated aryls are more chemically robust than their aryl analogs.

$Cp_2Ti(C_6F_5)OH$ can be prepared by base hydrolysis of $Cp_2Ti(C_6F_5)Cl$ (C23). The pentafluorophenyl groups in these compounds are inert toward alkalis; thus, $Cp_2Ti(C_6F_5)_2$ can withstand prolonged heating at 100° with 20% aqueous NaOH. A facile reaction occurs however, between $Cp_2Ti(C_6F_5)Cl$ and sodium ethoxide or methoxide (C22, C23). Both $Cp_2Ti(C_6F_5)_2$ and

$$Cp_2Ti(C_6F_5)Cl + NaOR \longrightarrow Cp_2Ti(C_6F_5)OR + NaCl$$

$Cp_2Ti(C_6F_5)Cl$ were shown to react with sodium methylmercaptide and the sodium derivatives of dithiols with cleavage of both C_6F_5 groups to give sulfur–titanium compounds (C22) (see Section B,4,a).

Recently, Gordon (G18) has prepared the fluorinated compounds, $Cp_2Ti(C_6F_5)Cl$, $Cp_2Ti(C_6F_5)C_6H_5$, $Cp_2Ti(C_6F_5)CH_3$, and $Cp_2Ti(C_6F_5)_2$ and has compared their PMR and mass spectra.

c. BIS(ALKYL)BIS(CYCLOPENTADIENYL)TITANIUM(IV)
DERIVATIVES

The first report of an alkyl group bonded to the Cp_2Ti entity was in 1956 when Piper and Wilkinson (P10) synthesized $Cp_2Ti(CH_3)_2$ in 1% yield from Cp_2TiCl_2 and a Grignard reagent in THF.

$$Cp_2TiCl_2 + 2CH_3MgI \xrightarrow{\text{THF}} Cp_2Ti(CH_3)_2 + 2MgClI$$

Later that year in the patent literature (F2, see also C38) the preparation of a range of mono- and bis(alkyl)titanium compounds containing Cp_2Ti

Table III-17

Bis(cyclopentadienyl)pentafluorophenyltitanium Derivatives

Compound	Yield (%)	Other data	Refs.
$Cp_2Ti(C_6F_5)_2$	44	M.p. 228°–230° (C23, T40); IR (C23); MW, monomer (C23); PMR (G18); MS (B95, G18); ^{19}F NMR (B57)	B57, B95, C23, G18, T40
$Cp_2Ti(C_6H_5)C_6F_5$	—	PMR; MS	G18
$Cp_2Ti(CH_3)C_6F_5$	—	PMR; MS	G18
$Cp_2Ti(F)C_6F_5$	24	M.p. 240°(dec) (C23, T40); sublimes 140°/10^{-3} mm (C23); IR (C23); MW, monomer (C23)	C23, T40
$Cp_2Ti(Cl)C_6F_5$	29	M.p. 187°–188° (R3), 201°–203° (C23, T40); MW, monomer (C23, R3); IR (C23); MS (B95, G18), PMR(CDCl$_3$) $\delta6.42(Cp)$ (G18, R3)	B95, C23, G18, R3, T40
$Cp_2Ti(Br)C_6F_5$	—	PMR(CDCl$_3$) $\delta6.48(t)(Cp)$	R22
$Cp_2Ti(OH)C_6F_5$	—	M.p. 183°–185°; MW, monomer; IR	C23
$Cp_2Ti(OCH_3)C_6F_5$	70	M.p. 144°–145°; IR	C22, T12
$Cp_2Ti(OC_2H_5)C_6F_5$	39	M.p. 117°; IR	C23

or similar moieties with substituted cyclopentadienyl groups was described, including that of $Cp_2Ti(CH_3)_2$ in 95% yield. The dimethyl derivative is stable to air, water, and alcohols, but sensitive to light and heat. It must be stored at low temperature (preferably $-78°$). It undergoes the expected reactions with halogens and hydrogen halides giving Cp_2TiX_2 (C38). Reactions with H_2 and CO give titanium(II) compounds (Chapter VII), while SO_2 and NO insert between titanium and one of the CH_3 groups (Section B,3,e). $Cp_2Ti(CH_3)_2$ has been claimed to be useful as a catalyst for the polymerization of olefins (B48, C38, D11), including dienes and vinyl monomers (R54).

The dissociation energy of the $Ti–CH_3$ bond was found to be 60 kcal/mole; so, although the bond is reactive, it is thermodynamically quite strong (T17). An electrochemical investigation (G36) showed that under the experimental conditions, reduction of the compound is best considered as a one-electron process involving a cleavage of the $Ti–CH_3$ bond.

The failure of a direct reduction of bis(pentamethylcyclopentadienyl)-titanium dichloride, $[\pi\text{-}(CH_3)_5C_5]_2TiCl_2$, to yield $[(CH_3)_5C_5]_2Ti$ prompted Brintzinger *et al.* (B43) to prepare $[\pi\text{-}(CH_3)_5C_5]_2Ti(CH_3)_2$ using methyllithium in diethyl ether. However, unlike $Cp_2Ti(CH_3)_2$, the yellow pentane solutions of $[\pi\text{-}(CH_3)_5C_5]_2Ti(CH_3)_2$ were found to show little or no reactivity toward H_2 even at 100 atm. pressure. $[\pi\text{-}(CH_3)_5C_5]_2Ti(CH_3)_2$ can be stored at room temperature for days without decomposition and is stable in solution to $90°$. On heating in toluene at $110°$ for 4 hr decomposition occurred with evolution of 1 mole of methane and formation of a diamagnetic turquoise compound whose structure as either **9** or **10** was inferred from PMR and mass spectra data. An observation consistent with structure **9** was the presence of an

CH₃ structures (9) and (10)

(9)   (10)

olefinic C–H stretch at 3040 cm^{-1} in the infrared spectrum. This turquoise compound reacts readily with H_2 in pentane at $0°$ to yield $[(CH_3)_5C_5]_2Ti$ (see Chapter VII), which could be converted to $[\pi\text{-}(CH_3)_5C_5]_2TiH_2$.

First mention of bis(benzyl)bis(cyclopentadienyl)titanium appeared in a review in 1970 (D14). Recently, Fachinetti and Floriani (F1) have investigated the thermal decomposition and the carbonylation of this compound (see Chapter VII, Sections A and C).

Rausch and co-workers (R4) have prepared bis(π-indenyl)titanium dimethyl and diphenyl derivatives in > 50% yield. A study of trends in the PMR spectra of these compounds has been conducted and their chemical reactivity is being examined.

d. MISCELLANEOUS σ-BONDED Ti–C COMPOUNDS

In two patents in 1962 Feay claimed the preparation of a new red organotitanium–aluminum complex by heating Cp_2TiCl_2 with trimethylaluminum (F7, F8). PMR data was in agreement with the structure,

$$
\begin{array}{c}
Cp \diagdown \quad \diagup CH_2 \!-\! Al \diagup ^{CH_3} \\
\quad Ti \quad \quad Cl \quad \quad Cl \\
Cp \diagup \quad \diagdown CH_2 \!-\! Al \diagdown _{CH_3}
\end{array}
$$

The molecular weight was high, indicating association.

A series of metal derivatives in which a barene nucleus and the metal atom are separated by a methylene group have been prepared (Z1) from the metal halides and the Grignard reagent, 1-bromomagnesiomethylbarene. With Cp_2TiCl_2 a high yield of the stable bis(σ-barenylmethyl) compound, Cp_2Ti-$(CH_2CB_{10}H_{10}CH$-$o)$ was obtained. In contrast to the σ-alkyl and σ-aryl compounds of Cp_2Ti^{IV}, the σ-barenyl methyl derivative is stable to air for long periods at 20°, decomposing only on heating at 185° and resembling $Cp_2Ti(C_6F_5)_2$ in its stability (C16). On prolonged heating in toluene, the compound decomposes with formation of methylbarene.

The polarographic reduction of this and other $Cp_2Ti(σ$-$R)_2$-type compounds has been investigated (G36).

Recently the novel alkyl derivatives obtained by the interaction of transition metal halides or complex halides with trimethylsilylmethyllithium have been reported (C42, Y1). A compound of interest was that formed from Cp_2TiCl_2.

$$Cp_2TiCl_2 + 2LiCH_2Si(CH_3)_3 \longrightarrow Cp_2T[CH_2Si(CH_3)_3]_2 + 2LiCl$$

The thermal stability of this compound has been attributed to the bulkiness of the ligands, which shields the metal from attack, and to the elimination of decomposition pathways, such as $β$ elimination of hydrogen (B78, M49). This work has been extended to include $Cp_2Ti[CH_2Ge(CH_3)_3]_2$ (C41a).

Bis(cyclopentadienyl)bis(σ-phenylethynyl)titanium was prepared by the reaction,

$$Cp_2TiCl_2 + 2NaC\equiv CC_6H_5 \longrightarrow Cp_2Ti(C\equiv CC_6H_5)_2 + 2NaCl$$

It can also be prepared from the organometallic amide, $Cp_2Ti[N(CH_3)_2]_2$, by reaction with phenylacetylene (J7).

Table III-18
Alkylbis(cyclopentadienyl)titanium Derivatives, Including Ethynyl Derivatives

Compound	Color	Highest yield (%)	Other data	Refs.
$Cp_2Ti(CH_3)_2$	Yellow-orange	95 (F2)	M.p. 97°(dec) (B48, C38, F2, P10); sublimes 40°/10⁻⁴ mm (P10); MW, monomer (C38); IR (P10); UV (P10); thermochemistry (T17); polarographic reduction (G38); PMR(CCl₄) δ5.90(Cp), 0.06(CH₃) (B21, P10)	B21, B48, C38, D11, F2, G36, L12, M23, P10, R27, R30, R31, T17, W7, W8
$[\pi\text{-}(CH_3)_5C_5]_2Ti(CH_3)_2$	Yellow	80	MS	B43
$Cp_2Ti(CH_2C_6H_5)_2$	—	—	M.p. 100°(dec) (D14); PMR δ5.6(Cp), 1.90(CH₂), 7.4-6.7(C₆H₅) (F1)	D14, F1, R21a
$(\pi\text{-}CH_3C_5H_4)_2Ti(CH_3)_2$	Yellow	65 (F2)	—	F2
$(\pi\text{-}Indenyl)_2Ti(CH_3)_2$	—	>50	PMR	R4
$Cp_2Ti\begin{bmatrix} CH_2\text{—}Al\text{—}CH_3 \\ Cl\ Cl \\ CH_2\text{—}Al\text{—}CH_3 \end{bmatrix}$	Dark red	Quantitative	PMR δ4.92(Cp), 2.47(AlCH₃), 0.37(Al-CH₂-Ti)	F7-F9
$Cp_2Ti\left[CH_2\text{—}C\text{—}CH\ \cdot\ B_{10}H_{10} \right]_2$	Orange	85	M.p. 185°(dec) (Z1); polarographic study (G36)	G36, Z1
$Cp_2Ti[CH_2Si(CH_3)_3]_2$	Yellow-orange	60	M.p. 185° (Y1); sublimes 50°/10⁻⁴ mm (Y1); MW (Y1); PMR(benzene) δ5.89(Cp), 0.90(CH₂), 0.03(CH₃)(W28); (C₆D₆) δ5.90(Cp), 0.96(CH₂), 0.10(CH₃), (CDCl₃) δ6.11(Cp), 0.89(CH₂), −0.06(CH₃); IR (W28)	C41a, C42, W28, Y1, Y3,
$Cp_2Ti[CH_2Ge(CH_3)_3]_2$	Orange	70	M.p. 136°-138°; IR; PMR(C₆D₆) δ5.87(Cp), 1.12(CH₂), 0.12(CH₃)	C41a
$Cp_2Ti(C{\equiv}CC_6H_5)_2$	Orange-brown	88	MW, monomer (K52); IR (J7, K52, T18); UV (K52, T18); PMR(CS₂) δ7.23(C₆H₅), 6.38(Cp) (J7, K52)	J7, K52, T18
$Cp_2Ti(C{\equiv}CCF_3)_2$	Orange	16	M.p. 125°(dec); IR; ¹⁹F NMR; PMR	B94

e. MIXED ALKYL (OR ARYL) HALIDO DERIVATIVES

The first characterization of this type of compound was reported in the patent literature (B48, F2). Orange-red $Cp_2Ti(CH_3)Cl$ was prepared in 90% yield by the reaction,

$$Cp_2TiCl_2 + Al(CH_3)_3 \xrightarrow[20°]{CH_2Cl_2} Cp_2Ti(CH_3)Cl + (CH_3)_2AlCl$$

A large range of alkyl (or aryl) bis(cyclopentadienyl)titanium chlorides, $Cp_2Ti(R)Cl$, has been synthesized by Waters and Mortimer (W13) and others (B48, F2, H5, L23) (see Table III-19) using a variety of alkylating agents. Beachell and Butter (B21) noted that to prepare pure $Cp_2Ti(CH_3)X$ (where X = Br or I), the corresponding Cp_2TiX_2 compound must be treated with the appropriate Grignard reagent (L20) since halogen exchange occurred during the reaction.

The compounds are normally orange to red and can be handled in air for short periods without appreciable hydrolysis. The methyl derivative dissolves in cold water without decomposition, but in acid the orange solution turns red with formation of methane. In the thermolysis of a series of these compounds, $Cp_2Ti(R)Cl$, it was noted (W14) that after an initial unsteady period, first-order loss of the R group occurred. The activation energy of bond breaking was approximately 25 kcal/mole for all compounds.

The alkyl derivatives have been studied mainly in connection with the olefin polymerization catalysts formed between Cp_2TiCl_2 and alkylaluminum derivatives (B3, B48, C38, F2, H5, L20). Electron spin resonance studies have been carried out (B13) on systems of the type $Cp_2Ti(CH_3)Cl + CH_3AlCl_2$ and $AlCl_3$.

More recently, the insertion reactions of $Cp_2Ti(CH_3)Cl$ with NO and SO_2 have been studied (W7, W8).

f. ACYL DERIVATIVES

Floriani and Fachinetti (F21) reported some oxidative addition reactions of $Cp_2Ti(CO)_2$ with acyl and alkyl halides and halogens, carried out at room temperature in toluene, e.g.,

$$Cp_2Ti(CO)_2 + RCOCl \xrightarrow{24\ hr} Cp_2Ti(Cl)\overset{\displaystyle O}{\overset{\displaystyle \|}{C}}R + 2CO$$

$$Cp_2Ti(CO)_2 + RI \longrightarrow Cp_2Ti(I)\overset{\displaystyle O}{\overset{\displaystyle \|}{C}}R + CO$$

The yellow-orange, air-stable acyl derivatives were obtained in 80% yield. They were insoluble in hydrocarbons, but soluble in chloroform in which the

Table III-19

Alkyl- or Arylbis(cyclopentadienyl)titanium Halides

Compound	Other data	Refs.
$Cp_2Ti(CH_3)Cl$	M.p. 168°–170°(dec) (B21, C36, F2); polarography (G36); IR (W13); M.p. 91.6° (W13, see comment); UV (W13); PMR(CDCl$_3$); δ6.34(Cp), 0.72(CH$_3$) (W13); (CCl$_4$) δ6.19(Cp), 0.68(CH$_3$) (B21); thermochemistry (W14)	B3, B13, B21, C38, F2, G36, H5, L20, W7, W8, W13, W14
$Cp_2Ti(CH_3)Br$	PMR(CDCl$_3$) δ6.28(Cp); 0.41(CH$_3$)	B21
$Cp_2Ti(CH_3)I$	PMR(CDCl$_3$) δ6.36(Cp), 0.2(CH$_3$)	B21
$Cp_2Ti(C_2H_5)Cl$	M.p. 89°–94° (L23, W13); thermochemistry (F30); IR (W13); UV (W13); PMR(CDCl$_3$) δ6.30(Cp), 1.64(CH$_2$), 1.23(CH$_3$) (W13)	F30, H5, L23, W13
$Cp_2Ti(n\text{-}C_3H_7)Cl$	M.p. 75° (W13), 160°(dec) (F2)	B50, F2, W13
$Cp_2Ti(n\text{-}C_4H_9)Cl$	M.p. 85°; IR; UV; PMR(CDCl$_3$) δ6.16(Cp), 1.58(CH$_2$), 0.79(CH$_3$)	W13
$Cp_2Ti(i\text{-}C_4H_9)Cl$	M.p. 71°; IR; UV; PMR(CDCl$_3$) δ6.23(Cp), 1.58(CH$_2$), 0.75(CH$_3$); thermochemistry (W14)	H5, W13, W14
$Cp_2Ti(n\text{-}C_5H_{11})Cl$	M.p. 52°; IR; UV; PMR(CDCl$_3$) δ6.20(Cp), 1.61(CH$_2$), 0.83(CH$_3$)	W13
$Cp_2Ti(neo\text{-}C_5H_{11})Cl$	M.p. 95°; IR; UV; PMR(CDCl$_3$) δ6.52(Cp), 2.31(CH$_2$), 0.91(CH$_3$)	W13
$Cp_2Ti(n\text{-}C_8H_{17})Cl$	—	H5
$Cp_2Ti(CH_2CH_2C_6H_5)Cl$	M.p. 139°; IR; UV; PMR(CDCl$_3$) δ6.26(Cp), 7.10(C$_6$H$_5$), 2.75, 1.85(CH$_2$); thermochemistry (W14)	W13, W14
$Cp_2Ti(CH_2C_6H_5)Cl$	M.p. 107°; IR; UV; PMR(CDCl$_3$) δ6.53(Cp), 7.18(C$_6$H$_5$), 2.92(CH$_2$); thermochemistry (W14)	W13, W14
$Cp_2Ti(C_6H_5)Cl$	M.p. 121°; IR; UV; PMR(CDCl$_3$) δ6.33(Cp), 6.85(C$_6$H$_5$)	W13, W14
$Cp_2Ti(CH=CH_2)Cl$	M.p. 153°(dec); IR; UV; PMR(CDCl$_3$) δ6.58(Cp)	W13

95

Table III-20

Acylbis(cyclopentadienyl)titanium Halides[a]

Compound	Physical data
$\underset{\parallel}{Cp_2Ti(Cl)\overset{O}{C}}-CH_3$	IR; MS; PMR(CDCl₃) δ5.8(Cp), 3.0(CH₃)
$Cp_2Ti(I)\overset{O}{\underset{\parallel}{C}}-CH_3$	IR; PMR(CDCl₃) δ5.8(Cp), 3.0(CH₃)
$Cp_2Ti(I)\overset{O}{\underset{\parallel}{C}}-C_2H_5$	IR; PMR(CDCl₃) δ5.8(Cp), 3.28(CH₂), 1.4(CH₃)
$Cp_2Ti(Cl)\overset{O}{\underset{\parallel}{C}}-C_6H_5$	IR; MW, monomer; PMR(CDCl₃) δ7.9(C₆H₅), 5.9(Cp)

[a] From ref. (F21).

benzoyl derivative was monomeric. Their structure was established by infra-red, PMR, and mass spectra.

g. METALLOCYCLES CONTAINING σ-BONDED CARBON
ATTACHED TO THE BIS(CYCLOPENTADIENYL)TITANIUM(IV)
MOIETY

From the reaction between $TiCl_4$ and excess sodium cyclopentadienide in the presence of diphenylacetylene, Vol'pin *et al.* isolated a black-green, air-stable crystalline solid (V19), which was considered to form from "nascent" titanocene. The compound can be prepared also from the di-

$$[(C_5H_5)_2Ti] + 2C_6H_5C{\equiv}CC_6H_5 \longrightarrow Cp_2Ti(C_6H_5C_2C_6H_5)_2$$

carbonyl, $Cp_2Ti(CO)_2$, and diphenylacetylene (S60). On the basis of PMR, infrared, and electronic absorption spectral data, the complex was formulated as a titanacyclopentadiene (S60). The structure was confirmed by direct synthesis (S42). The same compound was obtained by Watt and Drummond

(W18) from Cp_2TiCl_2, $C_6H_5C{\equiv}CC_6H_5$, and sodium, although it was origin-ally thought to be the titanium(II) complex, $Cp_2Ti(C_6H_5C{\equiv}CC_6H_5)_2$.

With phenylacetylene, $Cp_2Ti(CO)_2$ formed in 7% yield a green, air-stable complex, probably of similar structure (S60).

Masai *et al.* (M23) reacted $Cp_2Ti(C_6H_5)_2$ with diphenylacetylene in refluxing benzene under nitrogen and formed a dark green, air-stable organotitanium complex which was thought to be 1,1-bis(cyclopentadienyl)-2,3-diphenyltitanaindene.

$$Cp_2Ti \overset{C_6H_5 \; C_6H_5}{\diagup\diagdown}$$

A similar ring structure was obtained in 16% yield (R6) by condensation of the butyllithium–tolan dilithium reagent with Cp_2TiCl_2 in diethyl ether.

$$Cp_2TiCl_2 + \left[\begin{array}{c} C_4H_9 \\ C_6H_5 \end{array} \right]^{2-} \cdot 2Li^+ \xrightarrow{(C_2H_5)_2O} Cp_2Ti \overset{C_6H_5 \; C_4H_9}{} + 2LiCl$$

From 2,2′-dilithiobiphenyl and Cp_2TiCl_2 in diethyl ether a titanafluorene derivative was isolated as bright red crystals in 12% yield. The F_8 analog of

$$\underset{Cp \quad Cp}{Ti}$$

the titanafluorene has been made by Cohen and Massey (C39) by the same method. The compound showed high thermal stability, being decomposed *in vacuo* only after 24 hr at temperatures in excess of 330°.

Kolomnikov *et al.* (K38) investigated the reaction of $Cp_2Ti(C_6H_5)_2$ at 80°–90° in xylene with CO_2 at atmospheric pressure. From this reaction, the mechanism of which has been discussed in Section B,5,a, was isolated a diamagnetic, air-stable, red crystalline solid in 35% yield. Its identity as a metallocyclic carboxylate was confirmed by X-ray analysis (see Fig. III-12) (A4, K38).

6. Nitrogen-Bonded and Phosphorus-Bonded Compounds of Bis(cyclopentadienyl)titanium(IV)

Bis(cyclopentadienyl)bis(dimethylamido)titanium was isolated in low yield from the reaction between Cp_2TiCl_2 and lithium dimethylamide. The

$$Cp_2TiCl_2 + 2LiN(CH_3)_2 \longrightarrow Cp_2Ti[N(CH_3)_2]_2 + 2LiCl$$

Table III-21

Metallocyclic Compounds Containing Titanium

Compound	Color	Yield (%)	Other data[a]	Refs.
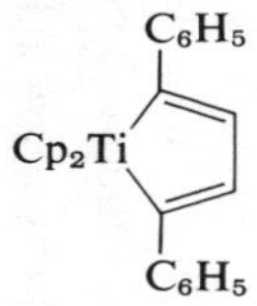 Cp₂Ti with ring bearing C₆H₅, C₆H₅	Green	7	M.p. 134°–136°; IR; UV; PMR δ6.2(Cp), 6.7(m)(C₆H₅)	S60
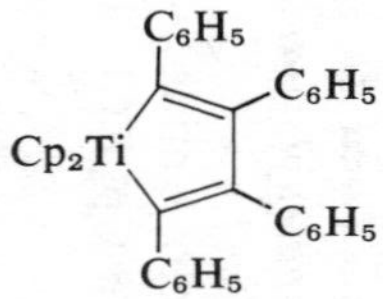 Cp₂Ti with ring bearing C₆H₅, C₆H₅, C₆H₅, C₆H₅	Green	16	M.p. 150°; IR (S42, S60); UV (S42, S60); MW, monomer-(benzene) (S42, V19, W18); X-ray diffraction (W18); PMR δ6.2(Cp), 6.8(m)(C₆H₅) (S42, S60)	S42, S60, V19, W18
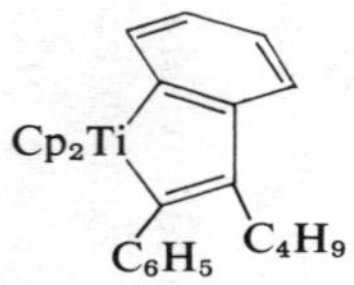 Cp₂Ti with indene ring bearing C₄H₉, C₆H₅	Burgundy	16	M.p. 143°–146°; MS; PMR(CDCl₃) δ7.38–6.57 (aromatic, 8H,m), 6.23(Cp, 10H, s) 5.98 (aromatic 1H, d of t), 1.93, 1.27, and 0.78(C₄H₉, 9H, br.t, br.m, and apparent d)	R6
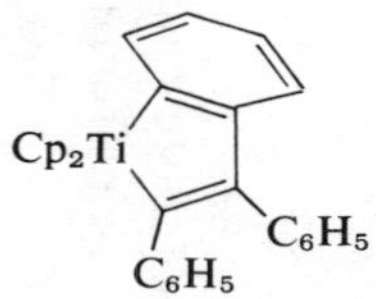 Cp₂Ti with indene ring bearing C₆H₅, C₆H₅	Green	46	M.p. 227°–228°; MW, monomer(benzene); IR; PMR(CCl₄) δ6.6 (C₆H₅), m) 6.1(Cp)	M23

98

Structure	No.	Color	Properties	Ref.
	12	Red	M.p. 143°–144°; MS; PMR(CDCl₃) δ6.24(Cp, 10H), δ7.48(2H, d of t)	R6
	25	—	M.p. 230°; IR	C39
	35	Red	M.p. 192°–195°(dec) (K38); IR (K38); PMR (K38) δ6.5–7.55(C₆H₄), 6.45(Cp) (K38); X-ray study (A4, K38)	A4, K38
	—	Green	Suggested intermediate (D40, K38, M23, S37); possible intermediate (R22–R24, R28)	D40, K38, M23, R22–R24, R28, S37

^{a} s, Singlet; d, doublet; t, triplet; m, multiplet; br, broad.

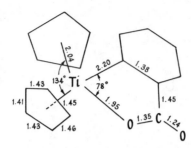

Fig. III-12.   Structure of Cp$_2$TiO$_2$CC$_6$H$_4$ (K38).

compound was obtained (C15) as a dark brown, viscous liquid boiling at 120°–128°/0.2 mm and showing a monomeric molecular weight in benzene.

Bis(amido) derivatives of Cp$_2$Ti^{IV} have been prepared by Issleib and Bätz (I3) from Cp$_2$TiCl$_2$ and an alkali metal derivative of pyrrole, indole, carbazole, or phthalimide (see Table III-22). Tille has also studied the reaction of lithium pyrrole with metal halides including Cp$_2$TiCl$_2$ (T35). The product, Cp$_2$Ti-(NC$_4$H$_5$)$_2$, was dark brown and sublimed at 185°/1 mm. PMR and infrared data led the author to conclude that the pyrrole was N-bonded to the metal.

Lappert and co-workers have prepared a range of metal ketimides (C40) including the organotitanium derivatives, Cp$_2$Ti(Cl)N=CR$_2$ (where R = C(CH$_3$)$_3$ or C$_6$H$_5$). These products were formed in high yield by a ligand exchange reaction between LiN=CR$_2$ and Cp$_2$TiCl$_2$. The bis(substituted) derivatives were not obtained from LiN=CR$_2$ [R=C(CH$_3$)$_3$ or C$_6$H$_5$], but with LiN=C(CF$_3$)$_2$ at 6° a significant amount of Cp$_2$Ti[N=C(CF$_3$)$_2$]$_2$ was obtained (C14). The mixed derivative, Cp$_2$Ti(Cl)N=C(CH$_3$)C$_6$H$_5$, was prepared in relatively low yield from LiN=C(CH$_3$)C$_6$H$_5$, formed *in situ* from lithium methyl and phenyl cyanide.

A chlorosilane elimination has been used to prepare Cp$_2$Ti[N=C(C$_6$H$_5$)$_2$]Cl in high yield (C41). No purification of the product, apart from washing with

$$Cp_2TiCl_2 + (CH_3)_3SiN{=}C(C_6H_5)_2 \xrightarrow[\text{10 hr, 120°}]{\text{xylene}} Cp_2Ti[N{=}C(C_6H_5)_2]Cl + (CH_3)_3SiCl$$

a small quantity of hexane, was required. The compounds were air-sensitive solids which could be sublimed at 150°/0.001 mm and could be handled in air for short periods without appreciable hydrolysis.

Anagnostopoulos and Nicholls (A12) studied the degree of solvolysis of Cp$_2$TiCl$_2$ by ammonia and monomethylamine under a variety of conditions. At −37° only one Ti–Cl bond was broken, giving mixtures of Cp$_2$TiCl(NHR)·nRNH$_2$ (where R = H or CH$_3$) and the amine hydrochloride.

Bis(cyclopentadienyl)titanium(III) chloride reacts with organic azides with the liberation of nitrogen and the formation of nitrogen-bridged titanium derivatives of varying thermal stability (C60). With anhydrous HN_3, Cp_2TiCl

$$2Cp_2TiCl + RN_3 \longrightarrow (Cp_2TiCl)_2NR + N_2$$

formed a similar nitrogen-bridged product, $(Cp_2TiCl)_2NH$, with liberation of nitrogen (C71).

Nitrogen-bonded pseudohalides of bis(cyclopentadienyl)titanium(IV) are discussed in Section B,2.

Issleib et al. (I5) have recently prepared the first bis(cyclopentadienyl)-titanium(IV) phosphido derivative, bis(cyclopentadienyl)(1,2,3-triphenyltri-phosphanato)titanium(IV), from the reaction.

$$M_2P_4(C_6H_5)_4 + Cp_2M'X_2 \longrightarrow Cp_2M' \overset{\displaystyle \overset{C_6H_5}{\underset{|}{P}}}{\underset{\underset{|}{C_6H_5}}{\overset{P}{\diagup}}} \hspace{-0.5em} P{-}C_6H_5 + 2MX + \frac{1}{n}(PC_6H_5)_n$$

$$M = Li \text{ or } Na; \quad M' = Ti \text{ or } Zr; \quad X = Cl \text{ or } Br$$

The titanium compound was dark violet, diamagnetic, and relatively stable to air in the solid state. The four-membered metal–phosphorus ring structure followed from the mass and ^{31}P NMR spectra. The complexes are monomeric in dioxane.

7. Bimetallic Complexes of Bis(cyclopentadienyl)titanium(IV)

Compounds which contain a direct metal–metal bond, as well as those which contain two metals, separated by bridging groups, are the subject of this section.

The first attempt to prepare compounds with titanium–metal bonds was by Dickson and West (D20) from the reaction of Cp_2TiCl_2 with $NaPb$-$(CH_2CH_3)_3$ or $LiSn(CH_2CH_3)_3$. The unstable bimetallic products postulated were $Cp_2Ti(NH_2)Pb(CH_2CH_3)_3$ and $Cp_2Ti(Cl)Sn(CH_2CH_3)_3$, but their characterization was incomplete.

It was not until 1967 that the first titanium–metal-bonded compounds were successfully isolated and characterized (C65). These were triphenyl-germyl and triphenylstannyl derivatives. The corresponding lead compound

$$Cp_2TiCl_2 + Na(Li)M(C_6H_5)_3 \xrightarrow{\text{THF}} Cp_2Ti\overset{\displaystyle \diagup M(C_6H_5)_3}{\underset{\diagdown Cl}{}} + Na(Li)Cl$$

Table III-22
Amido and Phosphido Derivatives of Bis(cyclopentadienyl)titanium(IV)

Compound	Color	Yield (%)	Other data	Refs.
Cp$_2$Ti[N(CH$_3$)$_2$]$_2$	Brown	—	B.p. 120°–128°/0.2 mm; MW, monomer(benzene); MS	C15, F4
Cp$_2$Ti(pyrrolyl)$_2$	Red-brown	97	M.p. 175°–177°(dec) (I3); MW (I3); IR (I3, T35); dipole moment (I3); PMR(C$_6$D$_6$) δ6.57(t), 6.37(t)(C$_4$H$_4$N), 5.52(Cp)	I3, T35
Cp$_2$Ti(indolyl)$_2$	Dark green	66	M.p. 170°–171°(dec); MW, IR; dipole moment	I3
Cp$_2$Ti(carbazolyl)$_2$	Dark green	25	M.p. 205°–208°; IR	I3
Cp$_2$Ti(phthalimido)$_2$	Yellow	33	M.p. 175°(dec); MW; IR	I3

Compound	Color	Yield (%)	Properties	Ref.
Cp$_2$Ti(Cl)N=C(C$_6$H$_5$)$_2$	Yellow	70	MS; IR; PMR(C$_6$D$_6$) δ5.2(Cp)	C40, C41
Cp$_2$Ti(Cl)N=C(p-CH$_3$C$_6$H$_4$)$_2$	Yellow	70	MS; IR; PMR(C$_6$D$_6$) δ7.3(C$_6$H$_4$), 5.95(Cp), 2.3(CH$_3$)	C41
Cp$_2$Ti(Cl)N=C(tert-C$_4$H$_9$)$_2$	Yellow	70	MS; IR; MW; PMR(C$_6$D$_6$) δ5.08(Cp), 0.68, 0.29(C$_4$H$_9$)	C40, C41
Cp$_2$Ti(Cl)N=C$\begin{smallmatrix}CH_3\\C_6H_5\end{smallmatrix}$	Yellow	Low	MS; IR; PMR(C$_6$D$_6$) δ5.88, 5.87(Cp) 2.27, 1.87(CH$_3$)	C41
Cp$_2$Ti(Cl)N=C$\begin{smallmatrix}CF_3\\CF_3\end{smallmatrix}$	—	60–80	M.p. 194°–195°(dec); sublimes <85°/10^{-3} mm; MS; IR	C14
(Cp$_2$TiCl)$_2$NC$_6$H$_5$	Brown	60–90	M.p. 185°(dec); IR; PMR	C60
(Cp$_2$TiCl)$_2$N(p-ClC$_6$H$_4$)	Red-brown	70	M.p. 150°(dec); IR; PMR	C60
(Cp$_2$TiCl)$_2$NH	Red-brown	>90	MW	C71
Cp$_2$Ti$\begin{smallmatrix}C_6H_5\\P\diagdown P-C_6H_5\\P\diagup\\C_6H_5\end{smallmatrix}$	Violet	40–60	M.p. 243°–245°; MW; MS; ^{31}P NMR	I5

was unstable; the green reaction solution decomposed even at low temperature giving metallic lead. The bis(triphenylsilyl) derivative, $Cp_2Ti[Si(C_6H_5)_3]_2$, was prepared by a similar method as a yellow sublimable solid (H7), but other attempts to prepare titanium–silicon compounds have failed (C9, C65, K31).

The burgundy colored germanium analog is less stable [m.p. 171°–173° (C72); 139°–141° (R21b)] while the green tin analog decomposes at 80° (C65). Recently some reactions of $Cp_2Ti[Ge(C_6H_5)_3]_2$ have been studied (R21d).

Silylation of Cp_2TiCl_2 with $KSiH_3$ yields olive green $(Cp_2TiSH_2)_2$, soluble in benzene (H6a). Crystal structure determination shows the molecule consists of two Cp_2Ti units linked by two SiH_2 bridges (see Table I-1).

Many bis(cyclopentadienyl)titanium(IV) compounds of general formula,

$$
\begin{array}{c}
R \\
| \\
Cp \diagdown \quad X \\
\quad Ti \diagup \diagdown MLn \\
Cp \diagup \quad X \\
| \\
R
\end{array}
$$

X = O, S or Se; M = Mo, W, Cr, Cu, Pd

have been prepared (B80, C6, D1, D7, J16, J17, K45). The question of metal–metal bonding within four-membered rings of this type has been discussed in Section B,4,b.

The trinuclear phenoxo-bridged complex, $Cp_2Ti(OC_6H_5)_2Mo(CO)_2$-$(OC_6H_5)_2TiCp_2$, was prepared (K48) as a yellow unstable solid from $Cp_2Ti(OC_6H_5)_2$ and $Mo(CO)_6$ or $(CH_3CN)_2Mo(CO)_4$. The cis structure of the complex follows from the appearance of two carbonyl IR bands at 1900 and 1950 cm^{-1} and two PMR signals ($\delta6.37$ and 6.44) caused by two pairs of Cp groups in different environments.

Carter has reported some preliminary results (C13) on the formation of titanium–boron-bonded compounds. Cp_2TiCl_2 was treated with $NaB_{10}H_{13}$ in THF giving a product formulated as $Cp_2Ti(B_{10}H_{13})_2$. The compound had limited stability and decomposed with the evolution of hydrogen.

Another novel route to a mixed transition–metal complex involved the reaction below (Y3).

$$Cp_2Ti(C{\equiv}CC_6H_5)_2 + Ni(CO)_4 \xrightarrow[\text{room temp.}]{\text{benzene}} Cp_2Ti \begin{array}{c} C{\equiv}C \diagup C_6H_5 \\ \diagup \quad \diagdown NiCO \\ C{\equiv}C \diagdown C_6H_5 \end{array}$$

Table III-23

Bimetallic Complexes of Bis(cyclopentadienyl)titanium(IV)

Compound	Color	Other data	Refs.
$Cp_2Ti(NH_2)Pb(C_2H_5)_3$	Buff	—	D20
$Cp_2Ti(Cl)Sn(C_2H_5)_3$	Yellow-brown	—	D20
$Cp_2Ti(Cl)Sn(C_6H_5)_3$	Green	M.p. 177°–180°(dec); PMR(CDCl$_3$) δ7.32(C$_6$H$_5$), 6.25(Cp)	C65
$Cp_2Ti(Cl)Ge(C_6H_5)_3$	Green	M.p. 193°–196°(dec); PMR(CDCl$_3$) δ7.30(C$_6$H$_5$), 6.15(Cp); MW, monomer (benzene) (R21b)	C65, R21b
$Cp_2Ti(Cl)Pb(C_6H_5)_3$	Green	Unstable	C65
$Cp_2Ti[Sn(C_6H_5)_3]_2$	Green	M.p. 80°(dec); PMR(CDCl$_3$) δ7.10(C$_6$H$_5$), 6.77(Cp)	C65
$Cp_2Ti[Ge(C_6H_5)_3]_2$	Burgundy	M.p. 171°–173° (C72), 139°–141° (R21b); IR (C72); PMR(CDCl$_3$) δ7.15(C$_6$H$_5$), 6.80(Cp); MW, monomer (benzene) (R21b)	C65, C72, R21b, R21d
$(Cp_2TiSiH_2)_2$	Olive green	Crystal structure	H6a
$Cp_2Ti[Si(C_6H_5)_3]_2$	Yellow	Sublimes at 280°	H7
$Cp_2Ti(SR)_2MLn$	—	For data on this type of compound, see Section B,4,b	—
$[Cp_2Ti(OC_6H_5)]_2Mo(CO)_2$	Yellow	PMR(CS$_2$) δ6.44, 6.37(Cp); IR	K48
$Cp_2Ti(B_{10}H_{13})_2$	—	—	C13
$[Cp_2Ti(C{\equiv}CC_6H_5)_2]_2NiCO$	Olive green	M.p. 141°–145°(dec); IR; MS; MW, monomer; PMR(CDCl$_3$) δ5.51(Cp), 7.75–7.20(C$_6$H$_5$)	Y3

105

C. Tris(cyclopentadienyl)titanium(IV) Chloride

This compound is the sole representative of this class of compounds. The synthesis (C5) involved the reaction of Cp_2TiCl_2 with cyclopentadienyl-magnesium chloride in methylene dichloride at $-78°$. After removal of unreacted Cp_2TiCl_2, a trace of an orange crystalline compound was obtained by extraction with toluene. The product, which was not characterized by analysis, was judged to be tris(cyclopentadienyl)titanium chloride on the basis of a single peak in the PMR spectrum in toluene at $25°$ at $\delta 5.75$, which remained a singlet down to $-100°$.

D. Tetracyclopentadienyltitanium(IV)

The first report of tetracyclopentadienyltitanium(IV) was in a patent by Breedeveld and Waterman (B84); however, no details of its properties or possible structure were given. The brown sample was probably impure and partly polymeric.

More recently, Siegert and de Liefde Meijer (S40) prepared the compound by reaction of Cp_2TiCl_2 with 2 moles of sodium cyclopentadienide. The violet-black, oxygen-sensitive product was isolated in 39% yield as a thermally stable crystalline solid melting at $128°$. It can be converted into $Cp_2Ti(h^2\text{-}C_5H_5)$ by heating *in vacuo*. With HCl it reacts to form Cp_2TiCl_2 and cyclopentadiene. This chemical evidence, coupled with infrared and PMR data, led Siegert and de Liefde Meijer (S40) to formulate tetracyclopentadienyltitanium as $Cp_2Ti(\sigma\text{-}C_5H_5)_2$.

The crystal and molecular structures of the compound (C4) show unambiguously that the molecule contains two *pentahapto-* and two *monohapto-* cyclopentadienyl rings (see Fig. III-13). The metal therefore has a sixteen-electron configuration.

The fluxional behavior has been fully reported (C5) in a study of the

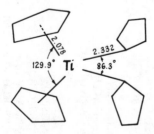

Fig. III-13.   Structure of tetra(cyclopentadienyl)titanium (C4).

temperature dependence of the PMR spectrum from $+80°$ to $-140°$. The observations were explained by assuming that the molecule has the same structure in solution as in the solid state and that it executes two types of intramolecular rearrangements. At the lowest temperature $(-140°)$ the σ-C_5H_5 rings rotate relatively slowly, but by about $-25°$ averaging is rapid enough to give a single line as narrow as that for the protons of the Cp rings. Between about $-20°$ and $+35°$ the σ-C_5H_5 and Cp lines broaden and coalesce to a single line which continues to become narrower as the temperature is raised further. The relatively ready occurrence of ring interchange has been attributed to the presence of an empty valence shell orbital on the titanium atom which makes a reasonably low-lying transition state accessible.

These findings lend considerable support to the claim (C3) that tetracyclo-pentadienyltitanium provides the first genuine example of the occurrence of rapid interchange of *monohapto*-cyclopentadienyl- and *pentahapto*-cyclopentadienyl rings as part of the fluxional behavior of the molecule.

E. Bis(cyclooctatetraene)titanium

This air-sensitive, violet-black compound was first made by the reaction (S66),

$$TiCl_4 + 2Na_2C_8H_8 \xrightarrow[\text{2 days, 70°}]{\text{benzene}} Ti(C_8H_8)_2 + 4NaCl$$

An 80% yield was obtained (B85) from $Ti(OC_4H_9)_4$, C_8H_8 and triethyl-aluminum, while electrochemical synthesis from $TiCl_4$ and C_8H_8 gave 45% (L14). Dietrich and Soltwisch (D23) have shown by X-ray diffraction that one C_8H_8 group is planar, the other being boat-shaped (Fig. III-14). The planar ligand is symmetrically attached to the metal; all Ti–C bond lengths are 2.32 ± 0.02 Å. However, the other C_8H_8 group is bound via only four carbons, giving rise to two pairs of Ti–C bond lengths, namely, 2.2 and 2.5 Å. Another form of the molecule is thought to exist in the crystal state in which the bent ring is bonded to the metal atom only through three carbon atoms.

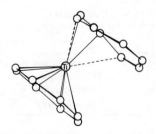

Fig. III-14. Structure of bis(cyclooctatetraene)titanium (D23).

The oxidation state of the metal in this complex is uncertain. If a bent C_8H_8 ligand is taken to imply that it is uncharged, one may consider the metal to be in its divalent state in this complex. Recently, Schwartz and Sadler (S19a) have shown by variable temperature PMR spectroscopy that the two C_8H_8 rings are fluxional, bending, and flattening in turn, and have hence suggested that this process involves an intramolecular redox reaction. The activation energy ($\Delta G^{\ddagger}$) for this process is 16.7 ± 0.2 kcal/mole at $63°$.

F. π-Pyrrolyl Compounds

These diamagnetic, liquid compounds which are unstable in solution, were made by Bradley and Chivers (B61) by the reaction,

$$Ti(NR_2)_4 + xC_4H_4NH \xrightarrow{\text{toluene}} (C_4H_4N)_xTi(NR_2)_{4-x} + xR_2NH$$
$$x = 1 \text{ or } 2$$

In general, in the absence of a vast excess of pyrrole, this reaction led exclusively to the bis(pyrrolyl) complexes. These compounds undergo trans-amination reactions with secondary amines in which the pyrrole group is not replaced. Derivatives of 2,5-dimethylpyrrole were also made; with tetrakis(dimethylamido)titanium the bis(pyrrolyl) complex was formed; with tetrakis(diethylamido)titanium only the monopyrrolyl complex was obtained, possibly for steric reasons. It was argued that the infrared spectra of the compounds were more in support of a π-bonded pyrrolyl structure than the alternate σ-bonded analog. The apparent greater stability of the complexes containing two pyrrolyl groups was also taken to support this contention, by analogy with cyclopentadienyltitanium chemistry.

With excess pyrrole under reflux ($120°$) all dialkylamido groups were removed from the titanium and a black, amorphous, insoluble, unreactive and involatile compound was formed, possibly containing polymerized pyrrole groups.

Table III-24
Bis(dialkylamido)bis(pyrrolyl)titanium Compounds[a]

Compound	Color	Volatility[b]
$(C_4H_4N)_2Ti[N(CH_3)_2]_2$	Viscous brown liquid	$120°/0.1$ mm
$(C_4H_4N)_2Ti[N(C_2H_5)_2]_2$	Golden liquid	$135°/0.1$ mm
$(C_4H_4N)_2Ti[N(n\text{-}C_3H_7)_2]_2$	Golden liquid	$155°/0.25$ mm
$(C_4H_4N)_2Ti[N(n\text{-}C_4H_9)_2]_2$	Golden liquid	$210°/0.1$ mm

[a] From ref. (B61).
[b] Heating bath temperature necessary for their distillation.

Chapter IV

Organometallic Compounds of Zirconium(IV) and Hafnium(IV)

The impetus which the advent of Ziegler–Natta catalysis has had on the organometallic chemistry of titanium is not evident in the chemistry of zirconium or hafnium. Although the first organometallic zirconium compounds were isolated in the early 1950's, it was not until the mid-1960's that significant advances occurred in our knowledge of complexes containing hydrocarbon groups bonded to zirconium or hafnium. By far the most attention has been given to π-cyclopentadienyl and related (indenyl, fluorenyl) compounds of zirconium(IV).

Our understanding of the organometallic chemistry of hafnium is quite sketchy. Although to date the chemistry of hafnium closely parallels that of zirconium, it would be a tenuous assumption to extrapolate that analogy past what is now known. One need only compare the chemistry of titanium and zirconium to be struck by both the similarities and gross differences between them.

A. Alkylzirconium(IV) and Alkylhafnium(IV) Compounds

The initial attempts to prepare organometallic zirconium compounds were made as long ago as 1887 and have been reviewed by Gilman (G16). It was not until 1966 that Berthold and Groh (B46) isolated the first alkylzirconium complex. (There still are no arylzirconium or arylhafnium compounds known.*) It was found that on reaction of methyllithium with zircon-

---

* The preparation of tetraphenylzirconium has been claimed recently by Razuvaev *et al.* (R30a), but it was not isolated. Ether solutions of the compound readily degrade at room temperature to $[(C_6H_5)_2Zr \cdot O(C_2H_5)_2]_2$.

ium tetrachloride in diethyl ether–toluene at $-45°$, a red liquid reaction product was obtained. This was purified by distillation between $-30°$ and $-15°$ and was formulated as a probable etherate of tetramethylzirconium. The compound decomposed in the solid state and in solution even below $-15°$, liberating methane.

Excess methyllithium with zirconium tetrachloride in ether–toluene gave the bright yellow complex, $Li_2[Zr(CH_3)_6]$, which was stable for several hours in the solid at $0°$ (B46, B47). Reaction with water was vigorous even at low temperatures, giving a mixture of methane, ethylene, acetylene, and hydrogen. Controlled alcoholysis to yield methane exclusively is possible with 2-ethylhexanol (B47).

Recently, several tetraalkylated species have been reported which are considerably more stable thermally. The off-white neopentyl derivative, $Zr[CH_2C(CH_3)_3]_4$ (M50, M50a), has a melting point of $103°$, whereas the pale pink tetrakis(1-norbornyl)zirconium or -hafnium compounds (B60) decompose slowly at $23°$. Several tetra(silylmethyl) compounds of zirconium and hafnium have also been prepared (C41a) and these are either oils or low-melting $(8°–11°)$ solids. The chemical shifts of the $ZrCH_2Si$ protons were $\delta 2.16 \pm 0.02$ whereas that in tetrakis(trimethylsilylmethyl)hafnium was $\delta 0.57$. The metal-carbon stretching frequencies were around $470 \ cm^{-1}$, some $40 \ cm^{-1}$ to lower energy than those observed for tetrakis(neopentyl)zirconium. The methyl groups in the neopentyl compound possess a chemical shift of $\delta 1.38$, the methylene group appearing at $\delta 1.66$.

Unlike titanium tetrahalides, the zirconium halides (X = Cl, Br), react with zinc alkyls in saturated hydrocarbon solvents. In toluene, reaction with dimethylzinc gave an unstable, red, insoluble methylzirconium trichloride complex, which was probably polymeric (T26). In diethyl ether, insoluble

$$2ZrX_4 + Zn(CH_3)_2 \xrightarrow[-10° \ to \ 0°]{toluene} 2CH_3ZrX_3 + ZnX_2$$

and pyrophoric but thermally more stable bis(etherates) were formed, which decomposed only above $80°$. The ethyl and propyl complexes were not of sufficient thermal stability to allow their isolation in the pure state (T26).

In pyridine, bis(alkyl) derivatives were formed (T26, T30). The dark brown

$$ZrX_4 + ZnR_2 \xrightarrow[0°]{pyridine} R_2ZrX_2 + ZnX_2$$

$$R = CH_3, C_2H_5; X = Cl, Br$$

compounds were not isolated, but were converted into the blue-black, crystalline mono-2,2′-bipyridyl adducts, which were stable up to $110°–120°$, and could be handled in the air for short periods. With protic reagents they liberated methane (or ethane). They were readily soluble in pyridine, slightly soluble in THF, and insoluble in hydrocarbons, ethers, and nitrobenzene. The higher solubility of the bis(butyl) analogs prevented their isolation (T26).

Benzyl Compounds

Much interest in benzyl compounds of zirconium and hafnium has been generated by the initial report of the activity of tetrabenzylzirconium (Z8) as a polymerization and hydrogenation catalyst.

All preparations of tetrabenzylzirconium and -hafnium compounds employ the Grignard reagent, $ClMgCH_2C_6H_5$ (F12, Z7, Z8).

$$ZrCl_4 + 4ClMgCH_2C_6H_5 \xrightarrow[(C_2H_5)_2O,\ 2\ hr]{-15° \text{ to } -20°} Zr(CH_2C_6H_5)_4 + 4MgCl_2$$

Numerous compounds in which the phenyl ring of the benzyl group bears substituents such as methoxy, methyl, fluoro, or chloro have been reported and used to initiate polymerization of ethylene, styrene, and propylene (P9). The tris- and bis(benzyl)zirconium chlorides are derived from the tetrabenzyl compound (Z7, Z8) by reaction with a stoichiometric amount of HCl in toluene at $-10°$. They are less thermally stable than the parent tetrabenzyl derivative. This is also the case with the compounds $(C_6H_5CH_2)_{4-n}ZrI_n$ ($n = 1, 2,$ or 3), made by reacting tetrabenzylzirconium with iodine (T25a). In the case of titanium, it has been shown that the higher the electronegativity of the halogen on the metal, the lower is the stability of the benzylmetal halide (Z7). The physical properties of several compounds are given in Table IV-1.

Compounds bearing acidic protons such as alcohols react with tetrabenzylzirconium to produce stoichiometric amounts of toluene (Z7, Z8). With 2 moles of oxygen followed by hydrolysis, benzyl alcohol was obtained in 50% yield. With hydrogen at $50°$, toluene and an unidentified low oxidation state compound which catalyzes the hydrogenation of toluene to methylcyclohexane are formed (Z7). Only in the presence of an olefin will reaction with hydrogen and subsequent hydrogenation proceed at $0°$ and atmospheric pressure. With carbon dioxide a ready reaction occurs under ambient conditions, in which 2 moles of carbon dioxide are absorbed to give a product which yields nearly equimolar amounts of phenylacetic acid and tribenzylcarbinol on hydrolysis. The greater reactivity of the zirconium compound compared with its titanium analog in reactions with carbon dioxide (and hydrogen) is ascribed to the higher polarity of the Zr–C bond (Z7).

Tetrabenzylzirconium is readily soluble in aromatic solvents and ethers, but only slightly soluble in aliphatic hydrocarbons. If kept in the dark below $25°$, it is stable for considerable periods of time in both the solid state and solution. In light it decomposes to a brown, diamagnetic hydride which is

Table IV-1
Physical Properties of Some Benzyl Compounds[a]

Compound	Color	M.p. (°C)	PMR chemical shift (δ)		Solvent	Temp. (°C)
			CH_2	C_6H_5		
$Zr(CH_2C_6H_5)_4$	Orange-yellow	133–134	1.44	6.36(o), 7.00 (m, p)	$C_6D_5CD_3$	30
			1.30	6.16	$C_6D_5CD_3$	−60
$Hf(CH_2C_6H_5)_4$	Pale-yellow	112–114	1.48	6.50, 6.97	C_6D_6	—
$Zr(CH_2C_6H_5)_3Cl$	Yellow	89	1.47	6.34, 7.00	C_6D_6	30
$Zr(CH_2C_6H_5)_2Cl_2$	Orange-yellow	60(dec)	—	—	—	—

[a] From refs. (F12, Z7, Z8).

unstable even at $0°$ and shows catalytic activity. Zucchini *et al.* (Z7), on the basis of its hydrolysis products, believe this complex to have the structure

$$C_6H_5CH_2{-}\underset{}{\overset{}{\bigcirc}}{-}CH_2{-}\underset{\underset{CH_2C_6H_5}{|}}{\overset{\overset{H}{|}}{Zr}}{-}CH_2C_6H_5$$

The thermal stability of tetrabenzylhafnium is greater than that of the zirconium or titanium analogs (F12). It can be kept for months under argon at room temperature. Thermal decomposition of a solution of tetrabenzyl-zirconium at $110°$ gives toluene and no coupled benzyl products. The hydrogens required for the formation of toluene are believed to originate in the methylenic groups (Z7). Thiele *et al.* (T25a) have recently made a more detailed study of this thermolysis reaction.

THF, trimethylphosphine, quinoline, 2,2′-bipyridyl, and pyridine form adducts in solution with both tetrabenzylzirconium and tetrabenzylhafnium (F12). No evidence of complex formation was observed (from the PMR chemical shifts of the CH_2 protons) with the bases NR_3 ($R = C_2H_5$, C_3H_7, C_4H_9, C_6H_5), N,N,N',N'-tetraethylethylenediamine, diethyl ether, dioxane, triphenylphosphine, or trimethylphosphite. In chlorobenzene solution at $40°$, the equilibrium constants for their mono(pyridine) adducts have been determined. The red 2,2′-bipyridyl adducts of tetrabenzylzirconium and -hafnium are sparingly soluble in benzene and are believed to have an octahedral configuration. The tetrabenzylzirconium compound decomposes at $120°$ and so here one does not see the enhancement in thermal stability found in other alkylzirconium (or titanium) compounds (T25a) on adduct formation.

On reaction with an equimolar amount of tribenzylaluminum at room temperature, tetrabenzylzirconium forms a red product which melts at $114°–115°$ and has been formulated as

$$[C_6H_5CH_2]_4Zr\underset{}{\overset{\overset{\textstyle C_6H_5}{|}}{\underset{\diagdown}{\overset{\diagup}{CH_2}}}}Al[CH_2C_6H_5]_2$$

In benzene it partially dissociates into the initial reactants. Ether destroys the complex, forming tetrabenzylzirconium and the etherate of tribenzyl-aluminum (G6, Z7).

The crystal structures of the tetrabenzyl derivatives of titanium, zirconium, and hafnium, determined by X-ray analyses (D5,D6), show that in all cases the coordination about the metal approaches the tetrahedral (see Fig. IV-1).

A comparison with the structure of tetrabenzyltin has led Davies and co-workers (D5, D6) to infer that the distortions in both the coordination

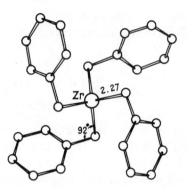

Fig. IV-1.   Structure of tetrabenzylzirconium (D5).

geometry and the ligand arise from the transition metal. Distortions are largest for zirconium and hafnium. Of particular interest are the distortions of the M–CH$_2$–C bond angles, which bring the aromatic rings into closer proximity to the metal (see Table IV-2). Davies and co-workers (D6) postulate that there is a weak interaction between the phenyl ring and the d orbitals of the metal. The enhanced stability of the tetrabenzyl compounds of zirconium, hafnium, and also titanium in comparison with other tetraalkyl compounds of the metals may arise from this interaction.

B. Cyclopentadienyl Compounds

In titanium(IV) chemistry and, in particular, with the mono(π-cyclopentadienyl)alkoxo halides, it was shown that the shift of the C$_5$H$_5$ proton resonance to higher frequencies in the PMR spectrum could be correlated with the

Table IV-2
Some Bond Lengths and Angles for Tetrabenzyl Compounds[a]

	Ti(CH$_2$C$_6$H$_5$)$_4$	Zr(CH$_2$C$_6$H$_5$)$_4$	Hf(CH$_2$C$_6$H$_5$)$_4$	Sn(CH$_2$C$_6$H$_5$)$_4$
M–C (Å)	2.04, 2.11	2.23, 2.26	2.23, 2.24	2.17, 2.17
	2.15, 2.21	2.28, 2.29	2.26, 2.27	2.18, 2.19
M–C (mean)	2.13(4)	2.27(1)	2.25(1)	2.18(1)
M–C–C (°)	92, 98, 109, 115	85, 90, 92, 101	88, 92, 93, 101	110, 110, 112, 114
M–C–C (mean)	103(3)	92(1)	94(1)	111(1)
C–M–C (°)	104, 105, 105	94, 96, 113	99, 100, 110	108, 108, 109
	109, 115, 119	116, 118, 122	115, 116, 118	110, 111, 111

[a] From ref. (D6).

basicity of the other groups (N29). The more basic the ligand, the lower is the frequency at which the resonance is seen. The same holds in zirconium chemistry (W7), namely, the shift to higher frequencies in chloroform is in line with the ability of other ligands to push electrons onto the zirconium and, hence, onto the π-cyclopentadienyl groups (see Table IV-3).

The chemical shift of the cyclopentadienyl protons in benzene differs appreciably from that found in other solvents. For example, while the chemical shift for $Cp_2Zr(CH_3)_2$ in benzene-d_6 is $\delta5.80$, in hexafluorobenzene, deuterochloroform, dichloromethane, cyclohexane, and pyridine-d_5 it is $\delta6.05$, 6.08, 6.08, 6.03, and 6.12, respectively (W8).

Table IV-3

PMR Chemical Shifts (δ) of the C_5H_5 Protons of Some Zirconium Compounds

Compound	$CDCl_3$	C_6D_6	Refs.
Cp_2ZrCl_2	6.52	5.83	W7
$Cp_2Zr(CH_3)Cl$	6.27	5.73	W7
$Cp_2Zr(CH_3)_2$	6.08	5.80	W8
$Cp_2Zr(OCH_3)CH_3$	6.22	—	W8
$Cp_2Zr(OCH_3)Cl$	6.35	—	W7
$Cp_2Zr(OCH_3)_2$	6.02	5.98	G28, W8
$Cp_2Zr(OCH_2CH_3)Cl$	6.32	5.90	G28, W7
$Cp_2Zr(OCH_2CH_3)_2$	—	6.00	G28
$Cp_2Zr[OCH(CH_3)_2]Cl$	6.33	6.00	W7
$Cp_2Zr[OCH(CH_3)_2]_2$	6.17	5.95	W7
$Cp_2Zr(OC_6H_5)_2$	6.35	5.88	W7
$(Cp_2ZrCl)_2O$	6.28	5.89	W7

Any slight alteration to the ionicity of the Zr–π-cyclopentadienyl bond from that found in Cp_2ZrCl_2 can significantly increase its reactivity toward reagents bearing acidic protons such as water, alcohols, phenols, carboxylic acids, and the like. As a working guide one may say that when the PMR chemical shift in $CDCl_3$ of the π-C_5H_5 protons in bis(cyclopentadienyl) compounds is further upfield than about $\delta6.20$, the above reactants can eliminate a cyclopentadienyl group. Such zirconium compounds are therefore often best treated as moisture-sensitive.

Besides being reflected in the chemical shift of the π-C_5H_5 protons, the basicity of the ligands on the metal is also said to affect the metal–ring stretching frequencies. The higher the basicity of the ligand, the lower the energy of these vibrations (G28).

Substituents on the cyclopentadienyl ring can influence the reactivity of bis(cyclopentadienyl)zirconium dihalide compounds toward water (S5).

Aromatic rings fused on to the cyclopentadienyl ring increase the lability of the other ligands, whereas the reverse is the case with aliphatic substituents. The variation is stability toward water with substituent on the cyclopentadienyl ring is in the order tetrahydroindenyl > methylcyclopentadienyl > cyclopentadienyl > indenyl > fluorenyl.

1. Mono- and Bis(cyclopentadienyl) Compounds of Zirconium(IV) and Hafnium(IV)

These compounds are the most numerous in the organometallic chemistry of zirconium and hafnium. The majority of cyclopentadienyl compounds are prepared from Cp_2ZrCl_2 (and Cp_2HfCl_2) by reaction with other ligands (see Fig. IV-3 below) which, in many cases, besides replacing the halides, will also remove one cyclopentadienyl group. Because of this it was considered desirable to discuss both the mono- and bis(cyclopentadienyl) compounds in this section.

a. CYCLOPENTADIENYL COMPOUNDS CONTAINING ONLY INORGANIC ANIONS

Since 1953 when Wilkinson *et al.* (W25) isolated the first cyclopentadienyl-zirconium compound, Cp_2ZrBr_2, all the bis(cyclopentadienyl)halido complexes of zirconium and hafnium have been synthesized, some only as recently as 1969. All compounds are colorless to yellow except $(\pi\text{-fluorenyl})_2ZrCl_2$, which is orange, and $CpZrI_3$, which is red.

i. BIS(CYCLOPENTADIENYL)METAL(IV) DIHALIDES. (a) *Methods of preparation.* The synthetic procedures most often employed involve reaction of the metal tetrahalide with either the cyclopentadienyl Grignard reagent, magnesium dicyclopentadienide, or an alkali metal cyclopentadienide. In their preparation of Cp_2ZrBr_2, Wilkinson *et al.* used the Grignard reagent (W24, W25) and obtained a 30% yield. The yield can be increased to $>60\%$

$$2(C_5H_5)MgBr + ZrBr_4 \longrightarrow Cp_2ZrBr_2 + MgBr_2$$

by using an alkali metal cyclopentadienide. The sodium compound is the one most often utilized (F25, S7). Cp_2HfCl_2 was first prepared by this method in

$$2NaC_5H_5 + ZrCl_4 \xrightarrow[\text{reflux}]{\text{THF}} Cp_2ZrCl_2 + 2NaCl$$

1957 (L24) and, recently, an excellent procedure for its preparation by this reaction has been described which leads to yields of up to 81% (D35). Thallium cyclopentadienide too can be used in these reactions (N31).

Magnesium cyclopentadienide in refluxing THF has been employed by

Reid and Wailes (R39) to make Cp_2ZrCl_2 in 63% yield. Many metal halides will react with molten magnesium cyclopentadienide to form cyclopentadienyl compounds (R40). However, whereas ZrF_4, $ZrCl_4$, and $ZrBr_4$ did not give the expected product, ZrI_4 surprisingly gave a 72% yield of Cp_2ZrI_2. Only the bis(cyclopentadienyl) compound was formed in this case, even in the presence of excess magnesium cyclopentadienide.

It is noteworthy that Cp_2ZrCl_2 has been obtained in up to 38% yield by reacting $ZrCl_4$ with cyclopentadiene at room temperature for 5 hr in the presence of an amine, in particular, diethylamine (M47, M48). A stipulation for this reaction is that the amine, which was generally also the reaction solvent, has a $pK_a > 9$.

When cyclopentadiene vapor was heated in excess of 270° in the presence of either $ZrCl_2$ or $ZrBr_2$, a novel reaction ensued in which the metal was oxidized and Cp_2ZrCl_2 or Cp_2ZrBr_2 was formed in 30% yield (R37). It was held that the hydrogen evolved reduced excess cyclopentadiene as no pressure build-up occurred in the reaction vessel. This reduction may well have been catalyzed by the zirconium(II) halide.

Several substituted bis(cyclopentadienyl)metal dichloride compounds are known (see Table IV-4). Bis(π-methylcyclopentadienyl)zirconium dichloride was the first made in 1959 by Reynolds and Wilkinson (R41). Only the sodium- or thallium-substituted cyclopentadienides have been used for these reactions. Interestingly, Samuel (S5) showed that the indenyl compound could only be synthesized with sodium indenide and not with the lithium or potassium salts. Reduction of bis(π-indenyl)zirconium dichloride with Adams catalyst in dimethoxyethane gave bis(π-tetrahydroindenyl)zirconium dichloride.

The alkyl-substituted π-cyclopentadienyl compounds are more soluble in organic solvents than is Cp_2ZrCl_2. Whereas Cp_2ZrCl_2 and the alkyl-substituted cyclopentadienyl compounds are air-stable, the fluorenyl and indenyl compounds are moisture-sensitive.

It should be pointed out that the preparation of these compounds is best carried out under an inert atmosphere. The importance of this lies in the ready hydrolysis of the Cp_2MX_2 compounds in the presence of base which may occur in the preparative reaction mixture from use of excess alkali metal cyclopentadienide or amine. The hydrolysis product is a μ-oxo compound of the type, $Cp_2Zr(Cl)OZr(Cl)Cp_2$. At times it has been found beneficial to extract the crude reaction mixtures with solvents containing dry HCl to convert the μ-oxo compound back to the dichloride.

Compounds containing halides other than chloride are best prepared from the commercially available Cp_2ZrCl_2. The yields obtainable from halide exchange reactions can be extremely good (see Table IV-4). The most satisfactory method for the preparation of the bromides and iodides of

Table IV-4
Cyclopentadienyl Compounds of Zirconium(IV) and Hafnium(IV) Containing Only Inorganic Anions (Other than Oxides or Hydrides)

Compound	M.p. (°C)	Sublimation (°C/mm pressure)	PMR shift of C_5 ring protons[a]	Refs.	Highest yield (%)	Refs.
Cp_2ZrF_2	>250	$135/2 \times 10^{-3}$	6.52(THF)	D35	79	D35
Cp_2HfF_2	>200	$135/10^{-3}$	6.48(THF)	D35	66	D35
Cp_2ZrCl_2	247–249	150–180	6.62(THF) 6.52($CDCl_3$) 5.83(C_6D_6)	D35 B74, S5, W7 W7	75	S7
Cp_2HfCl_2	>300	$155/10^{-2}$	6.52(THF) 6.46($CHCl_3$)	D35 B74, S5	81	D35
Cp_2ZrBr_2	259–260	150	6.69(THF)	D35	100	D35
Cp_2HfBr_2	265–267	—	6.59(THF)	D35	98	D35
Cp_2ZrI_2	299–300	180	6.82(THF)	D35, R37	100	D35
Cp_2HfI_2	303–305	—	6.69(THF)	D35	100	D35
$(\pi\text{-}CH_3C_5H_4)_2ZrCl_2$	180–181	—	6.20, 6.43, 6.50(acetone) 6.22, 6.31($CHCl_3$) 5.62, 5.75(C_6H_6)	S5 B74 B74	50	S5
$(\pi\text{-}CH_3C_5H_4)_2HfCl_2$	169	—	6.13, 6.38(acetone) 6.10, 6.22($CHCl_3$)	B74 B74	27.5	B74

Compound	M.p. (°C)	Conditions	NMR shift	Ref	J (Hz)	Ref
(π-tert-C₄H₉C₅H₄)₂ZrCl₂	189–190	—	6.31, 6.44(CHCl₃)	B74	76	N31
(π-tert-C₄H₉C₅H₄)₂HfCl₂	184.5–185.5	—	6.20, 6.32, 5.72(CHCl₃)	B74	31	B74
[(π-C₅H₄)₂(CH₂)₃]ZrCl₂	—	200/10⁻³–10⁻⁴	5.72, 6.27(C₆D₆)	H31	—	H31
[(π-C₅H₄)₂(CH₂)₃]HfCl₂	—	200/10⁻³–10⁻⁴	5.66, 6.18(C₆D₆)	H31	—	H31
(π-Indenyl)₂ZrCl₂	264	Sublimes	5.76d, 6.24t(CHCl₃)	S5	50	S9
(π-Tetrahydroindenyl)₂ZrCl₂	203–204	—	—	—	40	S5
(π-Fluorenyl)₂ZrCl₂	Decomposes below m.p.	—	—	—	15	S9
CpZrCl₃	237–238	Sublimes	6.65(THF)	R37	—	R37
CpZrBr₃	197–198	Sublimes	6.80(THF)	R37	72	R37
CpZrI₃	133–133.5	Sublimes	6.63(THF)	R37	—	R37
CpZrBrCl₂	190.5	—	—	—	66	G23
Cp₂Zr(NO₃)Cl	Explodes at 185°	—	—	—	67	F26
Cp₂Zr(NO₃)₂	Infusible	—	—	—	62	B72
Cp₂Zr(HSO₄)₂·4H₂O	—	—	—	—	—	F26
Cp₂Zr(NCS)₂	279–281	200/10⁻⁴	6.54(acetone)	C62	>95	C62
Cp₂Zr(NCO)₂	218	175/10⁻⁴	6.66(acetone)	C62	—	C62
Cp₂Zr(N₃)₂	123	140/10⁻³	6.53(THF)	C71	—	C71
Cp₂Hf(NCS)₂	257–259	—	6.35(CDCl₃)	C71	45	B100
Cp₂Hf(NCO)₂	185	—	—	—	66	B100
(π-tert-C₄H₉C₅H₄)₂Zr(NO₃)Cl	104–105	—	—	—	64	B74

[a] The coupling constants found in the substituted cyclopentadienyl rings were of the order 2.0 to 3.0 Hz.

both bis(cyclopentadienyl)zirconium and hafnium was developed by Lappert and co-workers (D35), namely,

$$3Cp_2MCl_2 + 2BX_3 \xrightarrow[\text{5 min}]{\text{CH}_2\text{Cl}_2} 3Cp_2MX_2 + 2BCl_3$$

The original method for the preparation of Cp_2ZrI_2 which involved refluxing Cp_2ZrCl_2 with excess potassium iodide in acetone (R37) suffers from the disadvantages of both lower yield and the possibility of the mixed halide product, $Cp_2Zr(Cl)I$, being formed. For this reason, the iodide compound so prepared is purified by sublimation. The equilibrium constant K for the redistribution reaction

$$Cp_2ZrCl_2 + Cp_2ZrBr_2 \rightleftharpoons 2Cp_2Zr(Cl)Br$$

in THF has been found to be 4.3 ± 0.2 liter mole^{-1} at $38°$. For the corresponding titanium system, $K = 4.1 \pm 0.6$ liter mole^{-1} at $38°$ in the same solvent (D35).

The bis(cyclopentadienyl)metal difluoride compounds were the last of the halide series to be prepared. The halide exchange reaction between Cp_2ZrCl_2 and an alkali metal fluoride cannot be utilized. However, Druce et al. (D35) have developed several methods which give the product in good yield. Of note is the rather novel reaction between $Cp_2M[N(CH_3)_2]_2$ and pentafluoro-

$$Cp_2M[N(CH_3)_2]_2 + 2C_6F_5CN \xrightarrow[\text{20°, 4 hr}]{\text{diethyl ether}} Cp_2MF_2 + 2p\text{-}C_6F_4(CN)N(CH_3)_2$$
$$\phantom{Cp_2M[N(CH_3)_2]_2 + 2C_6F_5CN \xrightarrow[\text{20°, 4 hr}]{\text{diethyl ether}}} 61\text{–}65\% \qquad\qquad 66\text{–}70\%$$

benzonitrile. The bis(cyclopentadienyl)bis(dimethylamido)metal complex also reacts with boron trifluoride–diethyl ether to form the difluoride (D35) in $66\text{–}79\%$ yield. With bis(cyclopentadienyl)bis(tert-butoxo)zirconium a 52% yield of Cp_2ZrF_2 was obtained.

(b) Structure and physical properties. Cp_2ZrCl_2 crystallizes in a monoclinic unit cell containing two molecules (B63). Its structure, determined in the vapor phase by electron diffraction, is similar to the distorted tetrahedral arrangement found for Cp_2TiCl_2 (R49). Similar structures for Cp_2ZrI_2 and Cp_2ZrF_2 have been deduced from X-ray diffraction studies (B102) (see Fig. IV-2).

Some selected bond lengths and angles found in these and other similar compounds are tabulated in Table IV-5. It is interesting to note that while the cyclopentadienyl groups in Cp_2ZrCl_2 are staggered, in $Cp_2ZrCl[Si(C_6H_5)_3]$ they are eclipsed.

The essentially tetrahedral arrangement of the ligands about the metal in Cp_2ZrCl_2 and its hafnium analog is also reflected in their dipole moments in benzene. For Cp_2ZrCl_2 the values found for the dipole moment were 4.87, 4.93 (K24), and 5.90 D (G14) and for Cp_2HfCl_2, 4.66 D (K24).

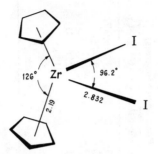

Fig. IV-2.   Structure of Cp_2ZrI_2 (B102).

The PMR chemical shifts experienced by the cyclopentadienyl protons (see Table IV-4) are indicative of π-bonded cyclopentadienyl groups. Large atoms such as iodine do not disturb the PMR equivalence of these protons. Splitting arising from a coupling of the ^{19}F atoms with the ring protons, found in the PMR spectrum of Cp_2TiF_2 (triplet, J = 1.7 Hz), was not observed in the spectra of the corresponding zirconium or hafnium compounds. The longer bond lengths in these latter compounds may account for this (D35). 1,1'-Trimethylenebis(π-cyclopentadienyl)zirconium and -hafnium dichloride have PMR (and infrared spectra) similar to the titanium analog (H31). The splitting of the ring protons into two multiplets (separation δ0.52 for Hf and δ0.55 for Zr) was ascribed to the tilting of the rings, the trimethylene bridge preventing their rotation and, thereby, the equilibration of the protons. The cyclopentadienyl ^{13}C NMR peaks for Cp_2ZrCl_2 and Cp_2HfCl_2 in $CDCl_3$ were found at δ115.7 and δ114.4 (relative to TMS), respectively (N28a).

The mass spectra of all bis(cyclopentadienyl)zirconium and -hafnium dihalides have been measured (D26, D35, R36). The breakdown pattern of these compounds proceeds via the preferential removal and fragmentation of the cyclopentadienyl ligands. This was thought to be indicative of the greater bond strength and higher ionic character of the metal–chloride bonds in these compounds, compared with that in $CpTiCl_3$ for example. $CpTiCl_3$ breaks down with loss of chloride ions rather than loss of the cyclopentadienyl group (D26). The ionization potential found for Cp_2ZrCl_2 was 9.4 $\pm$ 0.3 eV, while that for Cp_2TiCl_2 was 9.0 $\pm$ 0.2 eV. Dillard and Kiser (D26) have suggested that the electron removed came from a molecular orbital associated with the metal and the chlorine.

The methane chemical ionization mass spectra of Cp_2MCl_2 (where M = Ti, Zr, Hf) are unlike those of the cyclopentadienyl sandwich compounds of Fe, Ru, Os, Co, and Ni in that the parent ions are not formed in any abundance (H42). Here protonation occurs to form mainly Cp_2MCl^+, together with HCl.

Table IV-5
Bond Lengths and Bond Angles for Some Cyclopentadienylzirconium Compounds

Compound	Ref.	Bond lengths (Å) (standard deviation)				Bond angles (degrees)	
		Zr-halide	Zr-C(mean)	C-C(mean)	Zr-Cp($\perp$)	Halide-Zr-halide	Cp-Zr-Cp
Cp$_2$ZrF$_2$	B102	1.98(1)	2.50	1.37	2.21	96.2(3)	127.8(0.7)
Cp$_2$ZrCl$_2$	R49	2.309(5)	2.522(5)	1.42(1)	2.21(1)[b]	104(2)	134(5)
		2.441(5)[b]	—	—	—	97.1(2)[b]	—
Cp$_2$ZrI$_2$	B102	2.832(2)	2.48	1.35	2.19	96.2(1)	126(0.7)
Cp$_2$ZrCl[Si(C$_6$H$_5$)$_3$]	M51	2.430	2.49(1)	1.38(2)	—	93.9[c]	130.5
CpZr(acac)$_2$Cl	S63	2.50(1)	2.55(5)	1.43(5)	2.24(2)	—	—
CpZr(F$_6$-acac)$_3$[a]	E2	—	2.53(1)	1.40(1)	2.23(1)	—	—

[a] Enolate of hexafluoroacetylacetone.
[b] From reported unpublished work (G29).
[c] Cl-Zr-Si bond angle.

122

Heats of combustion for Cp_2ZrCl_2 and Cp_2TiCl_2 determined by bomb calorimetry (D26) are very similar; -1398 ± 5 kcal/mole for the zirconium compound and -1366 ± 5 kcal/mole for the titanium analog. The combustion reaction is

$$Cp_2ZrCl_2(s) + 13O_2(g) \longrightarrow ZrO_2(s) + 2HCl(aq) + 10CO_2(g) + 4H_2O(l)$$

On the basis of this reaction, the heats of formation of Cp_2ZrCl_2 and Cp_2TiCl_2 were evaluated at $-153(\pm 7)$ kcal/mole and $-145(\pm 7)$ kcal/mole, respectively.

The infrared spectra of the cyclopentadienylzirconium halides (B65, B100, D35, D36, F25, F27, M26, R35, R37, R41, S5, S8, S9, W24) and the corresponding hafnium compounds (B73, B100, D35, D36, S5) above 600 cm^{-1} are similar to those of the titanium compounds (see Chapter III). Assignments of the bands above 600 cm^{-1} for all Cp_2MX_2 (X = F, Cl, Br, I) compounds have been made (D35, F27). The near-infrared spectra have also been recorded for several compounds (R35). Little difference is generally observed between solution and mull spectra.

Many of the absorptions found in the infrared and Raman spectra below 600 cm^{-1} have recently been assigned as arising from the skeletal vibrations of the molecules (D36, M26, S6a). The assignments were confirmed by an approximate normal coordinate analysis (M26) (see Table IV-6). Metal–ring force constants were found to be indicative of the weakening of the metal–ring bond on tilting the rings from parallel. A steady decrease in bond strength with tilt has been predicted from metal–ring orbital overlap calculations (B8). The thesis that the intensity of the symmetric ring breathing mode near 1130 cm^{-1} can be correlated with the ionic character in the metal–ring bond (F27) is an oversimplification and does not appear to hold (B73, M26).

The infrared spectra of the methylcyclopentadienyl-, indenyl-, tetrahydro-indenyl-, and fluorenylzirconium dichloride compounds have been measured by Samuel (S5, S9).

(c) *Chemical properties.* All other cyclopentadienylzirconium derivatives, almost without exception, can be synthesized starting from Cp_2ZrCl_2. The reaction schemes in Fig. IV-3 outline some of its chemistry, which is discussed in greater detail elsewhere in this and subsequent chapters.

ii. BIS(CYCLOPENTADIENYL)METAL(IV) PSEUDOHALIDES, NITRATES, AND SUL-FATES. It is surprising that the first such compound (see Table IV-4) to find its way into the literature was the dinitrate, made by the reaction (B72)

$$(C_5H_5)_4Zr + 2HNO_3 \xrightarrow[-35°]{\text{dichloroethane}} Cp_2Zr(NO_3)_2 + 2C_5H_6$$

This benzene-soluble dinitrate explodes on heating to 185°. It is converted by HCl into the dichloride, Cp_2ZrCl_2. The mononitrate $Cp_2Zr(Cl)NO_3$ is more

Table IV-6

Observed Metal–Ligand Vibrations for Bis(cyclopentadienyl)Metal Dihalide Complexes[a]

Assignment	Symmetry	Cp_2MF_2			Cp_2MCl_2			Cp_2MBr_2			Cp_2MI_2		
		Ti	Zr	Hf	Ti	Zr	Hf	Ti	Zr	Hf	Ti	Zr	Hf
ν_s(M–Cp)	a_1	357	—	—	359	358	360	389	356	—	380	357	—
ν_a(M–Cp)	b_2	417	—	—	413	358	360	416	356	—	421	357	—
ν_s(M–X)	a_1	544	560[c]	562[c]	400	333	310	209	207	208[c]	197	178	170[c]
ν_a(M–X)	b_1	568	545[c]	536[c]	400	333	310	285	221	197[c]	231	194	153[c]
Tilt	a_1[b]	258	—	—	245	266	264	245	269	—	264	272	—
Tilt	b_1[b]	258	—	—	245	266	264	245	269	—	264	272	—
Tilt	a_2[b]	293	—	—	300	310	284	346	315	—	347	292	—
Tilt	b_2[b]	293	—	—	300	310	284	346	315	—	347	292	—
δ_s(X–M–X)	a_1	194	—	—	143	123	123	—	85	—	—	95	—
δ_a(X–M–X)	b_1	234	—	—	207	165	165	152	128	—	123	115	—
δ_a(Cp–M–X)	a_2	215	—	—	166	140	145	—	109	—	—	125	—
δ_s(Cp–M–Cp)	a_1	234	—	—	184	165	165	171	155	—	145	155	—
δ_a(Cp–M–Cp)	b_2	234	—	—	184	154	165	171	135	—	145	138	—

[a] Data in cm^{-1}. From refs. (D36, M26).
[b] Tentative assignments.
[c] These assignments from ref. (D36) of the symmetric and asymmetric metal–halogen stretching vibrations are opposite to those found in ref. (M26) from which the rest of the information in the table is taken.

124

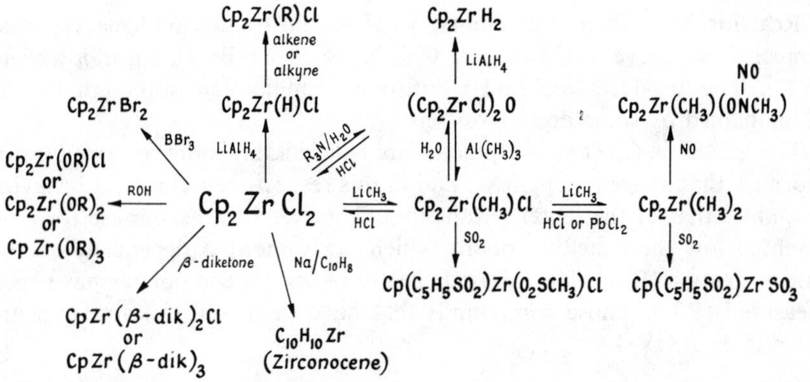

Fig. IV-3. Reaction schemes for Cp_2ZrCl_2, demonstrating its utility as a starting material.

heat-stable, melting at 190.5° (F26). It is readily soluble in the usual organic solvents and can be made from nitric acid and the dichloride or μ-oxo compound, $(Cp_2ZrCl)_2O$, in dichloroethane. The bisbisulfate, also made from the μ-oxo compound (F26), is an infusible hygroscopic substance insoluble in organic solvents.

The pseudohalide complexes, the thiocyanates and cyanates, are best prepared by exchange reactions from the dichloride in acetone, dichloromethane, benzene, or nitrobenzene (B100, C62, S5). All are nonelectrolytes

$$Cp_2MCl_2 + 2M'X \longrightarrow Cp_2MX_2 + 2M'Cl$$
$$M' = K \text{ or } Ag$$

in nitrobenzene (B100). The zirconium compounds were shown to be monomeric in acetone (C62). Although $Cp_2Zr(NCS)_2$ is reasonably air-stable, the cyanato complex, $Cp_2Zr(NCO)_2$, is very readily hydrolyzed to the μ-oxo compound, $[Cp_2Zr(NCO)]_2O$, even in the absence of a base (C62). While both Coutts and Wailes (C62) and Burmeister et $al.$ (B100) are in accord with an N-bonded thiocyanato group, the nature of the bonding between the metals and the cyanato ligand is open to conjecture. Any bonding assignment for the cyanates based solely on infrared data is rather tenuous. Dipole moment data (J10), however, tends to favor bonding through oxygen.

iii. MONO(CYCLOPENTADIENYL)METAL(IV) TRIHALIDES. The initial reports on cyclopentadienyl trihalide compounds are found in patents by Gorsich (G21, G23). Reid and Wailes (R37) carried out the first systematic preparations of the trihalides of zirconium utilizing the reaction

$$2ZrX_4 + Mg(C_5H_5)_2 \xrightarrow[\text{1.5 hr under } N_2]{\text{xylene, }100°-110°} 2CpZrX_3 + MgX_2$$

Reaction between a zirconium(III) halide and cyclopentadiene vapor at temperatures above 260° also gave $CpZrX_3$ (X = Cl, Br, I), but in low yield (R37). The products are best purified by sublimation, although this is accompanied by some decomposition.

The halides in $CpZrX_3$ compounds are considerably more reactive toward moisture than those in Cp_2ZrX_2, and in this respect their chemical behavior resembles that of the metal tetrahalides. They tend to decompose to some extent below their melting points which are somewhat dependent on the rate of heating (R37). The infrared spectra of several compounds have been measured (R37). Those compounds that have been well characterized are listed in Table IV-4.

iv. CYCLOPENTADIENYL μ-OXO AND HYDROXO COMPOUNDS. The reaction of Cp_2ZrCl_2 with amines in halogenated solvents was first observed by Samuel and Setton (S8). The nature of the reaction and of the zirconium product formed was established later by Brainina et al. (B69) and Reid and co-workers (R36). Thus, in the presence of water, often atmospheric, the reaction proceeds according to the equation (W2)

$$2Cp_2ZrCl_2 + 2C_6H_5NH_2 + H_2O \xrightarrow[\text{5–15 min}]{\text{room temp.}} \overset{\overset{\displaystyle Cl}{|}}{Cp_2Zr}-O-\overset{\overset{\displaystyle Cl}{|}}{ZrCp_2} + 2C_6H_5NH_2 \cdot HCl$$

Samuel (S5) found that on reaction of zirconium tetrachloride with 3 moles of sodium cyclopentadienide, $(Cp_2ZrCl)_2O$ was obtained in 50% yield. The assumption that the Cp_2ZrCl_2 initially formed was reduced by the excess sodium cyclopentadienide to $Cp_2Zr^{III}Cl$ and that subsequent oxidation gave the μ-oxo compound is erroneous. Cp_2ZrCl_2 is not reduced by sodium cyclopentadienide and the $(Cp_2ZrCl)_2O$ was most likely produced as a result of the basic conditions ensuing on hydrolysis of the excess sodium cyclopentadienide in the reaction mixture.

$(Cp_2ZrCl)_2O$ is a colorless compound subliming at 200°–220° in vacuo, it is stable in air and light for periods exceeding 1 month. It is soluble in all common organic solvents except saturated hydrocarbons, and forms mono-clinic crystals containing four molecules per unit cell (R36). Fragmentation in the mass spectrum proceeds by successive loss of cyclopentadienyl groups or chlorine atoms (R36). The intense band in the infrared at 749–755 cm^{-1} arises from the metal–O–metal stretching vibration. Bands at similar wave numbers are found in other μ-oxo compounds. The compounds of hafnium appear to be more soluble in organic solvents than the corresponding zirconium analogs.

Bis(cyclopentadienyl)-μ-oxo compounds, in which one or both chlorides are replaced by other ligands, have been made. When KOCN was added to an aqueous solution of Cp_2ZrCl_2, $[Cp_2Zr(OCN)]_2O$ precipitated (S5).

Similarly, addition of NaN_3 in water to Cp_2ZrCl_2 gave $(Cp_2ZrN_3)_2O$ (C71). This reaction was not duplicated with KSCN, as in this case $Cp_2Zr(SCN)_2$ was obtained (S5). The extremely facile hydrolysis of $Cp_2Zr(OCN)_2$ to form the μ-oxo compound has been noted by Coutts and Wailes (C71). KBr, KI, and $NaNO_3$ react in still a different way with bis(cyclopentadienyl)zirconium dihalides (S5). The hydroxoiodide compound and the analogous bromide,

$$Cp_2ZrX_2 + KI(excess) \xrightarrow{\text{H}_2\text{O}} Cp_2Zr(OH)I \cdot 3H_2O$$

$Cp_2Zr(OH)Br \cdot 2H_2O$, can also be prepared by dissolving the corresponding dihalide in a minimum amount of water (S5). These compounds are soluble in polar organic solvents, and on recrystallization from acetone or alcohol or on heating *in vacuo* to 150°, they are converted into the μ-oxo species.

The mixed halide–ethoxo compounds, $Cp_2Zr(X)OZr(OC_2H_5)Cp_2$ (X = Cl or Br), were formed on aerobic hydrolysis of $Cp_2Zr(OC_2H_5)X$ (B69). Reaction of Cp_2ZrH_2 with carboxylic acids can lead to a series of compounds containing both oxo and alkoxo bridges between zirconium atoms (W3).

Treatment of Cp_2ZrCl_2 with a stoichiometric amount of phenyl- or *p*-tolyllithium followed by hydrolysis of the reaction mixture gave [Cp_2Zr-$(C_6H_5)]_2O$ or the corresponding *p*-tolyl compound. It is of interest that here the aryl–zirconium bond is stable toward water. The phenyl product has also been prepared from $[Cp_2ZrCl]_2O$ and phenyllithium. These compounds are relatively unstable at room temperature (B65).

Several substituted cyclopentadienyl-μ-oxo compounds have been synthesized (see Table IV-7). The ease of reaction of bis(substituted-cyclopentadienyl)zirconium dichloride with water is dependent on the substituents borne by the cyclopentadienyl ring. The highest hydrolytic stability is shown by the tetrahydroindenyl compounds; the stability to water decreasing in the order methylcyclopentadienyl > cyclopentadienyl > indenyl > fluorenyl (S5). The asymmetry introduced by the μ-oxo group into the bis(π-methylcyclopentadienyl)- and bis(π-tetrahydroindenyl)zirconium chloride derivatives results in the removal of the PMR equivalence of several of the cyclopentadienyl protons (S5).

A general property of μ-oxo compounds is their reaction with protic reagents with the destruction of the oxygen bridge (B69, B73, B75, F26, M36).

$$[Cp_2ZrCl]_2O + 2HX \longrightarrow 2Cp_2Zr(Cl)X + H_2O$$

With β-diketones (B69, F26) or 8-quinolinol (F26) one cyclopentadienyl group per metal is also replaced, e.g.,

$$[Cp_2ZrCl]_2O + 4acacH \longrightarrow 2CpZr(acac)_2Cl + 2C_5H_6 + H_2O$$

These mono(cyclopentadienyl)-β-diketonates or 8-quinolinolates are converted to the corresponding μ-oxo complexes on treatment with water and

Table IV-7
μ-Oxo Compounds of Zirconium and Hafnium

Compound	PMR (Cp, δ)	M.p. (°C)	Yields (%)	Refs.
$(Cp_2ZrCl)_2O$	6.28($CDCl_3$)	300–309	97	B65, B69, R36, S5, W2, W7
$(Cp_2HfCl)_2O$	6.30($CHCl_3$)	265–269	63	B73, M36, S5
$(Cp_2ZrBr)_2O$	—	—	—	B65, S5
$(Cp_2HfBr)_2O$	—	267–282	48	B73
$(Cp_2ZrI)_2O$	—	—	—	S5
$(Cp_2ZrF)_2O$	—	232–234	21	M34
$[Cp_2Zr(OCN)]_2O$	6.46 (acetone)	275–277	—	C62, S5
$(Cp_2ZrN_3)_2O$	6.41(THF)	268(dec)	> 95	C71
$[Cp_2Zr(NO_3)]_2O^a$	—	—	—	S5
$Cp_2Zr(Cl)OZrCp_2(OC_2H_5)$	—	147–156	—	B69
$Cp_2Zr(Br)OZrCp_2(OC_2H_5)$	—	184–192	—	B69
$(Cp_2ZrH)_2O \cdot Cp_2ZrH_2$	—	—	25	W2
$[Cp_2Zr(C_6H_5)]_2O$	—	250–260	52.5	B65
$Cp_2Zr(p\text{-}CH_3C_6H_4)]_2O$	—	210–224	44	B65
$[\pi\text{-}CH_3C_5H_4)_2ZrCl]_2O$	5.91, 6.10, 6.2 (acetone)	—	—	S5
$[(\pi\text{-}tert\text{-}C_4H_9C_5H_4)_2ZrCl]_2O$	—	150–156	61	B74
$[(\pi\text{-}tert\text{-}C_4H_9C_5H_4)_2HfCl]_2O$	—	142–145	30	B74
$[(\pi\text{-}Tetrahydroindenyl)_2ZrCl]_2O$	5.54, 5.81, 6.13 (acetone)	—	—	S5

Compound		mp (°C)	Yield (%)	Ref.
$[(\pi\text{-}CH_2=CHC_5H_4)HfI_2]_2O$	6.09(CCl$_4$)	—	—	G22
$[CpZr(acac)_2]_2O$	—	164.5–165.5	55	B65, F26, M35
$[CpHf(acac)_2]_2O$	—	150–154	38	B73
$[CpZr(C_{10}H_9O_2)_2]_2O^b$	—	172–175	64	F26
$[CpZr(C_{15}H_{11}O_2)_2]_2O^c$	—	225–227	45	B66, F26
$[(\pi\text{-}CH_3C_5H_4)Zr(acac)_2]_2O$	—	118.5–119	27	M35
$[CpZr(C_9H_6NO)_2]_2O^d$	—	> 300	71	F26
$[CpHf(C_9H_6NO)_2]_2O$	—	> 300	46	B73

R = C$_2$H$_5$, $1 < n < 2$	6.08 (complex band)^e (C$_6$D$_6$)	—	—	W3
R = CD$_2$CH$_3$	6.08 (complex band)^e (C$_6$D$_6$)	—	—	W3
R = CH$_2$CF$_3$, $n = 1$	5.83 (complex band) (C$_6$H$_6$)	—	—	W3
R = CH$_2$C$_6$H$_5$, n is large	—	—	—	W3

^{a} Has not been obtained pure.
^{b} C$_{10}$H$_9$O$_2$ is the enolate of benzoylacetone.
^{c} C$_{15}$H$_{11}$O$_2$ is the enolate of dibenzoylmethane.
^{d} C$_9$H$_6$NO = 8-quinolinolato.
^{e} For R = C$_2$H$_5$, $\delta_{CH_3} = 1.17$, $\delta_{CH_2} = 3.97$; for R = CD$_2$CH$_3$, $\delta_{CH_3} = 1.15$.

triethylamine (B65). Both the μ-oxo group and the acetylacetonato groups

$$2CpZr(acac)_2Cl + 2N(C_2H_5)_3 + H_2O \longrightarrow [CpZr(acac)_2]_2O + 2(C_2H_5)_3N \cdot HCl$$

which occupy cis positions (M35) in this compound are replaceable by 8-quinolinol (B65).

$$[CpZr(acac)_2]_2O + 6C_9H_7NO \longrightarrow 2CpZr(ONC_9H_6)_3 + 4acacH + H_2O$$

The reaction of the Lewis acid, $Al_2(CH_3)_6$, with $[Cp_2ZrCl]_2O$ provides a very useful method for the synthesis of $Cp_2Zr(Cl)CH_3$ (S74, W7). The Zr–O–Zr linkage in $(Cp_2ZrCl)_2O$ is also reactive toward lithium aluminum hydrides (see Section B,1,c), whereas sodium naphthalene leaves the oxygen bridge intact while extracting the halides to give the μ-oxozirconium(III) compound (W4).

b. CYCLOPENTADIENYL COMPOUNDS CONTAINING OXYGEN-BONDED ORGANIC LIGANDS

The majority of cyclopentadienyl compounds fall into this class. The ligands include alkoxides, phenoxides, β-diketonates, carboxylates, sulfinates, and also those believed to be sulfonates.

i. ALKOXIDES AND PHENOXIDES. The formation of organometallic alkoxides and phenoxides of zirconium and hafnium was first mentioned in a 1960 U.S. Patent by Gorsich (G21).

Methods of preparation. Many suitable synthetic routes for the formation of π-cyclopentadienyl alkoxide and phenoxide complexes have been developed and, in general, these involve replacement of labile groups on the metal by an alkoxo or phenoxo group. For example,

$$Cp_2HfCl_2 + 2C_6H_5OH + 2(C_2H_5)_3N \longrightarrow Cp_2Hf(OC_6H_5)_2 + 2(C_2H_5)_3N \cdot HCl$$

$$CpZrCl_3 + 3C_9H_7NO \longrightarrow CpZr(C_9H_6NO)_3 + 3HCl$$

$$Cp_2ZrCl_2 + 2NaOC(CH_3)_3 \longrightarrow Cp_2Zr[OC(CH_3)_3]_2 + 2NaCl$$

These reactions can be conducted at room temperature over a period of 1–2 hr or under reflux for 25 min in ether or, more usually, benzene.

Other reactions such as the treatment of tetracyclopentadienylhafnium or -zirconium, bis(cyclopentadienyl)dimethylzirconium, or the bis(dimethylamido) compound, $Cp_2Zr[N(CH_3)_2]_2$, with alcohols or phenols have also been used to prepare alkoxides or phenoxides (B66, C16, M34, W8). Hydrido compounds, too, can be used. Thus, Cp_2ZrH_2 reacts with acetone to give the diisopropoxide, $Cp_2Zr[OCH(CH_3)_2]_2$, and with carboxylic acids to form polymers containing both μ-oxo and μ-alkoxo groups (see Section B,1,c). The more acidic reagents such as 8-quinolinol react readily with μ-oxo compounds to form cyclopentadienyltris(quinolinolato)metal derivatives. Tables IV-8 and IV-9 list the compounds reported in the literature.

Bis(cyclopentadienyl) alkoxides and phenoxides (not 8-quinolinolates) are white crystalline compounds readily soluble in most organic solvents and, in general, can be recrystallized from petroleum ethers. Although the phenoxides appear to be air-stable in the solid state, in solution they are considerably more sensitive, so that their preparation must be carried out under an inert atmosphere (A17). The alkoxo and phenoxo groups on both hafnium and zirconium are quite reactive toward protic reagents.

The C–O stretching vibrations in the infrared for the phenoxides are found in the 1280–1300 cm^{-1} region (A17) and in the 1090–1140 cm^{-1}

Table IV-8
Bis(cyclopentadienyl)zirconium(IV) and -hafnium(IV) Alkoxides and Phenoxides

Compound[a]	M.p. (°C)	Yield (%)	Refs.
Cp$_2$Zr(OCH$_3$)Cl	111–114.5	—	G28, W7
Cp$_2$Zr(OCH$_3$)CH$_3$	—	—	W8
Cp$_2$Zr(OCH$_3$)$_2$	63–68[b] (sublimes 85°/0.1 mm)	98	G27, G28, M34
Cp$_2$Zr(OCH$_2$CH$_3$)Cl	70–77	43	B69, G28
Cp$_2$Zr(OCH$_2$CH$_3$)Br	—	76	B69
Cp$_2$Zr(OCH$_2$CH$_3$)$_2$	52–57[b] (sublimes 55°/0.1 mm)	91	G27, G28, M34
Cp$_2$Zr[OCH(CH$_3$)$_2$]Cl	81–82.5 (sublimes 86°/0.1 mm)	23	B69, G28
Cp$_2$Zr[OCH(CH$_3$)$_2$]$_2$	110–116.5 (sublimes 68°)	—	G27, G28, W3
Cp$_2$Zr[OC(CH$_3$)$_3$]$_2$	148 (sublimes 110°)	86	C16, D35
Cp$_2$Hf[OC(CH$_3$)$_3$]$_2$	Sublimes 110°	58	C16
Cp$_2$Zr(OC$_6$H$_5$)$_2$	118–120	85	A16, A17, M34
Cp$_2$Hf(OC$_6$H$_5$)$_2$	122–123.5	71–76	B66
Cp$_2$Zr(O-2-ClC$_6$H$_4$)$_2$	104–106	72	A17
Cp$_2$Zr(O-4-ClC$_6$H$_4$)$_2$	66–68	72	A17
Cp$_2$Zr(O-2-CH$_3$C$_6$H$_4$)$_2$	132–134	88	A17
Cp$_2$Zr(O-3-CH$_3$C$_6$H$_4$)$_2$	150–152	92	A17
Cp$_2$Zr(O-4-CH$_3$C$_6$H$_4$)$_2$	164–166	88	A17
Cp$_2$Zr(1-naphtholato)$_2$	178–180	92	A17
Cp$_2$Zr(2-naphtholato)$_2$	173–175	85	A17
Cp$_2$Zr(O-4-C$_6$H$_5$C$_6$H$_4$)$_2$	156	70	A17
Cp$_2$Zr(O-2-NH$_2$C$_6$H$_4$)$_2$	160	70	A17
Cp$_2$Zr(O-4-NH$_2$C$_6$H$_4$)$_2$	184	20	A17
Cp$_2$Zr(benzene-1,2-diolato)	>180	25	A17
Cp$_2$Zr(biphenyl-2,2′-diolato)	>200	45	A17
Cp$_2$Zr(Cl)OZrCp$_2$(OC$_2$H$_5$)	147–156	—	B69
Cp$_2$Zr(Br)OZrCp$_2$(OC$_2$H$_5$)	184–192	—	B69

[a] Several μ-oxo–μ-alkoxo compounds are listed in Table IV-7.
[b] Ref. (M34) reports a m.p. of 112° for the methoxide and 140° for the ethoxide.

range for alkoxides (G28, W3). Bands between 730 and 765 cm^{-1} have been tentatively assigned to Zr–O bending modes, while those found in the 420–570 cm^{-1} region were considered to arise from Zr–O stretching vibrations (G28). The infrared spectra of several 8-quinolinolates have been recorded (B73).

Proton chemical shifts recorded for these compounds are given in Table IV-10. Attempts have been made to draw a correlation between the chemical

Table IV-9
Monocyclopentadienyl Phenoxides of Zirconium(IV) and Hafnium(IV)

Compound	M.p. (°C)	Yield (%)	Refs.
$CpZr(C_9H_6NO)_2Cl^a$	260–263	56–94	C17, F26
$CpHf(C_9H_6NO)_2Cl$	—	99	C17
$CpHf(C_9H_6NO)_2Cl \cdot 0.5C_6H_6$	248.5–251.5	68	B73
$CpZr(C_9H_6NO)_2Br$	—	98	C17
$CpZr(C_9H_6NO)_2Br \cdot 0.5C_6H_6$	245–246	90	B73
$CpHf(C_9H_6NO)_2Br \cdot 0.5C_6H_6$	236–237	66	B73
$[CpZr(C_9H_6NO)_2]_2O$	> 300	71	F26
$[CpHf(C_9H_6NO)_2]_2O$	> 300	46	B73
$CpZr(C_9H_6NO)_3$	357–360	51–90	B75
$CpHf(C_9H_6NO)_3$	> 250	34–58	M36
$CpZr(C_9H_6NO)_3 \cdot CHCl_3$	—	—	B75
$\pi\text{-}CH_3C_5H_4Zr(C_9H_6NO)_3$	252–260	60	B75
$\pi\text{-}tert\text{-}C_4H_9C_5H_4Zr(C_9H_6NO)_3 \cdot 0.5C_6H_6$	> 300	81	M35
$\pi\text{-}tert\text{-}C_4H_9C_5H_4Zr(C_9H_6NO)_3 \cdot CHCl_3$	> 300	—	M35
$\pi\text{-}tert\text{-}C_4H_9C_5H_4Hf(C_9H_6NO)_3 \cdot 0.5C_6H_6$	239–241	64	M35
$CpZr(acac)_2(OC_6H_5)$	62–63.5	27	B66
$CpHf(acac)_2(OC_6H_5)$	67.5–70	37	B66
$CpZr(C_{15}H_{11}O_2)_2(OC_6H_5)^b$	178–179	35	B66
$CpZr(C_{15}H_{11}O_2)_2(OC_6H_5) \cdot 0.5C_6H_6$	166–168	70	B66
$CpHf(C_{15}H_{11}O_2)_2(OC_6H_5)$	164–166.5	52	B66

^{a} C_9H_6NO is the 8-quinolinolate ion.
^{b} $C_{15}H_{11}O_2$ is the enolate of dibenzoylmethane.

shift of the cyclopentadienyl protons and the basicity of the other groups on the metal. The more basic the groups, the higher upfield is the cyclopentadienyl proton resonance peak expected to be. Meaningful trends in the chemical shift with basicity of the ligand have been found for the compounds $CpTi(OC_2H_5)_nCl_{3-n}$ ($n = 0$–3) in CCl_4 (N29) and $Cp_2Ti(SR)_nCl_{2-n}$ ($n = 0$–2) in $CDCl_3$ (C61). However, for a series of $Cp_2Zr(OR)_nCl_{2-n}$ compounds [where $n = 1$ or 2 and $R = CH_3$ or $CH(CH_3)_2$] no relationship between chemical shift in C_6D_6 and basicity was established (G28). This is undoubtedly due to the solvent used as was pointed out in Table IV-3. It should be recognized that a comparison of the chemical shifts in benzene is not meaningful, since the solvent effect is both large and variable for different alkoxides (W7). From Table IV-10 it can be seen that in $CDCl_3$ a relationship between basicity and the cyclopentadienyl proton chemical shifts exists in the compounds, $Cp_2Zr(OR)_nCl_{2-n}$ [where $n = 1$ or 2, R = CH_3 or $CH(CH_3)_2$].

The reaction between $Cp_2Zr[OC(CH_3)_3]_2$ and boron trifluoride–diethyl ether has been utilized to prepare Cp_2ZrF_2 in better than 50% yield (D35).

Table IV-10

Proton Chemical Shifts (δ) for Some π-Cyclopentadienylzirconium(IV) Alkoxides and Phenoxides[a]

Compound	Solvent	π-C_5H_5	O–R	Ref.
$Cp_2Zr(OCH_3)Cl$	C_6D_6	5.95	3.65	G28
	$CDCl_3$	6.35	—	W7
$Cp_2Zr(OCH_3)CH_3$	$CDCl_3$	6.22	3.78 [δ(Zr–CH_3] = 0.04]	W8
$Cp_2Zr(OCH_3)_2$	C_6D_6	5.98	3.77	G28
	$CDCl_3$	6.02	3.81	W8
$Cp_2Zr(OCH_2CH_3)Cl$	C_6D_6	5.90	—	W7
	C_6D_6	5.99	1.00 (CH_3); 3.86 (CH_2)	G28
	$CDCl_3$	6.32	—	W7
$Cp_2Zr(OCH_2CH_3)_2$	C_6D_6	6.00	1.11 (CH_3); 3.93 (CH_2)	G28
$Cp_2Zr[OCH(CH_3)_2]Cl$	C_6D_6	6.00	—	W7
	C_6D_6	5.95	1.00 (CH_3)	G28
	$CDCl_3$	6.33	—	W7
$Cp_2Zr[OCH(CH_3)_2]_2$	C_6D_6	6.01	1.08, 1.17 (CH_3); 4.11 (CH_2, sept.)	W3
	C_6D_6	5.99	1.08 (CH_3)	G28
	$CDCl_3$	6.17	—	W7
$Cp_2Zr(OC_6H_5)_2$	C_6D_6	5.88	—	W7
	$CDCl_3$	6.35	—	W7
$CpZr(C_9H_6NO)_2Cl$[b]	$CDCl_3$	6.52	[c]	C17
$CpZr(C_9H_6NO)_2Br$	$CDCl_3$	6.54	[c]	C17
$CpHf(C_9H_6NO)_2Cl$	$CDCl_3$	6.43	[c]	C17

[a] Several compounds containing an OC_6H_5 group on the metal are listed in Tables IV-11 and IV-12 and those containing μ-oxo–μ-alkoxy groups are in Table IV-7.

[b] C_9H_6NO is the 8-quinolinolato ligand.

[c] Complex multiplet lying between $\delta 6.8$ and 8.6.

As might be expected the bidentate 8-quinolinolato ligands are less labile than the monodentate alkoxides or phenoxides. In fact, the 8-quinolinolates are considered air-stable. $CpZr(C_9H_6NO)_2Cl$ reacts only slowly with ethanol under reflux (C17). $CpHf(C_9H_6NO)_2Cl$ reacts readily with water in the presence of an amine to form the μ-oxo complexes in which all the 8-quinolinolato ligands remain intact.

The 8-quinolinolates are yellow in color, soluble in chloroform and THF and to a lesser extent in benzene, and insoluble in ether or hexane. In nitrobenzene the compounds, $CpM(C_9H_6NO)_2X$ (where M = Zr or Hf, X = Cl or Br), are nonelectrolytes (C17). The metal–halogen stretching vibrations in these complexes are found at 321 cm^{-1} for Zr–Cl, 288 cm^{-1} for Hf–Cl, and 218 cm^{-1} for Zr–Br (C17).

ii. β-DIKETONATO COMPOUNDS. All the compounds fall into two classes, the bis(β-diketonates), $CpM(diketone)_2X$, and the tris(β-diketonates), $CpM(diketone)_3$.

(a) *Compounds of type* $CpM(diketone)_2X$. *Methods of preparation.* The reaction

$$Cp_2ZrCl_2 + 2\,acacH \longrightarrow CpZr(acac)_2Cl + HCl + C_5H_6$$

in which other β-diketones can be used (except tri- or hexafluorinated derivatives, see following section), is the one most often utilized for the synthesis of these bis(β-diketonates). The yield is generally high [94–98% for the acetylacetonate (F25, P6)], the reaction being driven to near-completion by removal of one or both of the by-products. When excess acetylacetone is used, this is accomplished by carrying out the reaction under reduced pressure to expel the volatile HCl and C_5H_6 (F25). If the acetylacetone is present in stoichiometric proportions, triethylamine is added to the reaction mixture to remove the acid as the triethylamine hydrochloride (P6).

Groups other than halides such as acetato, alkoxo, phenoxo, and μ-oxo can be replaced by β-diketones. The redistribution reaction

$$Cp_2ZrX_2 + Zr(acac)_4 \xrightarrow[\text{3 hr, 115°–120°}]{\text{toluene}} 2CpZr(acac)_2X$$

$$X = Cl \text{ or } Br$$

has also been used (B68, F25), but yields were low (19–36%). This, too, was the case (10%) for the reaction (B73)

$$(acac)_2 Zr(NO_3)_2 + NaC_5H_5 \xrightarrow[20°]{\text{benzene}} CpZr(acac)_2NO_3 + NaNO_3$$

Improved yields (51%) of this nitrate product were obtained from the reaction between the μ-oxo compound, $[CpZr(acac)_2]_2O$, and HNO_3 (F26).

All these compounds (see Table IV-11) are colorless crystalline substances, readily soluble in chloroform, somewhat less soluble in benzene and THF, and insoluble in petrol. The hafnium compounds are more soluble in the majority of organic solvents than are the corresponding zirconium derivatives (M36). $CpZr(acac)_2Cl$ is readily hydrolyzed in solution (P6), but in the solid

Table IV-11

Bis(β-diketonato)cyclopentadienyl Compounds of Zirconium(IV) and Hafnium(IV)

Compound	M.p. (°C)	Yield (%)	Refs.
CpZr(acac)$_2$Cl	189–190	19–98	B68, B69, F23, F25, P6
CpHf(acac)$_2$Cl	181–182	58	M36, P6
CpZr(acac)$_2$Br	202–204	39–67	B68, P6
CpHf(acac)$_2$Br	175–179	28	B70a
CpZr(acac)$_2$NO$_3$	142.5–143.5	45–47	B73, F26
CpHf(acac)$_2$NO$_3$	139–142	10–51	B73
CpZr(acac)$_2$(OCOCH$_3$)	128.5–131	70	B74, N32
CpZr(acac)$_2$(OC$_6$H$_5$)	62–63.5	27	B66
CpHf(acac)$_2$(OC$_6$H$_5$)	67.5–70	37	B66
CpZr(acac)$_2$(p-CH$_3$C$_6$H$_4$SO$_3$)·$\frac{1}{2}$THF	187.5–190	77	F26
(π-CH$_3$C$_5$H$_4$)Zr(acac)$_2$Cl	127–130	92	M35
(π-CH$_3$C$_5$H$_4$)Hf(acac)$_2$Cl	121	64	M35
(π-tert-C$_4$H$_9$C$_5$H$_4$)Zr(acac)$_2$Cl	170–171	54	M35
CpZr(C$_{10}$H$_9$O$_2$)$_2$Cl^a	173–185.5	69–97	F23, F25, M37, P6
CpHf(C$_{10}$H$_9$O$_2$)$_2$Cl	171.5–172.5	67	M37
CpZr(C$_{15}$H$_{11}$O$_2$)$_2$Cl^b	224–232.5	45–71	B66, F23, F26, M37
CpHf(C$_{15}$H$_{11}$O$_2$)$_2$Cl	216.5–219.5	70–75	B66, M37
CpZr(C$_{15}$H$_{11}$O$_2$)$_2$(OC$_6$H$_5$)	178–179	35–39	B66
CpZr(C$_{15}$H$_{11}$O$_2$)$_2$(OC$_6$H$_5$)·$\frac{1}{2}$C$_6$H$_6$	166–168	70	B66
CpHf(C$_{15}$H$_{11}$O$_2$)$_2$(OC$_6$H$_5$)	164–166.5	52	B66
[CpZr(acac)$_2$]$_2$O	164.5–165.5	55	B65
[CpHf(acac)$_2$]$_2$O	150–154	38	B73
[π-CH$_3$C$_5$H$_4$Zr(acac)$_2$]$_2$O	118.5–119	27	M35
[CpZr(C$_{10}$H$_9$O$_2$)$_2$]$_2$O	172–175	64	F26

^{a} C$_{10}$H$_9$O$_2$ is the enolate of benzoylacetone.

^{b} C$_{15}$H$_{11}$O$_2$ is the enolate of dibenzoylmethane.

state appears to be relatively stable (S63). In nitrobenzene it is a weak electrolyte, as are CpZr(acac)$_2$Br and CpHf(acac)$_2$Cl (P6).

Structural data. If the cyclopentadienyl group in CpZr(acac)$_2$X is assumed to occupy only one coordination site and the acetylacetonates to be bidentate, then the metal can be considered as being octahedrally coordinated and, hence, both cis and trans isomers are possible (F25). The bidentate nature of the acetylacetonato groups was demonstrated by the absence of free ketonic carbonyl stretching vibrations in the infrared spectrum (B73, F25, M35, P5). In the late 1960's it was firmly established, first by PMR spectroscopy (P5) and then by X-ray crystallography (S63), that CpZr(acac)$_2$Cl has the cis configuration to the complete exclusion of the trans, both in solution and in the solid state.

In the crystal, $CpZr(acac)_2Cl$ has C_1 symmetry. The plane containing the Cp and that containing the three oxygens and chlorine (see Fig. IV-4) are nearly parallel, the angle of intersection of the planes being only 3.0°. The zirconium is raised 0.45 Å above the four-coordinate plane. The mean Zr–O bond length (2.15 ± 0.04 Å) and the mean O–Zr–O bond angle (79.1° ± 1.3°) compare well with the values found for $Zr(acac)_4$. The average Zr–C distance is 2.55(5) Å.

Stezowski and Eick (S63) observed that the stereochemistry of CpZr-$(acac)_2Cl$ can be viewed in terms of eight-coordination about the zirconium and can best be fitted to D_{2d} dodecahedral geometry (see Fig. IV-4). The suggestion has been made (H33) that the B site atoms are more favorably

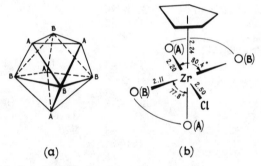

(a)　　　　　　　　　(b)

Fig. IV-4. (a) D_{2d} dodecahedron showing the A and B sites. (b) Coordination arrangement about the zirconium in $CpZr(acac)_2Cl$ (S63).

located for $d\pi–p\pi$ overlap with the zirconium. The cyclopentadienyl ring, which is located on the triangular AAB face is, in this scheme, bound by three bonding molecular orbitals to the metal.

The PMR spectra* for a series of $\pi\text{-}RC_5H_4Zr(acac)_2X$ compounds are tabulated in Table IV-12. The methyl groups, in general, give rise to four bands of equal intensity and the methine protons to two equiintense bands. It was inferred from the data that the acetylacetonato ligands were non-equivalent, probably cis to each other (M35, P5, P6). The benzoylacetone complexes, $CpZr(C_{10}H_9O_2)_2Cl$ and $CpHf(C_{10}H_9O_2)_2Cl$, occur in at least four isomeric forms in solution and these are interconvertible above 70°. This large number of isomers was ascribed to the asymmetry inherent in this diketone (M37, P5, P6).

Pinnavaia and Lott (P6) observed that the bands arising from the acetylacetonato ligands broaden on increasing the temperature above 70° and coalesce at 105° (Fig. IV-5). The rearrangement interconverting the non-

* The ^{13}C NMR spectrum of $CpZr(acac)_2Cl$ has also been measured (N28a); $\delta^{13}C_{Cp} = 116.47$ (relative to TMS).

equivalent acetylacetonates is first-order ($k_{25°} = 6.8 \times 10^{-3}$ sec^{-1}) for CpZr(acac)$_2$Cl, the activation energy being 23.1 $\pm$ 2.0 kcal/mole and the entropy 7.1 $\pm$ 5.4 eu at 25°. Similar values for these parameters are found for CpZr(acac)$_2$Br and CpHf(acac)$_2$Cl. A possible mechanism for the inter-conversion entails the rupture of one metal–oxygen bond leading to the formation of a symmetrical trans form by a rearrangement and metal–oxygen bond reformation, which then subsequently interconverts to one of the more stable cis isomers (P6).

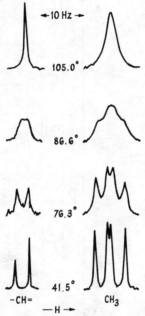

Fig. IV-5. Temperature dependence of the acetylacetonato methyl and methine proton resonance lines (60 MHz) for CpZr(acac)$_2$Cl in benzene (P6).

Recent dipole moment studies on CpZr(acac)$_2$Cl (5.71 D) and CpHf(acac)$_2$Cl (5.81 D) in benzene at 25° tend to confirm their cis configuration in solution (K24). In the solid state the compounds are isomorphous (P6). [CpZr(acac)$_2$]$_2$-O has a dipole moment of 3.62 D (K24).

(b) *Compounds of type* CpM(*diketone*)$_3$. *Methods of preparation.* While acetylacetone, benzoylacetone, and dibenzoylmethane react with Cp$_2$ZrCl$_2$ to give bis(diketonato) complexes, with either hexafluoro- or trifluoro-acetylacetone this reaction leads to tris(diketonato) complexes (H37).

$$Cp_2ZrCl_2 + 3F_6\text{-acacH} \longrightarrow CpZr(F_6\text{-acac})_3 + C_5H_6 + 2HCl$$

Table IV-12

Proton Chemical Shifts for π-RC$_5$H$_4$Zr(β-diketone)$_2$X Compounds[a]

Compound	Solvent	β-Diketonato protons		Cyclopentadienyl protons		Refs.
		$-CH_3$	$-CH=$	Aromatic	Alkyl	
CpZr(acac)$_2$Cl	Benzene[b]	1.55, 1.62, 1.63, 1.69	5.18, 5.25	6.50	—	P5, P6
	Benzene	1.48(3), 1.56(6), 1.63(3)	5.12(1), 5.19(1)	6.39(5)	—	M35
	CHCl$_3$	1.88(6), 1.95(3), 2.01(3)	5.51(2)	6.36(5)	—	M35
	CCl$_4$	1.89(6), 1.99(3), 2.05(3)	5.45(1), 5.49(1)	6.29(5)	—	M35
	C$_6$H$_5$NO$_2$[b]	1.93, 1.94, 1.96, 1.97	—	—	—	M37
CpHf(acac)$_2$Cl	Benzene[b]	1.56, 1.61, 1.63, 1.70	5.17, 5.21	6.44	—	P6
	Benzene	1.47(3), 1.54(3), 1.55(3), 1.63(3)	5.12(1), 5.17(1)	6.40(5)	—	M35
	CCl$_4$	1.91(6), 2.02(3), 2.08(3)	5.46(1), 5.48(1)	6.18(5)	—	M35
CpZr(acac)$_2$Br	Benzene[b]	1.53, 1.61, 1.63, 1.67	5.17, 5.27	6.55	—	P6
CpZr(acac)$_2$OCOCH$_3$	CCl$_4$	1.71(3), 1.80(3), 2.00(6)	5.26(1), 5.35(1)	6.09(5)	—	M35
CpZr(acac)$_2$OC$_6$H$_5$	CCl$_4$	1.60, 1.89, 2.02, 2.04	5.35, 5.41	6.22	—	B66
CpHf(acac)$_2$OC$_6$H$_5$	CCl$_4$	1.51, 1.88, 2.04, 2.07	5.31, 5.40	6.14	—	B66
[CpZr(acac)$_2$]$_2$O	Benzene	1.59(6), 1.73(3), 1.89(3)	5.05(1), 5.26(1)	6.59(5)	—	M35
	CCl$_4$	1.79(6), 1.86(3), 2.09(3)	5.17(1), 5.34(1)	6.09(5)	—	M35

Compound	Solvent					Ref.
$\pi\text{-}CH_3C_5H_4Zr(acac)_2Cl$	CCl_4	1.88(3), 1.89(3), 2.01(3), 2.07(3)	5.44(1), 5.49(1)	6.00(2), 6.16(2)	2.20(3)	M35
$\pi\text{-}CH_3C_5H_4Hf(acac)_2Cl$	Benzene	1.46(3), 1.54(6), 1.61(3)	5.07(1), 5.13(1)	—	2.27(3)	M35
	$CDCl_3$	1.92(6), 2.01(3), 2.08(3)	5.55(2)	6.04(2), 6.20(2)	2.27(3)	M35
$\pi\text{-}tert\text{-}C_4H_9C_5H_4Zr(acac)_2Cl$	CCl_4	1.91(6), 2.02(3), 2.09(3)	5.45(1), 5.47(1)	5.89(2), 6.06(2)	2.26(3)	M35
	Benzene	1.49(3), 1.56(3), 1.60(3), 1.65(3)	5.15(1), 5.19(1)	6.37(2), 6.42(2)	1.35(9)	M35
	$CDCl_3$	1.90(6), 1.99(3), 2.06(3)	5.58(2)	6.25(2), 6.29(2)	1.30(9)	M35
$CpZr(C_{10}H_9O_2)_2Cl^c$	Benzene	1.52, 1.56, 1.58, 1.61, 1.71, 1.73, 1.78, 1.83	—	—	—	M37
	$CHCl_3$	1.93, 1.94, 1.97, 2.01, 2.12, 2.13, 2.19, 2.21	—	—	—	M37
	CCl_4	1.98, 1.99, 2.02, 2.04, 2.17, 2.25	—	—	—	M37
	$C_6H_5NO_2{}^b$	1.92, 1.99, 2.01, 2.09, 2.14, 2.18, 2.31	—	—	—	M37
$CpHf(C_{10}H_9O_2)_2Cl$	$CHCl_3$	1.91, 1.93, 1.94, 1.99, 2.10, 2.11, 2.18, 2.21	—	—	—	M37
	CCl_4	1.93, 1.96, 2.00, 2.13, 2.19, 2.20	—	—	—	M37
	$C_6H_5NO_2{}^b$	1.98, 2.01, 2.03, 2.07, 2.09, 2.12, 2.17	—	—	—	M37
$CpZr(C_{15}H_{11}O_2)_2Cl^d$	$CHCl_3{}^b$	—	6.92	6.66	—	M37 / F23

^{a} δ, Relative to TMS.

^{b} Temperature variable spectrum measured.

^{c} $C_{10}H_9O_2$ is the enolate of benzoylacetone.

^{d} $C_{15}H_{11}O_2$ is the enolate of dibenzoylmethane.

139

Purportedly, the main hurdle in preparing the tris(diketonates) is the slow substitution by the third diketone (H37). The complexes of the less acidic diketones can be made if the metal reactant contains more readily replaceable phenoxo or alkyl groups.

Table IV-13

Cyclopentadienyltris(β-diketonato) Complexes of Zirconium(IV) and Hafnium(IV)

Compound	M.p. (°C)	Yield (%)	Molecular weight	Refs.
$CpZr(acac)_3 \cdot \frac{1}{2}C_6H_6$	Oil	Quantitative	—	B75
$CpZr(C_{10}H_9O_2)_3{}^a$	141–142	52	613	B70a, B75
$CpZr(C_{15}H_{11}O_2)_3{}^b$	186–187	63–72	792, 814	B66, B70a, B75
$CpHf(C_{15}H_{11}O_2)_3$	158–162	60	—	B66, B70a
$CpZr(C_{11}H_{19}O_2)_3{}^c$	230	62.4	720	H37
$CpZr(C_7H_6O_2)_3{}^d$	—	76	—	F23
$CpZr(acac)_2(C_{10}H_9O_2)$	—	80	—	B70a
$CpZr(acac)_2(C_{15}H_{11}O_2)$	116	61	—	B70a
$CpZr(C_{10}H_9O_2)_2(C_{15}H_{11}O_2)$	50–52	96	—	B70a
$CpZr(C_{15}H_{11}O_2)_2(C_{10}H_9O_2)$	84–87	79	—	B70a
$CpZr(F_3\text{-}acac)_3$	87–90	42.7	630	H37
$CpZr(F_6\text{-}acac)_3$	98–110	98.9	790	H37

[a] $C_{10}H_9O_2$ is the enolate of benzoylacetone.
[b] $C_{15}H_{11}O_2$ is the enolate of dibenzoylmethane.
[c] $C_{11}H_{19}O_2$ is the enolate of dipivalomethane.
[d] $C_7H_6O_2$ is the enolate of tropolone.

The tris(β-diketonato) complexes so prepared (see Table IV-13) are monomeric and range in color from white to yellow. In nitrobenzene the fluorinated acetylacetonato and dipivalomethanato compounds are very weak electrolytes (H37). They are very sensitive to hydrolysis, especially in solution, but are quite stable toward oxygen.

Structural data. The crystal structure of the hexafluoroacetylacetonate has been determined (E2, E3). The geometry of the molecule has been discussed in terms of a distorted icosahedron in which the cyclopentadienyl moiety occupies five coordination sites or in terms of a pentagonal bipyramid when the cyclopentadienyl group is considered monodentate (see Fig. IV-6). The Zr–C distances are similar, the mean value is 2.527(6) Å and the cyclopentadienyl ring is quite symmetrical. The zirconium atom is displaced 0.39 Å above the equatorial plane containing the five oxygen atoms and toward the cyclopentadienyl group. The equatorial oxygens do not quite lie in a plane, but deviations from planarity are small. Of interest is the axial oxygen–zirconium bond length which is appreciably shorter than the equatorial oxygen–zirconium bonds.

CpZr(β-diketone)$_3$ compounds in which the diketone is the hexa- or trifluoroacetylacetone or dipivalomethane have been shown to be stereochemically labile (E3, H37) (see Fig. IV-7). The spectra are explicable on the basis of the pentagonal-bipyramidal structure of the complex in which there are two equatorial diketones and one bridging an equatorial and an axial position. The stronger absorptions arise from the equatorial diketones, while the others are due to the unique diketone.

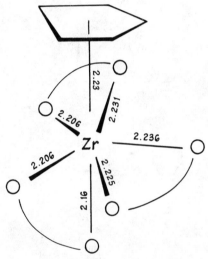

Fig. IV-6. Pentagonal-bipyramidal coordination arrangement in CpZr(F$_6$-acac)$_3$ (E2).

The more intense pair of fluorine bands broaden with increasing temperature and collapse into a singlet at about 50° owing to the intramolecular rearrangement interconverting the two equatorial diketones. The unique diketone is stereochemically rigid over this temperature range. At 25° the activation energy and entropy for this rearrangement is 15.7 ± 0.5 kcal/mole and 5.1 ± 1.7 eu. This compares with values of 21.3 ± 1.9 kcal/mole and 3.8 ± 5.5 eu for the corresponding dipivalomethane complex. Howe and Pinnavaia (H37) suggest that the isomerization proceeds via bond rupture or a digonal twist of an equatorial ligand.

Above 50°, the diketone spanning the axial and an equatorial position becomes labile and begins to exchange with the equatorial diketones. The activation energy and entropy for this exchange in CpZr(F$_6$-acac)$_3$ is 10.7 ±

2 kcal/mole and 12.1 $\pm$ 5.5 eu greater than the equatorial scrambling process. It has been suggested that this result concurs with the finding that the Zr–O axial bond length is appreciably shorter than the other Zr–O bonds.

CpZr(F$_3$-acac)$_3$ has six possible isomeric forms of which at least three have been detected by ^{1}H and ^{19}F NMR spectroscopy (H37).

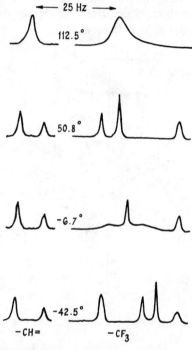

Fig. IV-7. Temperature dependence of the ^{19}F and the –CH= proton NMR lines for CpZr(F$_6$-acac)$_3$ in diisopropyl ether (H37).

iii. CARBOXYLATES. The reactions of carboxylic acids with cyclopentadienylzirconium compounds were first investigated in 1963 (B67, N32). When Cp$_2$ZrCl$_2$ was heated with a carboxylic acid, the hydrogen chloride formed was expelled and the product obtained was a mono(cyclopentadienyl)zirconium tris(carboxylate). With fluorinated acids such as CF$_3$CO$_2$H or

$$3RCO_2H + Cp_2ZrCl_2 \xrightarrow{\text{heat}} CpZr(OCOR)_3 + C_5H_6 + 2HCl$$

CF$_3$CF$_2$CO$_2$H the reaction stops at the bis(acylate) stage. The loss of a cyclopentadienyl group in the reaction of nonfluorinated carboxylic acids with Cp$_2$ZrCl$_2$ and other similar reactions has been rationalized (W7) on the basis of the ionicity of the Zr–Cp bond. The ionicity of this bond is highly dependent on the nature of the groups on the zirconium; the more electron-donating the group the greater the lability of the cyclopentadienyl. Wailes and Weigold (W3) have prepared $Cp_2Zr(O_2CCF_3)_2$ and $CpZr-(O_2CCH_3)_3$ by treating the dihydride Cp_2ZrH_2 with excess acid. King and Kapoor (K29) have prepared the volatile bis(trifluoroacetate) and bis(heptafluoropropionate) in over 70% yield by reaction of Cp_2ZrCl_2 with the silver salt of the acid. The monoacylates, $Cp_2Zr(OCOR)Cl$, have been obtained by treatment of the μ-oxo compound with a carboxylic acid at 20° for 1 hr (B74).

$$[Cp_2ZrCl]_2O + 2RCO_2H \longrightarrow 2Cp_2Zr(OCOR)Cl + H_2O$$

In contrast with cyclopentadienyl compounds, only the halides were replaced when bis(π-indenyl)zirconium dichloride was refluxed with either mandelic or p-aminosalicyclic acid in chloroform–dimethoxyethane solvent (G2).

Studies of the reactions undergone by these complexes (listed in Table IV-14) with HCl (B67, N32), acetylacetone (B67, N32), and H_2SO_4 (F24) have confirmed the lability of the carboxylato group by replacement with other ligands.

The infrared spectrum of $CpZr(O_2CCH_3)_3$ has three main bands centered at about 1510, 1450, and 1395 cm^{-1} in the carbonyl stretching region, suggesting that the acetato groups are in several nonequivalent environments. Furthermore, because of the relatively weak splitting ($< 120\ cm^{-1}$) between the asymmetric and symmetric stretching frequencies, the acetato ligands are bonded more or less symmetrically, be it inter- or intramolecularly (W3).

iv. SULFINATES. All compounds are O-sulfinates, as may be predicted from the affinity which zirconium shows for oxygen.

(a) *Substitution reactions.* Lindner and co-workers (L17) obtained $Cp_2Zr(O_2SCH_3)_2$ by the reaction

$$Cp_2ZrCl_2 + 2NaSO_2CH_3 \xrightarrow{\text{THF}} Cp_2Zr(O_2SCH_3)_2 + 2NaCl$$

In ethanol the reaction product was $Zr(O_2SCH_3)_4$. Here, presumably, the ethanol splits the Zr–Cp bond after its activation by the initial replacement of the halides by sulfinato groups. Subsequent replacement of the ethoxo ligands then leads to the formation of the tetrasulfinate.

(b) *Sulfur dioxide insertion.* The majority of the sulfinates have been made by this method (W7, W8). While Cp_2ZrCl_2 is completely inert to SO_2 and can be recovered unchanged from liquid SO_2 at room temperature after

Table IV-14
Carboxylato Complexes of Zirconium(IV)

Compound	Properties	Refs.
$Cp_2Zr(OCOCH_3)Cl$	M.p. 137°–138°	B74
$Cp_2Zr(OCOC_6H_5)Cl$	M.p. 160.5°–161.5°	B74
$Cp_2Zr(OCOCH_3)_2$	PMR(CDCl₃) δ6.32(Cp), δ2.03(CH₃)	W7
$Cp_2Zr(OCOCF_3)_2$	Colorless; hygroscopic; m.p. 117° [a]	B67, K29, M34, N32, W3
$Cp_2Zr(OCOC_3F_7)_2$	Colorless; m.p. 35°–36°	K29
$Cp_2Zr(OCOCCl_3)_2$	M.p. 129°	M34
$(\pi\text{-Indenyl})_2Zr[OCOCH(OH)C_6H_5]_2$	Colorless solid	G2
$(\pi\text{-Indenyl})_2Zr(O_2C\text{-}4\text{-}NH_2C_6H_4)_2$	M.p. 230°	G2
$CpZr(OCOCH_3)_3$	Colorless; hygroscopic; dec >170°; PMR(C₆D₆) δ6.26(Cp), 1.73(CH₃); (CHCl₃) δ6.38(Cp), 2.03(CH₃)	B67, B74, W3
$(\pi\text{-}tert\text{-}C_4H_9C_5H_4)Zr(OCOCH_3)_3$	M.p. 89°–91°	B74
$CpZr(OCOC_4H_9)_3$	Viscous liquid, gradually solidifies	B67, N32
$CpZr(OCOC_6H_{13})_3$	Liquid	B67, N32
$CpZr(OCOC_6H_5)_3 \cdot C_6H_5CH_3$	M.p. 209–210°°	N32
$CpZr(OCOCH_3)(acac)_2$	—	B74

[a] Ref. (K29) gives the m.p. as 93°–94°.

periods exceeding 1 week, $Cp_2Zr(CH_3)Cl$ reacts readily, inserting an SO_2 molecule into both the $Zr-CH_3$ and one $Zr-Cp$ linkage to form the pale yellow insoluble compound, $Cp(C_5H_5SO_2)Zr(O_2SCH_3)Cl$.

On passing SO_2 through a solution of the μ-oxo compound, $[Cp_2ZrCl]_2O$, a white insoluble μ-oxosulfinate is obtained. The $Zr-O-Zr$ bridges in this

$$n[Cp_2ZrCl]_2O + nSO_2 \longrightarrow [Cp(C_5H_5SO_2)ZrO]_n + nCp_2ZrCl_2$$

compound give rise to a strong band in the infrared at 708 cm^{-1}. The bands associated with the SO_2 group found at 1157 cm^{-1} and at 918 and 902 cm^{-1} were taken as indicative of an unsymmetrically bonded SO_2 group, possibly an S,O-sulfinate. This compound reacts further with liquid SO_2 at room temperature by insertion of an SO_2 molecule into the $Zr-O$ bond to form a sulfite. Another O-sulfinatozirconium sulfite is obtained as the end product of the reaction of $Cp_2Zr(CH_3)_2$ with SO_2. The reaction scheme is

$$Cp_2Zr(CH_3)_2 + 3SO_2 \xrightarrow{\ 10°\ } Cp(C_5H_5SO_2)Zr(O_2SCH_3)_2$$

$$\Big\downarrow \text{liquid } SO_2, 20°$$

$$(C_5H_5SO_2)_2ZrSO_3 + (CH_3)_2SO$$

This is the only reaction in which SO_2 inserts into all $Zr-Cp$ bonds. Tetra-(cyclopentadienyl)zirconium reacts with SO_2 to give $Cp(C_5H_5SO_2)_3Zr$.

v. SULFONATES. In 1961 Brainina et al. (B71) reported the preparation of $Cp(\pi-C_5H_4SO_3H)Zr(SO_4H)_2 \cdot (CH_3CO)_2O$ by reaction of Cp_2ZrCl_2 with concentrated sulfuric acid in acetic anhydride. This compound was nonmelting, hygroscopic, and insoluble in most organic solvents with the exception of alcohols and dimethylformamide. The acetic anhydride in the compound could be readily replaced by either ethanol or dimethylformamide, the number of molecules of solvent on the complex being variable (F24, W11). When $CpZr(SO_4H)_3$ in acetic anhydride–dichloroethane was treated with SO_3, the complex obtained was $(\pi-C_5H_4SO_3H)Zr(SO_4H)_3$ (F24).

Although Brainina and co-workers (B71, F24) considered that these complexes contained a π-bonded cyclopentadienylsulfonic acid group, an alternative structure in which the SO_3 moiety has inserted between the C_5H_5 ring and the metal has been proposed (W11). Evidence in favor of the SO_3 insertion product was provided by the PMR spectrum of the bis(bisulfate) complex in D_2O, in which solvent there were only two replaceable protons in the complex, in agreement solely with the cyclopentadienesulfonato structure.

In line with the insertion of SO_2 into bis(cyclopentadienyl)zirconium complexes, SO_3 from H_2SO_4–acetic anhydride reagent will insert only between one π-cyclopentadienyl group and the metal. With the stronger sulfonating agent, SO_3, insertion into the mono-π-cyclopentadienyl compound, $CpZr(SO_4H)_3$, does take place (F24).

When either the monoacetic anhydride or ethanol adduct of $Cp(C_5H_5SO_3)$-$Zr(SO_4H)_2$ is heated with acetylacetone, the Cp and a bisulfate group are replaced to give $(C_5H_5SO_3)Zr(acac)_2(SO_4H) \cdot \frac{1}{2}acacH$ from which $(C_5H_5SO_3)$-$Zr(acac)_2(SO_4H) \cdot acacH$ can also be obtained (F24).

c. CYCLOPENTADIENYL COMPOUNDS CONTAINING METAL–HYDROGEN BONDS

i. TETRAHYDROBORATES. Bis(cyclopentadienyl)zirconium bis(tetrahydroborate) was prepared by Nanda and Wallbridge (N1) from Cp_2ZrCl_2 and

$$Cp_2ZrCl_2 + 2LiBH_4 \xrightarrow[\text{8 hr}]{\text{ether}} Cp_2Zr(BH_4)_2 + 2LiCl$$

excess lithium borohydride. Using only a twofold excess of $LiBH_4$, $Cp_2Zr(Cl)BH_4$ was obtained. $Cp_2Zr(BH_4)_2$ is a pale yellow solid, melting at 155°, which is best purified by sublimation at 110°–115° *in vacuo*. On treatment with trimethylamine, BH_3 is removed as the amine complex, leaving the transition metal dihydride (N1). The mono(tetrahydroborate) also sublimes, but is less soluble in organic solvents.

In the case of hafnium, only the bis(tetrahydroborate) has been synthesized (D8). Its properties are similar to those of the corresponding zirconium compound; it is colorless, readily soluble in ethers and aromatic hydrocarbons, sublimes *in vacuo* at 110°, reacts only slowly with air, but rapidly with acids to release hydrogen quantitatively. The tetrahydroborate groups react with amines to form hydrido compounds.

The tetradeuteroborates have been made in a similar manner and their infrared spectra have been measured (D8). The bonding between the metal and the BH_4 moiety in $Cp_2Zr(BH_4)_2$ was considered from infrared evidence to be via two Zr–H–B bridges (D8, J4) and not triple hydride bridges as found in $Zr(BH_4)_4$ [$\nu(BH_2) = 2440$ (asym.) and 2386 (sym.) cm^{-1}; $\delta(BH_2) = 1123$ cm^{-1}]. The bridge expansion vibration appears at 2142 cm^{-1} (D8).

The PMR spectrum of $Cp_2Zr(BH_4)_2$ is in accord with bridge-bonded BH_4 groups in which there is a rapid exchange of hydrogens (D8, J4, V30, V31). In benzene, the quartet associated with the BH_4 group is centered at $\delta 0.78$. The ^{11}B NMR spectrum, consisting of a 1:4:6:4:1 quintet, also shows equivalent coupling between the boron and all the hydrogens (D8, J2). The bonding of the BH_4 groups in the hafnium compound is analogous to that in the zirconium complex. The crystal structure of Cp_2TiBH_4 also shows bonding via two bridging hydrido ligands (M32a).

ii. HYDRIDO COMPOUNDS. Although the bis(cyclopentadienyl)hydrido complexes of Re, Mo, W, and Ta have been known for quite some time, it was not until 1966 that the first organometallic zirconium hydride was reported. Since then, more have been isolated (see Table IV-15). As expected for zirconium(IV), all are diamagnetic and near white in color. They were initially synthesized from the tetrahydroborate complexes (J3). The hydrido tetra-

$$Cp_2Zr(BH_4)_2 + (CH_3)_3N \xrightarrow{\text{benzene}} Cp_2Zr(H)BH_4 + (CH_3)_3NBH_3$$

$$Cp_2Zr(BH_4)_2 + 2(CH_3)_3N \xrightarrow{\text{benzene}} Cp_2ZrH_2 + 2(CH_3)_3NBH_3$$

hydroborate is an air-reactive compound, subliming *in vacuo* at 60°; it is soluble in organic solvents. In contrast, Cp_2ZrH_2 is nonvolatile and completely insoluble in all solvents (J3, W2). It is hydrolyzed only slowly in air. The difference in properties is a result of the polymeric nature of the dihydride, the hydrido groups acting as bridging ligands in this compound, but not in the hydrido tetrahydroborate derivative. This was inferred, in part, from the infrared spectra, the dihydride having strong broad Zr–H bands in the 1300–1540 cm^{-1} region (J3, W2), whereas for $Cp_2Zr(H)BH_4$ the band is sharp and at 1945 cm^{-1}.

An improved method of synthesis of many zirconium hydride compounds employs LiAlH$_4$ (K15, W2). When the oxo-bridged compound, $[Cp_2ZrCl]_2O$, in THF is treated with an equimolar amount of LiAlH$_4$, the insoluble dihydride precipitates in over 60% yield. The by-product from the reaction is a

$$[Cp_2ZrCl]_2O + LiAlH_4 \longrightarrow 2Cp_2ZrH_2 + LiCl + \text{``}[AlOCl]_2\text{''}$$

soluble complex between Cp_2ZrH_2 and $[AlOCl]_2$. Cp_2ZrH_2 can also be obtained from $[Cp_2ZrCl]_2O$ with 4 moles of LiAl(O-*tert*-C$_4$H$_9$)$_3$H. However, with only 2 moles of the aluminum hydride, the insoluble white complex, $[(Cp_2ZrH)_2O \cdot Cp_2ZrH_2]$ was isolated (W2).

The hydrido chloride, $Cp_2Zr(H)Cl$, was synthesized in 90% yield (W2) by the reaction

$$Cp_2ZrCl_2 + LiAl(O\text{-}\textit{tert}\text{-}C_4H_9)_3H \xrightarrow{\text{THF}} Cp_2Zr(H)Cl + LiCl + Al(O\text{-}\textit{tert}\text{-}C_4H_9)_3$$

LiAlH$_4$ can also be used. An alternative preparation of this insoluble white hydride, although in lower yield (30%), is by reaction of Cp_2ZrCl_2 with magnesium in THF. Presumably, the hydrido group originates from the solvent (W2) as the presence of the THF is essential.

The halide in $Cp_2Zr(H)Cl$ can be replaced by other groups (W2), for example,

$$Cp_2Zr(H)Cl + LiAlH_4 \longrightarrow Cp_2Zr(H)AlH_4 + LiCl$$

Although $Cp_2Zr(H)BH_4$ sublimes *in vacuo* above 50°, $Cp_2Zr(H)AlH_4$ is not volatile and decomposes on heating at 60°. On thermolysis of the solid,

from a cyclopentadienyl ring. The suggested reaction is

$Cp_2Zr(H)CH_3$ is prepared (W2) from the reaction between $Cp_2Zr(Cl)CH_3$ and $LiAl(O\text{-}tert\text{-}C_4H_9)_3H$ or $LiAlH_4$ in THF. This compound and the hydrido chloride are both noticeably light-sensitive, changing from white to pink on exposure to daylight.

Of all the hydrido compounds of zirconium, only $Cp_2Zr(H)BH_4$ has a nonbridging Zr–H moiety. All other compounds show Zr–H stretching frequencies below 1540 cm^{-1} (see Table IV-15) indicative of Zr–H–Zr bridging groups (J1, W2, W9).

Table IV-15

Infrared Frequencies of Some Bis(cyclopentadienyl)zirconium Hydrides^a and Deuterides

Compound	Assignment	Frequency (cm^{-1})	
		Hydride	Deuteride
Cp_2ZrH_2	Zr–H–Zr	1520, 1300	1100, 960
$Cp_2Zr(H)Cl$	Zr–H–Zr	1390	1020
$Cp_2Zr(H)CH_3$	Zr–H–Zr	1500	1090, 965
$[(Cp_2ZrH)_2O \cdot Cp_2ZrH_2]$	Zr–H–Zr	1510, 1385, 1240	1075, 1025, 905
Cp_2Zr—H—$Al(CH_3)_3$	Zr–H–Zr	1350	980
	Zr–H–Al	1780	1290
$Cp_2Zr(H)AlH_4$	Zr–H–Zr	1425	1055
	AlH_4	1790, 1700	1310, 1260
$Cp_2Zr(H)BH_4$	Zr–H	1945	—

^{a} The only other hydride known is (COT)ZrH$_2$, in which the metal–hydride absorptions are at 1537 and 1310 cm^{-1}.

Only two hydrido compounds, $Cp_2Zr(H)BH_4$ and $[Cp_2ZrH_2 \cdot Al(CH_3)_3]_2$, possess sufficient solubility to permit the measurement of their PMR spectrum. In benzene, the cyclopentadienyl protons of these compounds showed resonance lines at $\delta5.70$ [for the borohydride (J4)] and $\delta5.50$ [for the aluminum complex (W9)]. The BH_4 group gave rise to a quartet centered at $\delta - 0.20$. In the dimer

$$Cp_2Zr\text{—}H\text{—}Al(CH_3)_3$$
$$\diagup \; \diagdown$$
$$H \quad H$$
$$\diagdown \; \diagup$$
$$Cp_2Zr\text{—}H\text{—}Al(CH_3)_3$$

the Zr–H–Zr triplet is at $\delta - 2.92$, while that from the Zr–H–Al group arises at $\delta - 0.92$.

The basic nature of the hydrido groups is exemplified by their reaction with acidic protons. Cp_2ZrH_2 reacts with isopropanol (W3) to give both $Cp_2Zr[OCH(CH_3)_2]_2$ and $Zr[OCH(CH_3)_2]_4$. The bis(isopropoxide) can be conveniently prepared from Cp_2ZrH_2 and acetone as here there is no danger of cleaving Zr–Cp bonds.

$$Cp_2ZrH_2 + 2(CH_3)_2CO \xrightarrow{\text{room temp.}} Cp_2Zr[OCH(CH_3)_2]_2$$

With carboxylic acids two modes of reaction are possible, namely, reduction of the acid and/or the elimination of the hydrido group as hydrogen. When Cp_2ZrH_2 is added to excess acetic acid, $CpZr(OCOCH_3)_3$ is formed

$$Cp_2ZrH_2 + 3CH_3CO_2H \longrightarrow CpZr(OCOCH_3)_3 + 2H_2 + C_5H_6$$

(W3). However, on slow addition of a dilute solution of acetic acid to a stirred slurry of the dihydride, reduction of the acid also occurs, and a polymeric μ-oxo–μ-ethoxo complex was obtained in which the basic unit is

$$\left[\begin{array}{c} C_2H_5 \quad C_2H_5 \\ | \qquad | \\ O \quad Cp \quad O \\ \diagup \quad | \quad \diagdown \\ Cp_2Zr \quad \quad Zr \quad \quad ZrCp_2 \\ \diagdown \quad | \quad \diagup \\ O \quad Cp \quad O \end{array} \right]_n$$

The μ-oxo atom arises by abstraction of an oxygen from the carboxylic group.

Hydrido compounds react readily with unsaturated compounds. Their reactions with alkenes and alkynes are discussed in the following section. Cp_2ZrH_2 and $Cp_2Zr(H)Cl$ have been used (W10) to reduce alkenes and alkynes catalytically and quantitatively to alkanes (see Table IV-16). The dihydride is the better catalyst. Zirconocene also catalyzes this reduction.

Compounds containing multiple bonds such as azobenzene and benzonitrile also react with Cp_2ZrH_2 (W11). With carbon disulfide Cp_2ZrH_2 forms the blue, insoluble sulfide, Cp_2ZrS (W11). The dihydride abstracts fluorine from

Table IV-16
Catalytic Hydrogenation of Some Hydrocarbons with Cp_2ZrH_2 and Hydrogen[a]

Unsaturated compound	Temperature (°C)	Product
Cyclohexene	80	Cyclohexane
trans-Stilbene	100	1,2-Diphenylethane
cis-Stilbene	100	1,2-Diphenylethane
Styrene	120	Ethylbenzene
Vinylcyclohexane	120	Ethylcyclohexane
2-Methyl-1-butene	120	2-Methylbutane
2-Methyl-2-butene	120	2-Methylbutane
cis-Hept-2-ene	120	Heptane
Phenylacetylene	120	Ethylbenzene
3-Hexyne	120	Hexane

[a] Pressure, 50–100 atm. From ref. (W10).

C_6F_5H and C_6F_6. With carbon monoxide at 100 atm pressure at 45° for 2 hr in benzene, a polymeric carbonyl is formed. The CO stretches occur at 1849 and 1948 cm^{-1} (W11). The reaction of the hydrido groups with CH_2Cl_2 proceeds smoothly and quantitatively, giving the metal chloride derivative and methyl chloride (W2).

d. CYCLOPENTADIENYL COMPOUNDS CONTAINING METAL–CARBON BONDS

The thermal instability introduced into metal alkyl compounds by the presence of β-hydrogen atoms on the alkyl groups is also evident in bis-(cyclopentadienyl)zirconium compounds. For example, although $Cp_2Zr(CH_3)_2$ can be sublimed at above 100°, $Cp_2Zr(C_2H_5)_2$ is not known.

i. ALKYL COMPOUNDS (INCLUDING ALKENYL AND ALKYNYL). Since 1961 when Braye et al. (B83) synthesized the orange, light-sensitive, 1,1-bis(π-cyclopentadienyl)-2,3,4,5-tetraphenylzirconacyclopentadiene from 1,4-dilithiotetraphenylbutadiene and Cp_2ZrCl_2, a number of other alkyl, alkenyl, and alkynyl compounds have been prepared. These are shown in Table IV-17. Several methods have been used for their preparation, the most common being alkylation of Cp_2ZrCl_2 with lithium or magnesium (Grignard) alkyls. The reaction

$$[Cp_2ZrCl]_2O + 2Al(CH_3)_3 \xrightarrow{CH_2Cl_2} 2Cp_2Zr(Cl)CH_3 + [(CH_3)_2Al]_2O$$

also provides a convenient route to the monomethyl derivative. With acidic hydrocarbons, displacement of amine from a dialkylamide can be used (J7), e.g.,

$$Cp_2Zr[N(CH_3)_2]_2 + 2HC \equiv CC_6H_5 \xrightarrow[1 \text{ hr}]{\text{reflux}} Cp_2Zr(C \equiv CC_6H_5)_2 + 2HN(CH_3)_2$$

Wailes, Weigold, and Bell (W6) have employed the insertion of unsaturated hydrocarbons into Zr–H bonds to prepare both alkyl and alkenyl derivatives of zirconium, e.g.,

$$Cp_2Zr(H)Cl + C_6H_{10} \xrightarrow{\text{benzene}} Cp_2Zr \overset{Cl}{\underset{C_6H_{11}}{\diagdown}}$$

Cyclohexene also reacts with Cp_2ZrH_2 at room temperature, but the bisalkyl derivative is of insufficient stability and the only product isolated was zirconocene.

Although acetylene itself reacts readily with Cp_2ZrH_2, the black solid compounds obtained were of variable composition. Substituted acetylenes led to some well-characterized insertion products (W6). The reaction of diphenylacetylenes with Cp_2ZrH_2 is interesting. When the reactants were brought together in refluxing benzene, the hydride dissolved with liberation of some hydrogen gas and a compound formulated as **11** was isolated. On

(11)   (12)

$$R = C_6H_5, \ p\text{-}CH_3C_6H_4$$

further refluxing, cyclopentadiene was eliminated together with the rest of the hydrogen and the product obtained was the dizirconabenzene (**12**). To satisfy the coordination requirements of the metal, there may be a multiple bond between the two zirconium atoms.

ii. REACTIONS AND PROPERTIES. With the exception of the alkynyl derivatives, all compounds of zirconium and hafnium containing a metal–carbon bond are monomeric and soluble in benzene and often in aliphatic hydrocarbons. The metal–carbon bonds are extremely reactive toward protic reagents. For example,

$$2Cp_2Zr(Cl)CH_3 + H_2O \longrightarrow [Cp_2ZrCl]_2O + 2CH_4$$

$$Cp_2Zr(CH_3)_2 + H_2O \longrightarrow [Cp_2ZrCH_3]_2O \xrightarrow{H_2O} [Cp_2ZrO]_n$$

$[Cp_2ZrO]_n$ loses cyclopentadienyl groups on further reaction with water. With methanol, $Cp_2Zr(CH_3)_2$ forms methoxides (W8). The monomethoxide

Table IV-17
Alkyl- and Arylcyclopentadienyl Compounds of Zirconium and Hafnium

Compound	Color	M.p. or sublimation temp (°C/mm)	PMR chemical shift (δ)	PMR solvent	Refs.
$Cp_2Zr(CH_3)Cl$	Pale yellow	193(dec)	−0.39(CH₃), 6.08(Cp)	CDCl₃[a]	S74, W7
$Cp_2Zr(CH_3)H$	Colorless	—	—	C₆D₆	W2
$Cp_2Zr(CH_3)COCH_3$	Colorless	—	0.48(ZrCH₃), 2.33(OCH₃), 5.40(Cp)	C₆D₆	W11
$Cp_2Zr(CH_3)[ON(NO)CH_3]$	Colorless	—	0.14(ZrCH₃), 3.83(NCH₃), 5.88(Cp)	CDCl₃	W8
$Cp_2Zr(CH_3)_2$	Colorless	$110/10^{-4}$	0.42(CH₃), 5.78(Cp)	C₆D₆[b]	W8
$Cp_2Hf(CH_3)_2$	—	—	—	—	R4
$(\pi\text{-indenyl})_2Zr(CH_3)_2$	—	—	—	—	R4
$(\pi\text{-indenyl})_2Hf(CH_3)_2$	—	—	—	—	R4
$Cp_2Zr(C_2H_5)Cl$	Yellow	—	—	—	S46
$Cp_2Zr(Cl)CH_2CH_2CH_2Zr(Cl)Cp_2$	Yellow	140	—	—	S45
$Cp_2Zr(CH_2C_6H_5)_2$	Yellow	—	1.84(CH₂), 5.46(Cp), 6.67–7.35(C₆H₅)	C₆D₆	F1
$Cp_2Zr[CH_2Si(CH_3)_3]_2$	Colorless	96–97	—	—	C42
$Cp_2Hf[CH_2Si(CH_3)_3]_2$	Colorless	83	—	—	C42
$Cp_2Zr(cyclo\text{-}C_6H_{11})Cl$	Red-brown	—	—	—	W10
$Cp_2Zr\text{-}Cl,\ C(H^\alpha)=C(H^\beta)(C_2H_5)$	Yellow-brown	Viscous liquid	6.74(H^α), 1.93(CH₂), 0.90(CH₃), $J_{H^\alpha H^\beta} = 18$ Hz	C₆D₆	W6
$Cp_2Zr\text{-}Cl,\ C(C_6H_5)=C(C_6H_5)(H)$	Orange	—	5.77, 5.90(Cp), 6.76(C₆H₅) cis and trans forms	C₆D₆	W6

152

Structure	Color		NMR	Solvent	Ref.
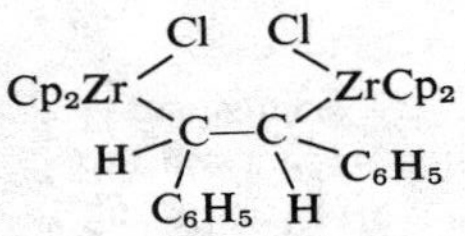 Cp₂Zr structure: Cp_2Zr with Cl, Cl, ZrCp₂, C₆H₅, C₆H₅, H	Crimson	—	5.78, 5.90(Cp), 6.76(C_6H_5) threo and erythro forms	C_6D_6	W6
$Cp_2Zr[HC{=}C(H)R]_2^c$					
R = C_2H_5	Brown-black	Oil	2.25(CH_2), 1.00(CH_3)	C_6D_6	W6
R = C_6H_5	Black	—	5.78(Cp), 6.98(C_6H_5)	C_6D_6	W6
$CpZr(C_8H_8)Cl$	Red	—	5.27(Cp), 5.94(C_8H_8)	$C_6D_5CD_3$	K4a
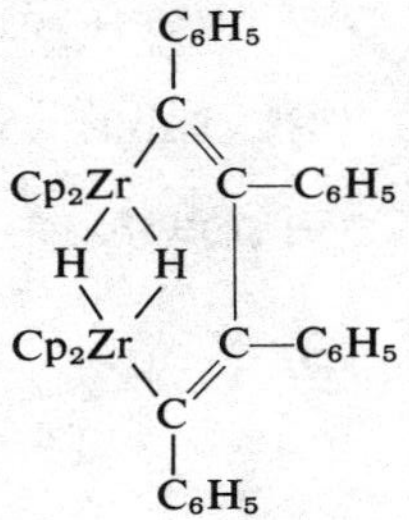 structure with C_6H_5, Cp_2Zr, H	Brown-black	—	—	—	W6
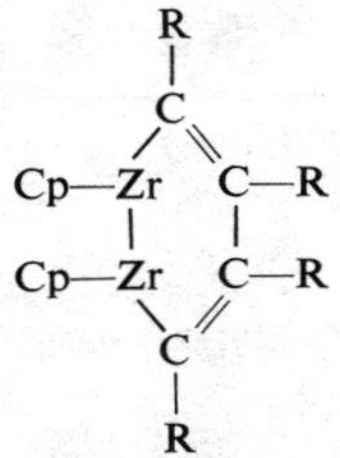 structure R = C_6H_5	Red-brown	—	6.00(Cp), 6.64–7.10(C_6H_5)	C_6D_6	W6
R = p-$CH_3C_6H_4$	Red-brown	—	5.94(Cp), 1.75, 1.96(CH_3)	C_6D_6	W6

153

(continued)

Table IV-17 (continued)

Alkyl- and Arylcyclopentadienyl Compounds of Zirconium and Hafnium

Compound	Color	M.p. or sublimation temp (°C/mm)	PMR chemical shift (δ)	PMR solvent	Refs.
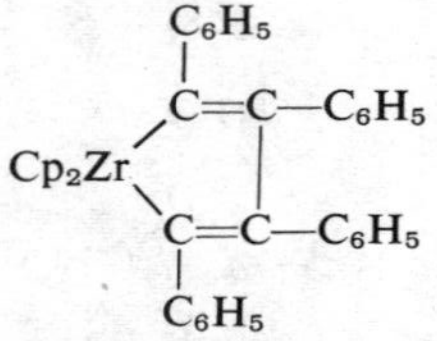 C_6H_5 structure with Cp_2Zr	Orange	140–170	—	—	B83, D38, H40
$Cp_2Zr(C{\equiv}CC_6H_5)_2$	Light brown	—	6.42(Cp), 7.27(C_6H_5) $\nu(C{\equiv}C)$ 2073 cm^{-1}(s)	CS_2	J7
$Cp_2Hf(C{\equiv}CC_6H_5)_2$	—	—	6.23(Cp), 7.17(C_6H_5) $\nu(C{\equiv}C)$ 2083 cm^{-1}(s)	CS_2	J7
$(\pi\text{-}CH_3C_6H_4)_2Zr(C{\equiv}CC_6H_5)_2$	Orange	73–75	6.12(Cp), 7.17(C_6H_5), 2.33(CH_3) $\nu(C{\equiv}C)$ 2078 cm^{-1}(s)	CS_2	J7
$Cp_2ZrCl\left[\begin{array}{c}-CHCH_2Al(C_2H_5)_2 \\ Al(C_2H_5)_2\end{array}\right]$	—	128	—	—	H5

[a] For other PMR solvents see (W8).
[b] For other PMR solvents see (W7).
[c] Trans isomers.

154

can be isolated in good yield, but since replacement of the second methyl group and the Cp ligands are mutually competitive, the bis(alkoxide) cannot be obtained satisfactorily by this method. Phenylacetylene is sufficiently acidic to react slowly with $Cp_2Zr(CH_3)_2$ in refluxing toluene (W8) to give mixtures of $Cp_2Zr(C\equiv CC_6H_5)_2$ and $Cp_2Zr(C\equiv CC_6H_5)CH_3$.

The Zr–C bonds, although readily degraded by acidic reagents, are unaffected by oxygen or carbon dioxide. Hydrogen does react (W8) with Cp_2Zr-$(CH_3)_2$ at 100°–120° in light petroleum to give the crimson-colored zirconium-(III) dimer, $[Cp_2ZrCH_3]_2$. Recently, Wailes, Weigold, and Bell have studied the insertion of hydrogen (W10), nitric oxide (W8), carbon monoxide (W11), and sulfur dioxide (W7, W8) into Zr–C bonds.

The ability of SO_2 to insert into metal–carbon bonds has been well documented in the literature and, not unexpectedly, the reaction occurs also with Zr–C-bonded compounds (W7, W8). However, with bis(cyclopentadienyl)-zirconium complexes there is the further interesting complication in that SO_2 reacts with the Zr–Cp linkage even though the cyclopentadienyl is symmetrically (pentahapto) bonded. The O-sulfinates so produced are discussed further in Section B,1,b,iv of this chapter.

On passing NO through a petrol solution of $Cp_2Zr(CH_3)_2$ at room temperature, the white insertion product $Cp_2Zr(CH_3)[ON(NO)CH_3]$, was formed (W8). The complex is soluble in hydrocarbons, halocarbons, and ethereal solvents. Although it is inert to further reaction with nitric oxide at atmospheric pressure and temperatures up to 100°, the monomethyl compound, $Cp_2Zr(Cl)CH_3$, reacts with nitric oxide in benzene at 20° to yield $Cp_2Zr(Cl)$-$[ON(NO)CH_3]$. In $CDCl_3$, the proton chemical shift of the N–CH_3 group was $\delta 3.84$, almost identical with that of the corresponding methylzirconium compound.

Carbon monoxide does not react (F1) with $Cp_2Zr(CH_2C_6H_5)_2$ even at 40 atm and 100°, but with $Cp_2Zr(CH_3)_2$, under milder conditions (W11) CO was inserted into a Zr–CH_3 bond. As in the nitric oxide reactions, insertion

$$Cp_2Zr(CH_3)_2 + CO \xrightarrow[\text{40–80 atm}]{\text{20°, petrol}} Cp_2Zr\begin{array}{c}CH_3\\ \diagdown\\ \diagup C \diagup O\\ |\\ CH_3\end{array}$$

occurs only into one of the Zr–CH_3 bonds. The low carbonyl stretching frequency of 1540 cm^{-1} suggests that the carbonyl bond order is somewhat below 2. The PMR chemical shift of the C–CH_3 group is $\delta 2.33$ in C_6D_6. The compound is unstable under an inert atmosphere, losing CO to reform $Cp_2Zr(CH_3)_2$.

iii. ARYL COMPOUNDS. It was not until 1964 that the first aryl compound,

$[Cp_2ZrC_6H_5]_2O$, was made (B65). Although stable in cold water, the $Zr-C_6H_5$ linkage is cleaved by acids or halides (B65).

Until recently, $Cp_2Zr(C_6H_5)_2$ was thought to lack sufficient stability to permit its preparation and isolation (C22). However, Rausch (R4) has now obtained this compound and the hafnium analog by reaction of Cp_2ZrCl_2 with phenyllithium in diethyl ether. The bis(π-indenyl)metal diphenyl compounds were also prepared in this way (R4).

Fluorinated derivatives of cyclopentadienylzirconium or -hafnium species have shown an enhanced thermal stability compared with the unsubstituted compounds (C22, R4). $Cp_2Zr(C_6F_5)_2$ is volatile, subliming at $120°/0.01$ mm; it explodes when heated in air above its melting point ($218°-219°$). The aryl group is readily lost through hydrolysis in moist solvents (C22). The penta-

$$Cp_2Zr(C_6F_5)_2 + H_2O \xrightarrow{\text{ether}} Cp_2Zr(C_6F_5)OH \xrightarrow{H_2O} Cp_2Zr(OH)_2$$

fluorophenyl groups are more prone to hydrolytic cleavage in the zirconium compounds than in the corresponding titanium analogs which require strong aqueous bases or protic acids for their decomposition (C22). The mono-hydroxide species sublimes at $190°/0.01$ mm and may explode on heating in air above $260°$.

Metallocyclic compounds of the type, $Cp_2M(C_6F_4C_6F_4)$ (M = Zr, Hf), containing the metallofluorene system have been obtained as pale yellow crystals in low yield ($8-23\%$) from the dilithioaryl and Cp_2MCl_2 (R4). Although hydrolytically unstable, the compounds exhibit an extremely high thermal stability, decomposing only above $300°$.

iv. ALLYL COMPOUNDS. Martin, Lemaire, and Jellinek (M16) synthesized a series of allylbis(cyclopentadienyl)zirconium compounds by the reaction

$$Cp_2ZrCl_2 + 2C_3H_5MgCl \xrightarrow{0°} Cp_2Zr(C_3H_5)_2 + 2MgCl_2$$

The allyl compounds are quite unstable, decomposing after several hours at room temperature in the light. When stored in the dark they may be kept for several months at $-18°$. With aqueous HCl or water, $Cp_2Zr(C_3H_5)_2$ evolves propene and then slowly decomposes further, releasing all the cyclopentadienyl groups. With dry HCl, Cp_2ZrCl_2 is formed, while reaction with CH_2Cl_2 at $0°$ replaces only one allyl group by chloride.

$$Cp_2Zr(C_3H_5)_2 \xrightarrow{<1 \text{ hr}} [Cp_2Zr(C_3H_5)Cl \cdot Cp_2Zr(C_3H_5)_2] \xrightarrow{6 \text{ hr}} Cp_2Zr(C_3H_5)Cl$$

The bonding between the allyl groups and metal was established from the infrared spectra, mainly in the C–C stretch region ($1610-1520$ cm^{-1}) (see Table IV-18). All compounds contain at least one σ-bonded allyl group. PMR spectroscopy shows the allyl ligands to be dynamic, even at $-90°$. The bond-

Table IV-18

Allyl and 2-Methylallyl Compounds of Zirconium[a]

Compound	Color	$\nu(C=C)(cm^{-1})$	Zr-allyl bonding types[b]	Chemical shift (δ)[c]
$Cp_2Zr(C_3H_5)_2$	Cream	1533(s), 1589(s),	π, σ	5.13(Cp), 2.89(CH$_2$), 5.63(CH)
$Cp_2Zr(C_3H_5)Cl$	Yellow	1598(s)	σ	5.74(Cp), 3.31(CH$_2$), 6.11(CH)
$[Cp_2Zr(C_3H_5)_2 \cdot Cp_2Zr(C_3H_5)Cl]$	Yellow	1533(m), 1589(s), 1598(s)	π, σ, σ	—
$Cp_2Zr(C_4H_7)_2$	Yellow oil (m.p. $-15°$)	1520(m), 1603(s)	π, σ	5.59(Cp), 2.86(CH$_2$), 1.61(CH$_3$)
$Cp_2Zr(C_4H_7)Cl$	Red oil	1606(s)	σ	6.33(Cp), 3.09(CH$_2$), 1.66(CH$_3$)

[a] From ref. (M16).

[b] Assignments made on basis: $\nu(C=C) \geqq 1589$ cm^{-1}, σ-bonded allyl; $\nu(C=C) \leqq 1533$ cm^{-1}, π-bonded allyl.

[c] In C_6D_6 except for $Cp_2Zr(C_4H_7)Cl$ which is in CDCl$_3$.

157

ing in these compounds has been discussed (D14, S41) in terms of the model proposed by Ballhausen and Dahl (B8).

e. CYCLOPENTADIENYL COMPOUNDS CONTAINING METAL–SULFUR AND METAL–SELENIUM BONDS

All the compounds known (see Table IV-19), with the exception of $[Cp_2ZrS]_n$, are alkyl or arylthio (or seleno) derivatives of bis(cyclopentadienyl)-zirconium(IV) or -hafnium(IV).

Several methods which are used in the preparation of alkoxides and phenoxides of Cp_2Zr^{IV} and Cp_2Hf^{IV} (see Section B,1,b,i) are suitable also for the synthesis of the corresponding thiolato and selenolato complexes. The most convenient method is

$$Cp_2ZrCl_2 + 2HSC_6H_5 + 2N(CH_2CH_3)_3 \xrightarrow{\text{benzene}}$$
$$Cp_2Zr(SC_6H_5)_2 + 2(CH_3CH_2)_3N \cdot HCl$$

as it uses Cp_2ZrCl_2 as starting material and gives high yields of product (K39). The thiolato or selenolato complex is obtained as a glassy solid which is extremely air- and moisture-sensitive, especially in solution. The sparingly soluble maleonitriledithiolato complex, obtained (K39) from Cp_2ZrCl_2 and $Na_2S_2C_2(CN)_2$ by NaCl elimination, is stable toward water.

While in titanium(IV) chemistry the bis(thiolato) compounds are more stable than the corresponding selenolato derivatives, the reverse is true in the case of zirconium (K39). Most of the compounds are monomers when freshly prepared, but in solution or on aging in the solid state, they tend to polymerize with subsequent loss of solubility. The chelate complexes formed from the dithiolates are more stable than the bis(monothiolates). They are all yellow in color.

The PMR chemical shifts for the aromatic thiolato and selenolato protons are in the range $\delta 7.0$ to 7.5 (multiplets), while their π-cyclopentadienyl protons occur as singlets at about $\delta 6.00$ (K39). A comparison of the PMR spectrum of bis(cyclopentadienyl)ethylene-1,2-dithiolatozirconium(IV) with those of the corresponding tungsten, molybdenum, and titanium compounds has led Köpf (K40) to propose that in the zirconium (and titanium) complex the sulfur ligand has partial thioaldehyde character. The contribution of the

dithioglyoxal form to the structure is small (K28, K40). In CS_2 the dithiolato protons of this compound show a PMR chemical shift of $\delta 7.13$, while the Cp protons are evident at $\delta 6.00$.

Table IV-19
Bis(cyclopentadienyl)zirconium(IV) and -hafnium(IV) Compounds Containing Metal–Sulfur or Metal–Selenium Bonds

Compound	M.p. (°C) or sublimation temp.	Yield (%)	MW	Refs.
$Cp_2Zr[SC(CH_3)_3]_2$	160–180/0.05 mm	81	412	C16
$Cp_2Hf[SC(CH_3)_3]_2$	140–160/0.02 mm	86	500	C16
$Cp_2Zr(SC_6H_5)_2$	147–154 (dec)	72	459	K39
$Cp_2Zr(SeC_6H_5)_2$	103–108 (dec)	81	548	K39
Cp₂Zr benzenedithiolate structure	187–190	92	378	K39
Cp₂Zr toluenedithiolate (CH₃) structure	179–185	89	383	K39
Cp₂Zr dithiolate with two CN groups structure	200 (dec)	28	—	K39
Cp₂Zr ethylenedithiolate (H, H) structure	180–186, 170/0.2 mm	12–21	337	K28, K40

f. CYCLOPENTADIENYL COMPOUNDS CONTAINING METAL–NITROGEN BONDS (AMIDO AND KETIMIDO COMPOUNDS)

The organometallic chemistry of the amido complexes of titanium, zirconium, and hafnium has been developed to a large extent by Lappert and co-workers (C16, C41, D35, J7, J8, L4). The utility of amido compounds in organometallic synthesis lies in their behavior with weakly acidic hydrocarbons (pK_A up to about 20) such as cyclopentadiene and phenylacetylene and with compounds such as alcohols and thiols.

Mono(cyclopentadienyl) derivatives are formed on reaction of a tetrakis-(dialkylamido)metal compound with a stoichiometric amount of cyclopentadiene (C16), whereas use of excess cyclopentadiene leads to the bis(cyclopentadienyl)metal derivatives. The bis(amido) compounds can alternatively

$$M(NR_2)_4 + 2C_5H_6 \longrightarrow Cp_2M(NR_2)_2 + 2HNR_2$$

be made from the reaction between Cp_2ZrX_2 (X = Cl, Br) and a lithium or potassium amide (C16, I3). Ketimido compounds can also be prepared by this

Table IV-20

Amido and Ketimido Complexes of Zirconium and Hafnium

Compound	Color	B.p. or sublimation temp. (°C)	ν(Zr–N)(cm^{-1})	PMR chemical shifts (δ) (in C_6D_6)	Yield (%)	Ref.
CpZr[N(CH$_3$)$_2$]$_3$	Yellow, slightly viscous liquid	94–96/0.05 mm	—	—	33	C16
Cp$_2$Zr[N(CH$_3$)$_2$]$_2$	Yellow shining crystals	110–120/0.05 mm	—	—	54	C16
Cp$_2$Hf[N(CH$_3$)$_2$]$_2$	Yellow solid	120–125/0.02 mm	—	—	60	C16
Cp$_2$Zr[N(C$_2$H$_5$)$_2$]$_2$	Orange-yellow solid	120–130/0.03 mm	—	—	64	C16
Cp$_2$Hf[N(C$_2$H$_5$)$_2$]$_2$	Yellow crystals	120–130/0.03 mm	—	—	72	C16
Cp$_2$Zr[N(CH$_2$)$_2$]$_2$	Colorless solid	100/0.001 mm	—	6.05(Cp), 2.05(CH$_2$)	73	J7
Cp$_2$Zr[N(C$_6$H$_5$)$_2$]$_2$	Orange crystals	M.p. 180	507	—	56	I3
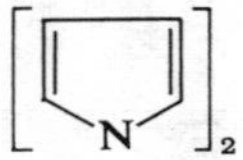	Yellow-green solid	M.p. 195(dec)	640	—	68	I3
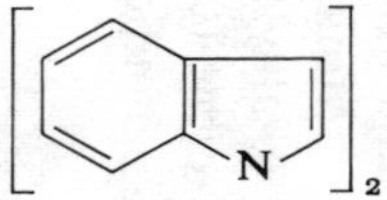	Orange solid	M.p. 59–62	550	—	38	I3

160

Compound		M.p./B.p.		NMR		Ref.
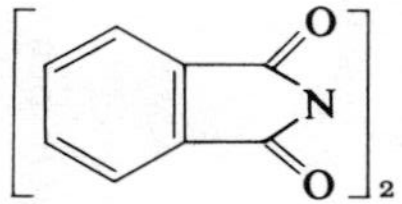 Cp₂Zr[structure]₂	Colorless solid	M.p. 200(dec)	—	—	61	I3
Cp₂Zr(C₁₂H₈N)₂[a]	Orange-red solid	M.p. 216–219	530	—	31	I3
(π-C₅H₄CH₃)₂Zr[N(CH₃)₂]₂	Reddish brown waxy solid	125–135/0.1 mm	—	—	67	C16
(π-Indenyl)Zr[N(CH₃)₂]₃	Orange viscous liquid	144/0.15 mm	—	—	20	C16
Cp₂ZrCl[N=C(C₆H₅)₂]	Yellow crystals	150/0.001 mm	ν(C=N) 1640	5.70(Cp), 7.02(C₆H₅)	—	C41
Cp₂Zr[N=C(C₆H₅)₂]₂	Red crystals	150/0.001 mm	ν(C=N) 1660(sh), 1640(br)	5.80(Cp), 7.1(C₆H₅)	—	C41
Cp₂Hf[N=C(C₆H₅)₂]₂	Red-orange powder	150/0.001 mm	ν(C=N) 1665(sh), 1643(br)	5.88(Cp)	—	C41
Cp₂ZrCl{N=C[C(CH₃)₃]₂}	Yellow powder	150/0.001 mm	ν(C=N) 1640	—	—	C41

^{a} C₁₂H₈N =

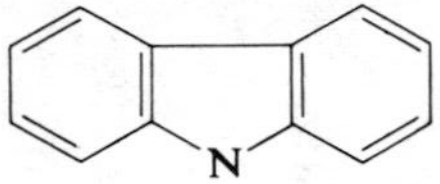

161

means (C41). The reactions are solvent-dependent. Under mild conditions (0°) the monoketimido complex is obtained. On refluxing with 2 moles of

$$Cp_2ZrCl_2 + LiN{=}CR_2 \xrightarrow[0°]{\text{ether}} Cp_2Zr(Cl)N{=}CR_2 + LiCl$$

lithium ketimide, both halogens of Cp_2ZrCl_2 are replaced and the bis-(ketimido) derivative is formed. Transamination reactions have also been employed to prepare these compounds (C41).

$$Cp_2M(NR_2)_2 + 2HN{=}CR_2' \xrightarrow[\text{reflux 2–3 hr}]{\text{ether}} Cp_2M(N{=}CR_2')_2 + 2HNR_2$$

All these complexes (see Table IV-20) are air-sensitive, some being pyrophoric. Issleib and Bätz (I3) established from both infrared and dipole moment measurements that those amido groups containing pyrrole rings are bound to the zirconium via the nitrogen rather than through a Zr–C linkage. An analogous bonding mode exists in the corresponding titanium compounds.

The $Cp_2M(NR_2)_2$ compounds react readily with alcohols or thiols (C16). Thus, with a stoichiometric amount of *tert*-butanol, $Cp_2M(O\text{-}tert\text{-}C_4H_9)_2$ was formed. Excess 2-methylpropane-2-thiol was necessary to obtain the corresponding thio compound.

The amido complexes, $Cp_2M[N(CH_3)_2]_2$ (M = Zr or Hf), have been used (D35) to prepare the fluoride compounds, Cp_2MF_2 (see Section B,1,a,i). $Cp_2Zr[N(CH_3)_2]_2$ initiates the polymerization of acrylonitrile (J8). It does not react with C_6F_5H even on heating to 100° for 1 hr (J7).

g. CYCLOPENTADIENYL COMPOUNDS CONTAINING METAL–PHOSPHORUS BONDS (PHOSPHIDES)

Organometallic compounds containing a zirconium–phosphorus bond are rare (E4, I4, I5) and hafnium–phosphorus-bonded compounds are unknown.

When the sodium phosphide, $C[CH_2P(Na)C_6H_5]_4$, was added to Cp_2ZrCl_2 in the molar ratio 1:2, the pale yellow spiro compound,

was formed in near quantitative yield. This monomeric complex melts with decomposition at 270°, and is soluble in THF and ethanol, but not in benzene or light petroleum. No organometallic zirconium compound was isolated when the potassium (rather than the sodium) phosphide was used as in this case all the cyclopentadienyl ligands were removed.

Whereas simple phosphides of the type, $LiPR_2$ ($R = C_2H_5$, $n\text{-}C_4H_9$), reduce the metal in Cp_2ZrBr_2 to the trivalent state (I4) with the concomitant formation of $[Cp_2ZrPR_2]_2$, the reaction of 1,4-dialkali-1,2,3,4-tetraphenyl-tetraphosphanes with Cp_2ZrX_2 ($X = Cl$, Br) results in the formation of an orange, monomeric, diamagnetic, 1,2,3-triphenylphosphanato complex of zirconium(IV). The solid complex (m.p. $233°\text{–}235°$) is stable in the air, but is

$$Cp_2ZrX_2 + M_2P(C_6H_5)_4 \longrightarrow Cp_2Zr \underset{P}{\overset{P}{\bigvee}} P\text{—}C_6H_5 + 2MX + \frac{1}{n}(PC_6H_5)_n$$

readily oxidized in solution (I5). It dissolves easily in most organic solvents except aliphatic hydrocarbons; it is not stable in water or alcohol. It reacts readily with HCl, and with iodine is converted quantitatively to Cp_2ZrI_2 and $C_6H_5PI_4$. The triphosphanato ring structure of the compound was deduced from its mass spectral fragmentation pattern and the ^{31}P NMR which was of the AX_2 type. The chemical shifts in THF, employing 85% H_3PO_4 as external standard, were $\delta - 181.4$ (P_A) and $\delta - 90.7$ (P_X) with $J_{PP} = 338$ Hz.

h. CYCLOPENTADIENYL COMPOUNDS CONTAINING METAL–METAL BONDS

The only compounds of this type known are those in which the dissimilar metal moiety is either the triphenylsilyl, triphenylgermyl, or triphenylstannyl group (see Table IV-21). They are made by alkali salt (NaCl or LiCl) elimination reactions. For example,

$$Cp_2ZrCl_2 + LiSi(C_6H_5)_3 \xrightarrow[-40° \text{ to } -78°]{THF} Cp_2Zr(Cl)[Si(C_6H_5)_3] + LiCl$$

Although the reaction is carried out at low temperatures, the compounds are all thermally stable (K30, K31). The second chlorine could not be replaced by another dissimilar metal group (C9). Dry HCl breaks the metal–metal bond (C9), as does $CDCl_3$ over a period of 24 hr (K31).

The crystal structure of $Cp_2ZrCl[Si(C_6H_5)_3]$ has been determined (M51) and some of the bond lengths and angles are given in Fig. IV-8. The complex is best viewed as having a pseudotetrahedral configuration in which the π-cyclopentadienyls are considered monodentate.

There is an approximate plane of symmetry through the Si, Zr, and Cl atoms with the consequence that the cyclopentadienyls are in an eclipsed conformation. Of interest in this structure is the long Zr–Si bond (2.813(2) Å).

Table IV-21

Properties of Complexes of Zirconium and Hafnium Containing Metal–Metal Bonds

Compound	Color	Sublimation temp. ($°C/5 \times 10^{-4}$ mm) or m.p.	PMR chemical shift (δ)[b]		Refs.
			π-C_5H_5	C_6H_5	
$Cp_2ZrCl[Si(C_6H_5)_3]$	Orange	175°–178°(m.p.)	6.29	7.44	C9
$Cp_2HfCl[Si(C_6H_5)_3]$	Yellow	180	6.24	7.45	K30, K31
$Cp_2ZrCl[Ge(C_6H_5)_3]$	Orange	190	6.43	7.45	K30, K31
$Cp_2HfCl[Ge(C_6H_5)_3]$	Yellow	200	6.15($CDCl_3$)	7.40($CDCl_3$)	C65
			6.36	7.45	K30, K31
$Cp_2ZrCl[Sn(C_6H_5)_3]$	Orange	a	6.62	7.45	K30, K31
$Cp_2HfCl[Sn(C_6H_5)_3]$	Yellow	a	6.32($CDCl_3$)	7.32($CDCl_3$)	C65
			6.52	7.45	K30, K31

[a] Not completely pure.
[b] In THF.

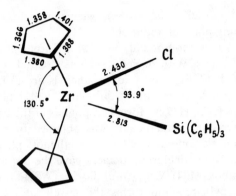

Fig. IV-8. Structure of $Cp_2ZrCl[Si(C_6H_5)_3]$ (M51).

2. Tris(cyclopentadienyl) Compounds of Zirconium(IV) and Hafnium(IV)

In parallel with titanium chemistry, where comparatively little is known about tris(cyclopentadienyl) compounds, this area in zirconium and hafnium chemistry has also been neglected. The only compound isolated which falls into this class is tris(methylcyclopentadienyl)zirconium chloride. It was made in 38% yield by stirring $ZrCl_4$ with sodium methylcyclopentadienide in toluene at room temperature for 3 hr under argon (B75). $(CH_3C_5H_4)_3ZrCl$ is yellow in color and melts at 168°–170°. It is extremely hygroscopic and reacts with 8-quinolinol to give a 60% yield of $(\pi\text{-}CH_3C_5H_4)Zr(C_9H_6NO)_3$.

3. Tetra(cyclopentadienyl)zirconium and -hafnium

Tetra(cyclopentadienyl)zirconium and -hafnium were first mentioned in the literature in 1964 by Fritz (F27) who, on the basis of their infrared spectra, listed these with other compounds containing σ-bonded cyclopentadienyl groups. The initial reports on the preparation of tetra(cyclopentadienyl)-zirconium, also in 1964 (M32), and the hafnium analog (M36) in 1966, entailed the reaction of excess sodium cyclopentadienide with the metal tetrahalide in an inert solvent under nitrogen. The zirconium compound is, however, best made from the readily available Cp_2ZrCl_2 (B64, B72). Both

$$Cp_2ZrCl_2 + 2NaC_5H_5 \xrightarrow[\text{room temp., 5 hr}]{\text{THF, ether or toluene}} \underset{54\text{–}69\%}{Zr(C_5H_5)_4} + 2NaCl$$
(excess)

compounds are monomeric in freezing benzene. The hafnium derivative melts with decomposition at 207°–208° (M36), whereas the zirconium analog has an indefinite melting point of about 360° (K24).

All reactions of these compounds involve replacement of two or more cyclopentadienyl groups. With reagents such as $ZrCl_4$, $FeCl_2$, HCl, HNO_3, and Br_2 (B72, M34, M36), two cyclopentadienyls are lost. Acidic organic ligands such as acetylacetone and 8-quinolinol will replace three cyclopentadienyl groups (B75, H37) to form compounds of the type, $CpM(acac)_3$. With ethanol, zirconium tetraethoxide was obtained in 71% yield. When used in stoichiometric amounts, ethanol, methanol, phenol, thiophenol, trifluoroacetic acid, and trichloroacetic acid gave the bis(cyclopentadienyl)-zirconium compounds (M34). With sulfur dioxide (W8) at room temperature, tetra(cyclopentadienyl)zirconium forms the O-sulfinate, $CpZr(O_2SC_5H_5)_3$. An insertion product was also believed to form when the compound was reacted with nitric oxide (W8).

The infrared spectra of tetra(cyclopentadienyl)zirconium and -hafnium are substantially richer in bands than the spectra of bis(cyclopentadienyl) compounds. Bands assignable to C–C double bonds (B73, B74, F27, L19, M36) suggested that not all of the cyclopentadienyls are pentahapto-bonded. The PMR spectrum of the zirconium compound (one peak at $\delta 5.75$ in $CDCl_3$) and that of the hafnium analog (one peak at $\delta 5.92$ in $CDCl_3$) did not confirm this. In the case of zirconium the singlet remained unsplit between 25° and $-52°$ (B64), undoubtedly owing to a rapid interchange of the ligands (B64, B70, B74, M36).

The asymmetry inherent in the molecules can also be ascertained from their dipole moments. In benzene at 25°, tetra(cyclopentadienyl)zirconium has a dipole moment of 3.41 D which increases to 3.57 D at 50°. The dipole moment of the hafnium analog at 25° is near 3.77 D. This similarity in dipole moments had led to the inference that both compounds have like structures (K24). However, a comparison of their molecular structures, determined by X-ray analysis, did not confirm this (K61, K63) (see Fig. IV-9).

The four rings in tetra(cyclopentadienyl)zirconium are disposed approximately tetrahedrally about the metal. From their structural results, Kulishov and co-workers (K61, K62) deduced that three of the cyclopentadienyls (A, B, and C in Fig. IV-9) are pentahapto-bonded while the fourth (ring D) was shown to be a *monohapto*-cyclopentadienyl group. The plane of the *monohapto*-cyclopentadienyl ring makes an angle of 52° with the Zr–C σ bond. The Zr–C bond lengths for rings A, B, and C vary over the range 2.50 to 2.74 Å.

Cotton *et al.* (C4) contend that the structural determination is not of sufficient quality to allow an unequivocal interpretation of the bonding between the metal and the rings A, B, and C. A bonding scheme incorporating one *monohapto*- and one *pentahapto*-cyclopentadienyl ring, together with two tilted

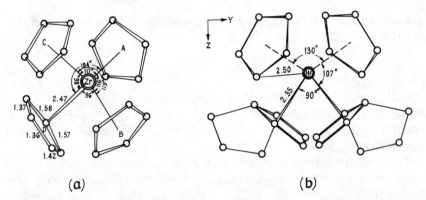

(a)   (b)

Fig. IV-9.   Structure of (a) tetra(cyclopentadienyl)zirconium (K62) and (b) tetra-(cyclopentadienyl)hafnium (K63) showing the alternate h^1-C_5H_5 ring positions.

rings considered as four-electron donors (so leading to an 18-electron system), was considered to be more acceptable.

The corresponding tetra(cyclopentadienyl)hafnium compound has two *pentahapto-* and two *monohapto*-cyclopentadienyl groups (K61a, K63). Such a molecular configuration was also found for the titanium compound (C4). The monohapto rings in $Cp_2Hf(h^1$-$C_5H_5)_2$ can occupy two symmetry-related positions as shown in Fig. IV-9. The angle between the Cp rings is 130° and that between the h^1-C_5H_5 rings 90°, values very reminiscent of those found in other Cp_2MX_2 (M = Zr, Ti) compounds (see Tables I-1 and IV-5).

C. Miscellaneous π Derivatives of Zirconium(IV) and Hafnium(IV)

Until recently, the only such compounds known were the tetraallyl derivatives. This was somewhat surprising in that the tetra(allyl)titanium(IV) compounds have not been isolated (T25). In 1970, Sharma *et al.* (S28, S29) observed that the reaction of $ZrCl_4$, $HfCl_4$, and $TiCl_4$ with cycloheptatriene or cyclooctatetraene gave the thermally stable compounds, $(C_7H_7)_2MCl_2$ and $(C_8H_7)_2MCl_2$, respectively. The mode of bonding in these compounds was not ascertained.

1. π-Allyl Compounds

Since the initial report of the preparation of an allylzirconium compound in 1963 (S65), many others have been synthesized. The red tetra(π-allyl)-

zirconium (B22, W21) and hafnium compounds (B22) have been made by the reaction

$$ZrCl_4 + 4(C_3H_5)MgCl \xrightarrow[-78°]{ether} Zr(\pi\text{-}C_3H_5)_4 + 4MgCl_2$$
$$45\%$$

Although they can be sublimed at room temperature into tubes cooled in liquid nitrogen, the compounds must be stored at low temperatures. The stability of the allyl compounds tends to increase with the atomic weight of the metal (B22, W21).

The compounds react vigorously with protic reagents. With stoichiometric amounts of dry HCl, tetra(π-allyl)zirconium will form either the bis- or red-brown tris(π-allyl) chloride (S65, W21). Job and Pioli (J13) in the patent literature reported the preparation of a large number of π-allyl and π-methyl-allyl halide compounds of zirconium, hafnium, and titanium. These were derived from the tetra(π-allyl) compound with a halide source such as tri-phenylmethyl chloride, *tert*-butyl chloride, *tert*-butyl iodide, benzyl chloride, 3-chloroprop-1-ene, etc.

Reactions of tetra(π-allyl)zirconium with ketones and esters have also been studied (B5) in detail. With acetone the reaction is

$$Zr[\pi\text{-}C_3H_5]_4 + 4(CH_3)_2CO \xrightarrow[-10°]{pentane} Zr \left[\begin{array}{c} CH_3 \\ | \\ -OCCH_2CH=CH_2 \\ | \\ CH_3 \end{array} \right]_4$$

(red)   (colorless)

The structures of this alkoxide and analogous derivatives from acetone-d_6, acetophenone, and benzophenone, formed quantitatively by this reaction, have been confirmed by PMR and infrared spectroscopy.

The fluxional behavior of $Zr(\pi\text{-}C_3H_5)_4$ in the PMR spectrum was first observed by Wilke and co-workers (W21). Recently, Becconsall, Job, and O'Brien (B22, B23) investigated this phenomenon in more detail (see Table IV-22) and concluded that interconversion of the syn and anti terminal hydrogens proceeded via the internal rotation of the CH_2 groups about the bonds linking them to the central carbon atom.* At low temperatures, namely, $-74°$, all the allyl ligands are symmetrically π-bonded. Calculations have indicated that the scrambling process requires an activation energy of 10.5 ± 1.0 kcal/mole (B22, W21). With tetra(π-allyl)hafnium in both $CDCl_3$ and $CFCl_3$, no fluxional behavior is evident even at $-72°$, the PMR spectrum at all temperatures being of the AX_4 type (B22). It has been postulated that

* A thorough investigation of this fluxional behavior has recently been made by Krieger *et al.* (K58a). They concluded that the interchange of the hydrogens occurs at one end of the alkyl group at a time rather than a simultaneous interchange at both ends.

Table IV-22

PMR Chemical Shifts for Tetra(π-allyl)zirconium and -hafnium[a]

Compound	Solvent	Temperature (°C)	Spectrum type	Chemical shifts (δ)[b]	J(Hz)
Zr(π-C$_3$H$_5$)$_4$	CFCl$_3$	−74	AM$_2$X$_2$	H(1) 5.18(n)[c] H(2) 3.28(d) H(3) 1.90(d)	$J_{12} = 8.5; J_{13} = 15.5; J_{23} < 0.5$
	CFCl$_3$	−10	AX$_4$	H(1) 5.19(quin) H(2), H(3) 2.63(d)	$J_{12} = J_{13} = 12.5$
Hf(π-C$_3$H$_5$)$_4$	CDCl$_3$	−20	AX$_4$	H(1) 5.64(quin) H(2), H(3) 2.82(d)	$J_{12} = J_{13} = 12.5$

[a] From ref. (B22).
[b] Protons are numbered thus

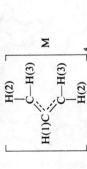

[c] d, Doublet; quin, quintet; n, nonet.

169

the dynamic nature of the allyl ligands results from an absence of d electrons available for back-bonding (B23, W21).

The stretching vibration of the coordinated C=C bond is found at about 1520 cm^{-1} in tetra(π-allyl)zirconium (B5).

2. Cyclooctatetraene Compounds

The chemistry of cyclooctatetraene complexes of both zirconium and hafnium is very recent and almost entirely due to the work of Wilke *et al.* (K4, K4a). The reaction between disodium cyclooctatetraene and the metal tetrahalides is the most general route (70–80% yields) to the red-brown bis-(cyclooctatetraene) complexes. Although the reaction of triethylaluminum with the metal tetraalkoxides in the presence of cyclooctatetraene (COT) gave good yields of the zirconium and titanium derivatives, this was not the case with hafnium, as not all the alkoxo groups could be replaced. The reaction between cyclooctatetraene and tetraallylzirconium gave $(COT)_2Zr$ in 98% yield.

Both the zirconium and hafnium bis(cyclooctatetraene) compounds are nearly insoluble. However, in toluene containing $HAl(C_2H_5)_2$ they dissolve readily forming complexes containing metal–H–Al bridges. Both the bis and monoadducts have been isolated in the case of zirconium, the bis adduct probably being tri- or tetrameric. Other monoadducts, in particular those of hexamethylphosphoramide and THF, have been isolated. The molecular structure of the oxygen- and moisture-sensitive THF adduct, $(COT)_2Zr \cdot THF$, determined by X-ray analyses (B82) is shown in Fig. IV-10. One cyclooctatetraene ring is planar and symmetrically bound to the metal, while the other is attached in butadiene fashion via four carbons. The compound is best described as h^8-cyclooctatetraene-h^4-cyclooctatetraene-tetrahydrofuranzirconium. The coordination planes of the cyclooctatetraene ligands are at an

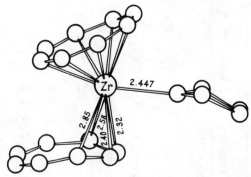

Fig. IV-10.   Structure of bis(cyclooctatetraene)tetrahydrofuranzirconium (B82).

angle of 146.2° to each other. The variations in the bond lengths between the metal and the tetrahapto ring were rationalized by Brauer and Krüger (B82) by proposing a bonding scheme involving the resonance forms

In view of the long Zr–O bond length found in the structure, it is not surprising that the THF molecule can be readily expelled from the molecule on warming. The C=C bonds not involved in bonding absorb at 1509 cm^{-1} in the infrared. The COT ligands in $(COT)_2Zr$ are considered to be aromatic (K4a).

With anhydrous HCl in THF, both $(COT)_2Zr$ and $(COT)_2Hf$ gave the species, $(COT)MCl_2 \cdot THF$, from which the THF can be driven off on heating *in vacuo*. The zirconium dichloride complex has also been prepared directly (W11) from $ZrCl_4$ and disodium cyclooctatetraene. The cyclooctatetraene ligands are also replaced on reaction of the compounds with alcohols.

Both of the chlorides in $(COT)ZrCl_2$ can readily be exchanged by alkyl (CH_3, C_2H_5) or allyl (C_3H_5, C_4H_7) groups on reaction with the corresponding Grignard reagent (K4a). With sodium cyclopentadienide only one Cp ligand enters the molecule to form $(COT)Zr(Cp)Cl$, possibly because of steric factors.

The compounds $(COT)Zr(CH_3)_2$ and $(COT)Zr(C_2H_5)_2$, isolated as the monoetherates, have moderate thermal stability. It has been suggested (K4a) that the β-elimination decomposition mechanism may be impeded to some degree in the ethyl compound through the lack of a suitable vacant coordination site on the metal.

The allyl groups in $(COT)Zr(C_3H_5)_2$, $(COT)Hf(C_3H_5)_2$, and $(COT)Zr$-$(C_4H_7)_2$ all exhibit dynamic behavior in the PMR spectra. $(COT)Zr(C_3H_5)_2$ catalyzes the stereospecific dimerization of butadiene at 65° to octatriene-(1,3-*trans*-6-*cis*). The intermediate in this reaction, $(COT)ZrC_8H_{12}$, contains the C_8H_{12} group bonded to the metal via the two terminal allyl groups. It also shows dynamic behavior in the PMR spectrum.

The dihydride $(COT)ZrH_2$ has been obtained (K4a) in 45% yield as small, dark green crystals by reaction of the metal tetrabutoxide with cycloocta-tetraene and $HAl(C_2H_5)_2$ at 60°. The metal–hydride vibrations are found at 1537 and 1310 cm^{-1}. On reaction with alcohols, the compound gives alkoxides and with iodine in toluene, 1-phenyl-1-tolylethane was formed.

Chapter V

Complexes of Titanium and Aluminum (Including Related Systems)

The importance of organic derivatives of the transition elements and of organotitanium compounds, in particular, has been highlighted in recent years by the voluminous work on the reactions of various titanium compounds with alkylaluminum derivatives. The products of such reactions, generally involving a trialkylaluminum or an alkylaluminum halide with a titanium halide, alkoxide, or β-diketonate, have been shown to be active in the polymerization of olefins, vinyl ethers, vinyl chloride, etc., and also in the fixation of nitrogen. The function of the alkylaluminum is to alkylate and reduce the titanium compound and generally, but not always, to complex with it.

Many titanium–aluminum complexes have resulted from the search for new polymerization catalysts. Those used in industry generally are prepared from $TiCl_4$ or $TiCl_3$ and triethylaluminum and, undoubtedly, contain alkyltitanium compounds, but their heterogeneous nature means that the structures and mechanism involved in the polymerization process are difficult to study and are not well understood.

Soluble complexes for the polymerization of ethylene were reported as early as 1957 by Breslow and Newburg (B88, B89) and Natta and co-workers (N22, N25). These were prepared by treatment of bis(cyclopentadienyl)-titanium dichloride with a trialkylaluminum or an alkylaluminum halide. Although less active and less specific than the best heterogeneous catalysts (N24), these soluble complexes, nevertheless, proved to be more amenable to studies of reaction mechanisms and kinetics. Numerous similar catalyst systems have been reported in the intervening years and many of these are tabulated in Table V-3. Almost all are based on bis(cyclopentadienyl)-titanium(IV) compounds, either halo, alkyl, or aryl or alkoxo derivatives together with a trialkyl- or triarylaluminum or alkylaluminum halide. Several monocyclopentadienyltitanium complexes are also known.

A. μ-Halo Complexes

In the reaction of organotitanium halides with alkylaluminum compounds in nonpolar solvents, the first step is probably alkylation of titanium. The stability of the alkyltitanium complex, now removed from crystal lattice effects by solubility, is conferred by complexation with the aluminum compound. In a hydrocarbon solvent such as benzene or hexane, reaction of Cp_2TiCl_2 with $(C_2H_5)_2AlCl$, for example, gives at first a red diamagnetic complex, probably with bridging groups between the two metals. Reduction now occurs by elimination of ethane and ethylene to give the blue titanium(III) complex (14) (B88, B89, L23, M43, N14, N22, Z3).

$$
\begin{array}{ccc}
\underset{Cp}{\overset{Cp}{\diagdown}}Ti\underset{Cl}{\overset{Cl}{\diagdown}}Al\underset{C_2H_5}{\overset{C_2H_5}{\diagdown}} & \underset{Cp}{\overset{Cp}{\diagdown}}Ti\underset{Cl}{\overset{Cl}{\diagdown}}Al\underset{Cl}{\overset{C_2H_5}{\diagdown}} & \underset{Cp}{\overset{Cp}{\diagdown}}Ti\underset{Cl}{\overset{Cl}{\diagdown}}Al\underset{Cl}{\overset{Cl}{\diagdown}} \\
(13) & (14) & (15)
\end{array}
$$

The yield of ethylene from this reduction is lower than that of ethane, a result which can be explained by polymerization. The mass spectrum of this complex has been measured by Takegami and co-workers and shows the parent ion (T8). Electron spin resonance evidence indicates that the diethyl complex (13) is also formed in this reaction and is then converted to the monoethyl derivative (H8).

With triethylaluminum reduction is considerably faster giving the diethyl complex (13)* (B89, M43, N14, N20–N22, Z3), while ethylaluminum dichloride reduces Cp_2TiCl_2 somewhat more slowly to the tetrachloro compound (15) (N14, N22).

Intermediate in this reaction is the monoethyl derivative, Cp_2TiCl_2Al-$(C_2H_5)Cl$, which is slowly converted into the final product (H8). Equilibration of alkyl and chloride ligands between the titanium and aluminum compounds is so facile that the positions of these ligands at the start of the reaction is immaterial. The course of the reaction is determined by the relative numbers of alkyl and chloride groups.

In all the above reactions, reduction of the titanium is paralleled by a loss in ability to polymerize olefins, although still apparently active with vinyl

* On the basis of ^{27}Al NMR results, the structure of this complex as a dichloride-bridged compound has been disputed by DiCarlo and Swift (D19), unless a very strong Ti^{III} exchange interaction is operating in the bimetallic complex.

ethers (M42, M45, N14, N16). Thus, $Cp_2TiCl_2 + R_3Al$ is a poor catalyst mixture because of rapid alkylation and reduction to a terminal complex. Addition of certain alkyl halides activates the catalyst, possibly because of *in situ* formation of R_2AlCl, a weaker alkylating and reducing agent (B32, K1). However, recent results (G33a) indicate that a more complex mechanism is operating.

Differences in the rates of reduction of Cp_2TiCl_2 by various ethylaluminum halides are illustrated in Fig. V-1.

Reduction by $(C_2H_5)_3Al$ is virtually complete in 1 hr, whereas with $C_2H_5AlCl_2$, complete reduction takes 1000 hr (H8). The rate for $(C_2H_5)_2AlCl$ lies between the two. Similar curves have been determined (B29) for $TiCl_4$ with ethylaluminum halides.

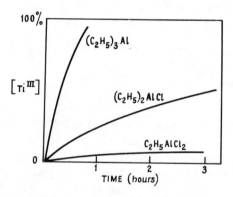

Fig. V-1.   Rates of reduction of Cp_2TiCl_2 by ethylaluminum compounds (H18a).

With triethylaluminum, Ti:Al ratios of 1:2 or 1:3 are ideal, provided that the blue diethyl complex, $Cp_2TiCl_2Al(C_2H_5)_2$, is removed as soon as it is formed. Otherwise, and particularly in the presence of a large excess of $(C_2H_5)_3Al$, further reaction occurs giving a violet complex. Similar behavior was noted with $(iso\text{-}C_4H_9)_3Al$ and Cp_2TiCl_2, a violet-red complex resulting with excess aluminum reagent even at room temperature (A9) (see later).

The ESR spectrum of $Cp_2TiCl_2Al(C_2H_5)_2$ shows only a single strong resonance at concentrations of 10 mmoles/liter or lower (A9, C28, M5, N47). With substitution of one terminal chlorine for an ethyl group, i.e., $Cp_2TiCl_2Al\text{-}(C_2H_5)Cl$, line width increases and hyperfine interaction becomes evident, although the spectrum cannot be fully resolved until the aluminum is fully chlorinated (H8, H19, M5). Thus, the spectrum of $Cp_2TiCl_2AlCl_2$ shows six broad lines due to interaction of the unpaired electron of the titanium(III) with the magnetic nucleus of the aluminum atom ($I = \frac{5}{2}$) (H8, H13, H18a, M5) (see Fig. V-2).

This effect is a measure of the delocalization of the unpaired electron, which spends less time on titanium and more time on aluminum as the number of chlorines on the aluminum is increased. It has been used as an indication of the Lewis acidity of the Group III halides (H16).

The ESR spectrum of the solution from Cp_2TiCl_2 and $(iso\text{-}C_4H_9)_3Al$ shows a singlet at a 1:1 ratio ($g = 1.976$) unchanged with time, whereas with higher Al:Ti ratios the spectrum changes with time, possibly owing to the presence of hydrides (A9, Z4).

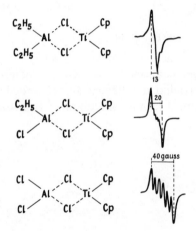

Fig. V-2.   ESR spectra of titanium–aluminum chloride complexes (H18a).

Reaction of Cp_2TiCl_2 with methylaluminum compounds follows the same course as with the ethyl derivatives, although reduction of titanium occurs less readily. A similar pattern of spectra has been observed, depending on the degree of alkylation of the aluminum (B13, M5). The systems Cp_2TiCl_2 + $(CH_3)_2AlCl$ and $Cp_2Ti(CH_3)Cl$ + CH_3AlCl_2 at an Al:Ti ratio of 1.0 rapidly become identical owing to ligand migration (B13, B14, G34).

From Cp_2TiCl_2 and trimethylaluminum in methylene chloride at 20°, a 90% yield of the monomethyltitanium chloride, $Cp_2Ti(CH_3)Cl$, has been obtained (C38, F2) after decomposition of the reaction mixture with diethyl ether. Further alkylation of this compound occurred with trimethylaluminum to give the interesting complex, $Cp_2Ti(CH_3)ClAl(CH_3)_2$, which has been reported briefly by Natta and co-workers (N17). The complex apparently has a bridging methyl group.

Alternative methods of preparation of the ethyl complexes have been claimed in a German patent, using $TiCl_3$ with $C_5H_5Al(C_2H_5)_2$ or $(C_5H_5)_2NaAl(C_2H_5)_2$ to give $Cp_2TiCl_2Al(C_2H_5)_2$ or $TiCl_3$ with $(C_5H_5)_2Al(C_2H_5)$ to give $Cp_2TiCl_2Al(C_2H_5)Cl$ (S20).

Titanium–aluminum complexes of this type are split by Lewis bases such as diethyl ether or trimethylamine, giving $(Cp_2TiCl)_2$ and the etherate or amine complex of the aluminum compound (B89, C38). This constitutes a useful preparation of the titanium(III) halide (see Chapter VI). The reaction is reversible since in hydrocarbon solvents the bimetallic complexes are reformed (L23, M45, N14).

$$Cp\diagdown \begin{matrix} Cl \\ Ti \\ Cp \diagup \quad Cl \end{matrix} \diagup \begin{matrix} R \\ Al \\ R' \end{matrix} \rightleftharpoons Cp_2TiCl + ClAl\diagup \begin{matrix} R \\ R' \end{matrix}$$

R, R' = alkyl or Cl, etc.

The structure of the diethyl complex, $Cp_2TiCl_2Al(C_2H_5)_2$, as a chlorine-bridged dimer has been confirmed by an X-ray structure determination carried out by Natta, Corradini, and Bassi (C44, N12) and later by Kocman et al. (K33a). The arrangement around the titanium is distorted tetrahedral (see Fig. V-3).

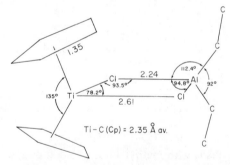

Fig. V-3. Structure of $Cp_2TiCl_2Al(C_2H_5)_2$ (K33a).

Complexes of type, $Cp_2TiCl_2MCl_2$ [where M = B(H11, H16), Ga (H12, H16), or In (H16)], have been identified in solution by ESR spectroscopy after reduction of $Cp_2Ti(C_3H_7)Cl$ with Na or K in the presence of MCl_3.

Bimetallic complexes similar to these have been obtained from Cp_2TiCl and beryllium chloride or zinc chloride (H2), although the latter is more easily prepared by reduction of Cp_2TiCl_2 with zinc metal in toluene (S2).

$$Cp\diagdown \begin{matrix} Cl \\ Ti \\ Cp \diagup \quad Cl \end{matrix} \diagup \begin{matrix} Cl \\ M \\ Cl \end{matrix} \diagup \begin{matrix} Cp \\ Ti \\ Cp \end{matrix}$$

M = Be or Zn

The magnetic moment of the zinc complex shows the spin-only value (1.71 BM), independent of temperature, with no magnetic interaction between titanium atoms (S2, W11). The ESR spectrum of a powdered sample showed a single resonance with a line width of ~ 20 gauss and a g value of 1.977 (S2). The bridging chlorides in the zinc complex can be replaced by cyanate, thiocyanate, cyanide, nitrate, etc., if an excess of these anions is present during the reduction with zinc in organic solvent (W11).

Complexes of lower valent titanium with aluminum compounds have resulted when Friedel–Crafts reductive conditions have been employed with titanium(IV) compounds. Thus, from Cp_2TiCl_2, $AlCl_3$, and aluminum the titanium(III) complex, $Cp_2TiCl_2AlCl_2$, was obtained (V14), while $CpTiCl_3$ under the same conditions afforded the trimetallic titanium(III) compound,

apparently containing a π-bonded cyclopentadienyl group. The same compound is believed to be formed by the action of moisture or halogenated solvents on the mixture, Cp_2TiCl_2, $AlCl_3$, and Al (H13). It was identified by an ESR signal of eleven lines. Removal of a cyclopentadienyl group under these conditions is now well established; in fact, the deliberate introduction of water during the reduction of Cp_2TiCl_2 with aluminum and aluminum chloride in tetrahydrofuran affords a useful preparation of $CpTiCl_2$ (W11) (see Chapter VI, Section B).

Arene complexes of titanium have been prepared from titanium tetrachloride, aluminum chloride, and aluminum in the presence of the aromatic hydrocarbon, namely, $C_6H_6 \cdot TiCl_2 \cdot 2AlCl_3$ from benzene (M18, N19, V14) and $\{Ti_3[C_6(CH_3)_6]_3Cl_6\}Cl$ from hexamethylbenzene (F18) (see Chapter VII). From the latter arene, blue crystalline $C_6(CH_3)_6Ti(OH)Cl$ has also been obtained under slightly different conditions (W19).

Titanium–aluminum indenyl derivatives have been prepared by reduction of the bis(indenyl)titanium dihalides with triethylaluminum in light petroleum (M8).

X = Cl or Br

Treatment of these complexes with diethyl ether or trimethylamine leads to the bis(indenyl)titanium(III) halides (M8). When mixed with Grignard reagents, these polymerize olefins (M8, M9).

The reactions of titanium tetrahalides with methylaluminum compounds generally lead to alkylated titanium species of type $(CH_3)_nTiCl_{4-n}$ which are described in the appropriate section. Exchange of methyl groups between $Ti(CH_3)_4$ and $Al(CH_3)_3$ in ether has been studied by PMR using the deuterated titanium compound (K23). After 20 min at $-75°$, 25% of the CD_3 groups had been transferred to the alkylaluminum via the etherates.

Ethylaluminum compounds (and higher alkyl), generally bring about reduction of the titanium halide to the trivalent state, and their mixtures are active as Ziegler–Natta olefin polymerization catalysts (see Section E).

A novel titanium–aluminum complex, believed to be the chlorine-bridged complex below, was formed as a red-brown anodic deposit on electrolysis of a mixture of titanium tetrachloride and triisobutylaluminum in heptane or ethyl chloride (M7). Unfortunately, the compound does not appear to have been studied further.

$$\begin{array}{c}
\textit{iso-}C_4H_9\diagdown\qquad\diagup Cl\diagdown\qquad\diagup\textit{iso-}C_4H_9 \\
\qquad Ti\qquad Al \\
Cl\diagup\quad\diagdown Cl\diagup\quad\diagdown\textit{iso-}C_4H_9
\end{array}$$

As in the case of aluminum, treatment of Cp_2TiCl_2 and Cp_2TiBr_2 with triethylgallium leads to reduction of the titanium and formation of halogen-bridged complexes.

$$\begin{array}{c}
Cp\diagdown\qquad\diagup X\diagdown\qquad\diagup C_2H_5 \\
\qquad Ti\qquad Ga \\
Cp\diagup\quad\diagdown X\diagup\quad\diagdown C_2H_5
\end{array}$$

$$X = Cl,\ Br$$

Reduction is considerably slower than with aluminum alkyls, but the final complex is blue in color, showing an asymmetric singlet in the ESR spectrum ($g = 1.974$) (D49). With a large excess of $(C_2H_5)_3Ga$ the parameters of the ESR spectrum change, possibly owing to replacement of X by C_2H_5.

B. μ-Alkoxo Complexes

During the search for new polymerization catalysts the reactions of titanium alkoxides with alkylaluminums have been studied by several groups of workers. Generally, the reactions which occur are complex, leading to more than one product. These have been studied by ESR spectroscopy, but have never been isolated.

From $CpTi(OC_2H_5)_3$ and $(C_2H_5)_3Al$ or $(C_2H_5)_nAlCl_{3-n}$, complexes believed to be of the type

$$
\begin{array}{c}
\text{C}_2\text{H}_5 \diagdown \quad \diagup \text{X} \diagdown \;\; \overset{\displaystyle \text{R}}{\underset{\displaystyle |}{\;}} \;\; \diagup \text{X} \diagdown \quad \diagup \text{C}_2\text{H}_5 \\
\text{Al} \qquad \text{Ti} \qquad \text{Al} \\
\text{C}_2\text{H}_5 \diagup \quad \diagdown \text{X} \diagup \quad \diagdown \text{X} \diagup \quad \diagdown \text{C}_2\text{H}_5
\end{array}
$$

R = Cp, X = OC_2H_5 or Cl

have been detected (D48). Similar compounds in which R = C_2H_5 were shown to be formed from $Ti(OC_4H_9)_4$ and $(C_2H_5)_3Al$ (D47, H32, T5).

The system $Ti(OC_4H_9)_4$ + $(C_2H_5)_3Al$ is active for polymerization of several monomers depending on the Al:Ti ratio, and in certain cases is stereoregulating (T5) (see Table V-3). The difference in activities with different Al:Ti ratios seems a further indication of the complexity of the reaction.

The proportion of titanium(III) in these reaction solutions was found to be small [10% (H32)] so that other active polymerizing species must be present. Natta has suggested that these catalysts are not soluble, but are colloidally dispersed, which could explain their stereoregulating ability. The evidence has been presented inconclusively by Boor (B55).

In the presence of cyclooctatetraene the system $Ti(OC_4H_9)_4$ + $(C_2H_5)_3Al$ has produced the derivatives $Ti(COT)_2$ and $Ti_2(COT)_3$ (see Chapter VI, Section A).

The reaction of titanium tetra(n-butoxide) with triethylgallium is slower than that with the corresponding aluminum compound, but gives similar complexes (D49).

C. μ-Hydrido Complexes

Although titanium hydrides have been predicted as intermediates in the formation of allyltitanium(III) compounds (M15), it was not until electron spin resonance studies were carried out that the presence of hydrides was established. Work by several groups has indicated the formation of hydride-bridged bimetallic complexes in the reduction of Cp_2TiCl_2 with lithium naphthalene or sodium naphthalene (H10, H14), sodium metal (H17, K18), alkyllithium (B91), alkylmagnesium halides (B91, M25), magnesium plus magnesium iodide (V17, V18), sodium metal in the presence of aluminum trichloride (H17), or sodium naphthalene followed by lithium aluminum hydride (H20). Although none of these complexes has been isolated and characterized, there is reasonable evidence that they all conform to the general structure (K18),

$$
\begin{array}{c}
\diagdown \qquad \overset{\displaystyle \text{H}}{\diagup \quad \diagdown} \\
\diagup \text{Ti} \diagdown \qquad \diagup \text{M} \\
\qquad \diagdown \text{H} \diagup
\end{array}
$$

Table V-1
Bimetallic Complexes with Bridging Groups (Excluding Hydrides)

Compound[a]	Color	Isolated	Other data	Refs.
$Cp_2Ti(\mu\text{-}Cl)_2Al(C_2H_5)_2$	Blue	Yes	M.p. 126°–130° (N14), 124° (A9); μ_{eff} 1.70; ESR (M5)	A9, B89, K33a, M5, M43, N12, N14, N21, N22, Z3
$Cp_2Ti(\mu\text{-}Cl)_2Al(C_2H_5)(Cl)$	Blue	Yes	M.p. 88°–92° (N14); μ_{eff} 1.57; ESR (H19, M5)	B88, B89, H19, L23, M5, M43, N14, N22, Z3
$Cp_2Ti(\mu\text{-}Cl)_2Al(Cl)_2$	Blue	Yes	M.p. 155°–160° (M45, N14); μ_{eff} 1.72; ESR (H19, M5)	H19, M5, M45, N14, N22
$Cp_2Ti(\mu\text{-}Cl)_2B(Cl)_2$	Blue	No	ESR	H11, H16
$Cp_2Ti(\mu\text{-}Cl)_2Ga(Cl)_2$	Blue	No	ESR	H12, H16
$Cp_2Ti(\mu\text{-}Cl)_2In(Cl)_2$	Blue	No	ESR	H16
$Cp_2Ti(\mu\text{-}Cl)_2Zn(\mu\text{-}Cl)_2TiCp_2$	Green	Yes	μ_{eff} 1.71 BM; ESR	H2, S2
$Cp_2Ti(\mu\text{-}Cl)_2Be(\mu\text{-}Cl)_2TiCp_2$	Blue	Yes	—	H2

Structure	Color		Properties	Refs
Cp$\diagdown$ $\diagup$Cl$\diagdown$ $\diagup$CH$_3$ $\qquad$Ti$\qquad$Al Cp$\diagup$ $\diagup$Cl$\diagdown$ $\diagdown$CH$_3$	—	No	—	N17
π-C$_9$H$_7\diagdown$ $\diagup$Cl$\diagdown$ $\diagup$C$_2$H$_5$ $\qquad$Ti$\qquad$Al π-C$_9$H$_7\diagup$ $\diagup$Cl$\diagdown$ $\diagdown$C$_2$H$_5$	Green	Yes	M.p. 80°–82°(dec)	M8
π-C$_9$H$_7\diagdown$ $\diagup$Br$\diagdown$ $\diagup$C$_2$H$_5$ $\qquad$Ti$\qquad$Al π-C$_9$H$_7\diagup$ $\diagup$Br$\diagdown$ $\diagdown$C$_2$H$_5$	Green	Yes	M.p. 88°–90°(dec); μ_{eff} 1.68	M8
Cl$\diagdown$ $\diagup$Cl$\diagdown$ $\diagup$R$\diagdown$ $\diagup$Cl$\diagdown$ $\diagup$Cl $\quad$Al$\quad$Ti$\quad$Al Cl$\diagup$ $\diagup$Cl$\diagdown$ $\diagdown$Cl$\diagup$ $\diagdown$Cl R = benzene, cyclopentadiene, or mesitylene	Red-violet	Yes	ESR	H13, V14, V20, V21
C$_2$H$_5\diagdown$ $\diagup$X$\diagdown$ $\diagup$R$\diagdown$ $\diagup$X$\diagdown$ $\diagup$C$_2$H$_5$ $\quad$M$\quad$Ti$\quad$M C$_2$H$_5\diagup$ $\diagup$X$\diagdown$ $\diagdown$X$\diagup$ $\diagdown$C$_2$H$_5$ M = Al; X = OC$_2$H$_5$; R = Cp	—	No	ESR	D48
M = Al; X = OC$_4$H$_9$; R = C$_2$H$_5$	Red-brown	No	ESR	D47, T5
M = Ga; X = OC$_4$H$_9$; R = C$_2$H$_5$	Dark green	No	ESR	D49
iso-C$_4$H$_9\diagdown$ $\diagup$Cl$\diagdown$ $\diagup$iso-C$_4$H$_9$ $\qquad$Ti$\qquad$Al iso-C$_4$H$_9\diagup$ $\diagup$Cl$\diagdown$ $\diagdown$iso-C$_4$H$_9$	Red-brown	Yes	—	M7
bipy$\diagdown$ $\diagup$Cl$\diagdown$ $\diagup$Cl C$_2$H$_5$$-Ti\qquad$Al $\quad$Cl$\diagup$ $\diagdown$Cl$\diagup$ $\diagdown$Cl $\quad$bipy	Violet	Yes	—	P11

^{a} C$_9$H$_7$, Indenyl.

181

(where M = Na, Li, MgBr, AlCl$_2$, or AlH$_2$), and that the H–Ti–H angle steadily increases in the order Na < Mg < Al, etc.

Their ESR spectra are characterized by interaction of the unpaired electron of the titanium(III) atom with the magnetic nucleus of the metal in the counter-ion and splitting of each of these lines into 1:2:1 triplets by the two equivalent bridging hydrogens. In some cases further hyperfine structure caused by interaction with the hydrogens of the cyclopentadienyl groups is observed (B91, H10, N48). Evidence from deuteration experiments (V18) indicates that the hydrido groups originated from the Cp ligands, so that both of the latter may now no longer be pentahapto. Indeed, inequivalence of the protons on the C$_5$ rings has been indicated (K18).

Aluminum complexes in which M = AlR$_2$ (where R is alkyl), have been detected in the reaction between Cp$_2$TiCl$_2$ and excess trialkylaluminum (A9, Z4). The effect is not observed at low Al:Ti ratios and may be due to alkyl-aluminum hydride impurities in the trialkylaluminum, since such compounds have been detected by ESR from Cp$_2$TiCl$_2$ + HAlR$_2$ (where R = iso-C$_4$H$_9$) (A8, Z4).

Of the titanium hydrides which have been isolated and characterized, the first was the tetrahydroborate, Cp$_2$TiBH$_4$, which was obtained as a violet-colored, volatile solid by reduction of Cp$_2$TiCl$_2$ with LiBH$_4$ in ether (N44). With HCl, HBr, HI, BCl$_3$, or BBr$_3$ the tetrahydroborate was converted to the corresponding halide, (Cp$_2$TiX)$_2$, while boron trifluoride afforded the tetra-fluoroborate, Cp$_2$TiBF$_4$, with two bridging fluorine atoms (N44). The ESR data for the tetrahydroborate have been reported (N46).

Although it has been pointed out that the infrared spectrum of Cp$_2$TiBH$_4$ differs from that expected for a compound with a double hydride bridge (D8), a recent single-crystal X-ray diffraction study (M32a) has confirmed the pseudotetrahedral arrangement suggested originally (N44), namely,

$$\text{Cp} \diagdown \quad \overset{H}{\diagup} \diagdown \overset{H}{\diagup}$$
$$\text{Ti} \quad \text{B}$$
$$\text{Cp} \diagup \quad \diagup H \diagdown \quad H$$

Bond lengths and angles are shown in Table I-1. The hydridoborohydride, CpTi(H)BH$_4$, has also been reported briefly (J5).

Closely related is the dark blue complex, Cp$_2$TiB$_3$H$_8$, prepared by Klanberg et al. (K32) from Cp$_2$TiCl$_2$ and two moles of CsB$_3$H$_8$. Bonding is believed to be through two three-center Ti–H–B systems.

The first hydride of titanium uncomplexed with other elements was obtained as a purple solid from the reaction of hydrogen with Cp$_2$Ti(CH$_3$)$_2$ in the solid state (B41). It is apparently a dimer with bridging hydrogens,

$$\text{Cp} \diagdown \quad \overset{H}{\diagup} \diagdown \quad \diagup \text{Cp}$$
$$\text{Ti} \quad \quad \text{Ti}$$
$$\text{Cp} \diagup \quad \diagdown \overset{H}{\diagup} \quad \diagdown \text{Cp}$$

and of marginal stability. In the infrared spectrum a band characteristic of bridging hydride was present at 1450 cm^{-1}. In solution this purple hydride is cleaved by tetrahydrofuran or triphenylphosphine to give monomeric species of low stability, for example, $Cp_2Ti(H)P(C_6H_5)_3$.

In solution $(Cp_2TiH)_2$ loses hydrogen forming "titanocene" (B44, C38) and, even in the solid state, it is slowly converted to a gray-green isomeric form (M21) of higher molecular weight, similar to that isolated by Martin and de Jongh from the hydrogenation of bis(cyclopentadienyl)methylallyl-titanium(III), (M11, M12). Decomposition in toluene or ether proceeds further to give an unstable species, $(Cp_2Ti)_2$, which rearranges to the final stable "titanocene," containing an hydridic hydrogen originating from one of the cyclopentadienyl rings (B43).

$$Cp_2Ti(CH_3)_2 \xrightarrow[\text{solid}]{H_2} (Cp_2TiH)_2 \xrightarrow{\text{rearrange}} (Cp_2TiH)_x$$
$$\text{(violet)} \qquad\qquad \text{(gray-green)}$$

$$\Big\downarrow \text{toluene} \Big| \text{ or ether, room temp.}$$

$$[(C_5H_5)(C_5H_4)TiH]_2 \xleftarrow[100°, \text{2 hr}]{\text{heat}} (Cp_2Ti)_2$$

Similar types of hydridic compounds have been detected by van Tamelen and co-workers by visible and infrared spectroscopy during the reduction of Cp_2TiCl_2 with sodium under argon (V8). Four sequential products were observed: $(Cp_2TiCl)_2$, $(Cp_2Ti)_n$ ($n = 1$–2), $[(C_5H_5)(C_5H_4)TiH]_x$, and $[(C_5H_5)(C_5H_4)TiH]_2$ ("stable titanocene"). The second of these reacts rapidly and reversibly with nitrogen (see Chapter VIII).

Bis(pentamethylcyclopentadienyl)titanium forms a hydride species by absorption of gaseous hydrogen in pentane or toluene (B42, B43). The orange $[\pi\text{-}(CH_3)_5C_5]_2TiH_2$ loses hydrogen at room temperature under vacuum.

Generally speaking, hydrides of titanium are much less stable, i.e., more reactive, than their zirconium analogs. For this reason they are promising intermediates in many processes and will, undoubtedly, assume more importance in the future. The use of the gray-green hydride, $(Cp_2TiH)_x$, as a hydrogenation and isomerization catalyst has been investigated (M11, M12).

From 1,3-pentadiene and the purple hydride, bis(cyclopentadienyl)1,3-dimethylallyltitanium(III) was obtained in almost quantitative yield (B41).

In 1959 Natta and co-workers reported (N15) the isolation of a purple titanium–aluminum complex which was formed from the titanium(III) compound, $(Cp_2TiCl)_2$, and triethylaluminum in boiling benzene. Although originally formulated (M44, M46, N11, N18) as the dimer, $[(C_5H_5)_2TiAl(C_2H_5)_2]_2$, it was later proposed (W4) that this compound was identical with that prepared by Wailes and Weigold from titanocene and triethylaluminum. A structure containing a bridging C_5H_4 group was proposed on the basis of loss of ethane without participation of solvent and which was not at variance with the X-ray crystal structure of Corradini and Sirigu (C45).

$$
\begin{array}{c}
\mathrm{C_2H_5} \quad \mathrm{C_2H_5} \\
\diagdown \diagup \\
\mathrm{C_5H_4 {-} Al} \\
| \qquad | \\
\mathrm{C_5H_5 {-} Ti} \qquad \mathrm{Ti {-} C_5H_5} \\
| \qquad | \\
\mathrm{Al {-\!-\!-} C_5H_4} \\
\diagup \diagdown \\
\mathrm{C_2H_5} \quad \mathrm{C_2H_5}
\end{array}
$$

More recently, Tebbe and Guggenberger (T15) have indicated that these complexes are different; they have prepared the Natta complex by the action of triethylaluminum on $Cp_2Ti(C_6H_5)_2$, Cp_2TiCl_2, or Cp_2TiCl in benzene at 70° for 12 hr. Proton magnetic resonance (220 MHz), molecular weight, and analytical data defined the composition as $[(C_5H_5)(C_5H_4)TiHAl(C_2H_5)_2]_2$, the hydrido ligand bridging between titanium and aluminum atoms, although no absorption in the infrared above 1500 cm^{-1} was observed. The structure from an X-ray crystal study is shown in Fig. V-4 (T15).

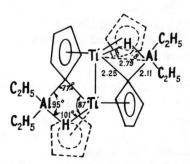

Fig. V-4.   Structure of $[(C_5H_5)(C_5H_4)TiHAl(C_2H_5)_2]_2$ (T15).

Tebbe and Guggenberger believe (T14) that the Wailes and Weigold complex is also a hydride and contains two C_5H_4 rings per titanium atom, i.e., $[(C_5H_4)_2TiHAl(C_2H_5)_2]_2$. They suggest that the two C_5H_5 rings in the dimeric molecule have each lost one proton and have joined.*

The Natta complex is diamagnetic and when mixed with α-$TiCl_3$ is claimed to be a more stereospecific olefin polymerization catalyst than the common system, $TiCl_3$-$Al(C_2H_5)_3$ (N17).

With trimethylaluminum under the same conditions, titanocene forms the analogous methyl compound, presumably, $[(C_5H_4)_2TiHAl(CH_3)_2]_2$ with loss of methane (W4).

---

* This suggestion has now been confirmed by an X-ray crystal structure determination (G36a) which shows the presence of the fulvalenide ligand $(C_5H_4{-}C_5H_4)^{2-}$.

In the same class of reactions is that between $TiCl_2$, sodium cyclopenta-dienide, and alkylaluminum halides patented by Hamprecht *et al.* for BASF whereby red $(C_5H_5)_2TiAl(C_2H_5)_2Cl$ or $(C_5H_5)_2TiAl(CH_3)_2Cl$ is formed (H1). These are apparently olefin polymerization catalysts.

D. Reactions of Alkylaluminum Derivatives with Zirconium Compounds

In contrast to the corresponding reactions with titanium derivatives, the action of aluminum alkyls on zirconium compounds is characterized by alkylation without reduction in oxidation state. The first stable alkylzirconium derivative was prepared in this way in 1965 by Surtees from trimethyl-aluminum and an oxygen-bridged zirconium complex in benzene (S74). It was later shown that methylene dichloride was a much better solvent for this preparation (W7).

$$(Cp_2ZrCl)_2O + 2(CH_3)_3Al \longrightarrow 2Cp_2Zr(CH_3)Cl + [(CH_3)_2Al]_2O$$

From Cp_2ZrCl_2 and triethylaluminum, Sinn and Kolk obtained a dimeth-ylene-bridged complex, $Cp_2Zr(Cl)CH_2CH_2Zr(Cl)Cp_2$, with evolution of ethane (S45). Reaction of the product with DCl gave 1,2-dideuteroethane. From a large excess of $(C_2H_5)_3Al$ and Cp_2ZrCl_2 in heptane at 35°, an unusual zirconium–aluminum complex was isolated (H5) containing two aluminums per zirconium. On the basis of the formation of 80% monodeuteroethane and 20% trideuteroethane on treatment with DCl, the product was formulated as

$$Cp_2Zr(Cl)CHCH_2Al(C_2H_5)_2$$
$$|$$
$$Al(C_2H_5)_2$$

Tetrabenzylzirconium forms a red crystalline complex with tribenzyl-aluminum (Z7) in hydrocarbon solvents at room temperature. From molec-ular weight, conductivity, and PMR data, it was postulated that in solution the complex is in equilibrium with its precursors.

$$Al(CH_2C_6H_5)_3 + Zr(CH_2C_6H_5)_4 \rightleftharpoons AlZr(CH_2C_6H_5)_7$$

Bimetallic complexes of zirconium and aluminum involving hydrides are also known. In particular, the hydride, $Cp_2Zr(H)Cl$, will react with $LiAlH_4$ (K15, W2) to give $Cp_2Zr(H)AlH_4$. Hydrido groups bridging both Zr–Zr and Zr–Al are probably present. In addition, the zirconium dihydride, Cp_2ZrH_2, will dissolve readily in trimethylaluminum giving the faintly blue complex,

$$Cp_2Zr-H-Al(CH_3)_3$$
$$/ \quad \backslash$$
$$H \quad H$$
$$\backslash \quad /$$
$$Cp_2Zr-H-Al(CH_3)_3$$

Table V-2.
Bimetallic Complexes with Bridging Hydrogens

Compound	Isolated	Method of preparation	Other data	Refs.
$\left[\begin{array}{c} Cp\ \ H \\ \diagdown Ti \diagdown \\ Cp\ \ H \end{array} \right]^{-} M^{+\,a}$ M = Na, Li, MgBr	No	Cp_2TiCl_2 + Na/naphthalene, Li/naphthalene, C_2H_5MgBr	ESR, g 1.99	B91, H10, K18
$Cp\text{–}Ti(H)_2\text{–}Al(Cl)_2\text{–}Cp$ (bridging H, Cl, Cl)	No	Cp_2TiCl_2 + Na + $AlCl_3$	ESR, g 1.992	H17, H20, K18
$Cp\text{–}Ti(H)_2\text{–}Al(H)_2\text{–}Cp$	No	Cp_2TiCl_2 + $LiAlH_4$	ESR, g 1.991	H20, K18
$Cp\text{–}Ti(H)_2\text{–}Al(R)_2\text{–}Cp$ $R = C_2H_5, C_3H_7, C_4H_9$	No	Cp_2TiCl_2 + R_2AlH (excess)	Violet-blue	A6, Z4
$Cp\text{–}Ti(H)_2\text{–}B(H)_2\text{–}Cp$	Yes	Cp_2TiCl_2 + $2LiBH_4$	Violet solid; IR; MW; crystal structure	M32a, N44
$CpTi(H)BH_4$	No	$CpTi(BH_4)_2$ + amine	—	J5
$Cp\text{–}Ti(H)_2\text{–}Ti(Cp)\text{–}Cp$	Yes	$Cp_2Ti(CH_3)_2$ + H_2 (no solvent)	Violet solid; IR	B41
$[Cp(C_5H_4)TiHAl(C_2H_5)_2]_2$	Yes	$(C_2H_5)_3Al$ + $Cp_2Ti(C_6H_5)_2$, Cp_2TiCl_2, or Cp_2TiCl	PMR (T15); MW; IR	M46, N18, T15
$[(C_5H_4)_2TiHAl(C_2H_5)_2]_2$	Yes	$(C_2H_5)_3Al$ + titanocene	MW	G36a, W4

a It is doubtful whether both of the C_5 rings in these compounds are Cp (see text).

in near quantitative yield (W9). The hydridic hydrogens are coupled in the PMR spectrum, but do not exchange, so that both types can be distinguished as triplets (Zr–H–Al, $\delta - 0.92$; Zr–H–Zr, $\delta - 2.92$) (see Chapter IV, Section B,1,d).

The zirconium–aluminum complexes are not particularly active as Ziegler–Natta-type olefin polymerization catalysts (see, for example, T13), perhaps a consequence of the difficulty with which zirconium is reduced (but see ref. A20a).

E. The Role of Organotitanium in Olefin Polymerization (Ziegler–Natta Catalysis)

It is not the intention to review here the field of olefin polymerization, but any coverage of organotitanium and organozirconium chemistry would not be complete without mention of some of the data which have accumulated during the search for new polymerization catalysts. This evidence clearly underlines the role of organotransition metal derivatives in polymerization and for this reason it is an important contribution to this field of chemistry. In fact, polymerization is possible only because of the lability of alkyl groups on transition metals, a lability which is responsible for much of the unique chemical behavior of this class of compounds.

The literature on olefin polymerization is voluminous and confusing. There is no point in repeating here evidence which has been critically discussed elsewhere, particularly since few definite conclusions have been reached from such evidence. The reader is referred to the most recent comprehensive review by Boor (B55) and the multitude of references therein.

1. The Catalyst System

Ziegler–Natta-type catalysts are, in general, composed of a transition metal compound and an organometallic reducing agent containing a Group I–III metal. The transition metal derivative is usually a halide, but sometimes an alkoxide or β-diketonate, of titanium or for some purposes vanadium, zirconium, cobalt, or chromium. The base metal compound is normally a trialkylaluminum or alkylaluminum halide, although organolithium, organomagnesium, and organozinc compounds have found some use. The purpose of the base metal alkyl is to alkylate and (sometimes) reduce the transition metal and possibly to complex with it. Not all combinations of these two components have the required activity and it is necessary to choose the catalyst to suit the monomer to be polymerized.

As Carrick has pointed out (C10), the various catalysts in the Ziegler–Natta group show great diversity in their general physical character, including solubility and structure and the nature of the polymers produced. This is to be expected when dealing with binary combinations of different metal derivatives. It is difficult, therefore, to describe the behavior of all these catalysts in one unified theory. However, similarities do exist if only the propagation reaction is considered.

2. The Nature of the Active Metal Site

The catalyst system most commonly used in industry is based on $TiCl_3$ and $(C_2H_5)_3Al$ or $(C_2H_5)_2AlCl$. Alkylation of titanium undoubtedly occurs and has been demonstrated by several groups (for a discussion, see B54, R43; also C26, K25, T6, T7). Considerable discussion has occurred in the literature over whether the resulting alkyltitanium is complexed with an aluminum derivative, presumably through bridging ligands such as chloride, or whether it is aluminum-free, but stabilized by crystal lattice effects on the surface of the $TiCl_3$ crystal.

Whatever its form, it is generally agreed that polymerization is initiated at alkyltitanium sites so that it is not difficult to realize that the stereoregulating effects of heterogeneous catalysts are dependent on the crystal form of the original halide as well as on the alkylating agent and reaction conditions (R43).

3. Oxidation Number of the Active Site

Much work has gone into efforts to establish the oxidation number of the titanium in the belief that this could help to determine the configuration at the active site. Unfortunately, the number of active sites is relatively small (R43) so that any measurement made on the bulk of the catalyst may bear no relationship to the situation at the active sites. As a result, much of the evidence is inconsistent and inconclusive (B55). Titanium(III) and titanium(IV) sites have been proposed as well as those of mixed oxidation states (e.g., M10, O5, W26).

On the basis of increased activity of various catalyst systems on addition of oxygen (B89) or halogenated solvents (B9, B10, F31, F32) which are known to oxidize titanium(III) to titanium(IV), tetravalent active sites were assumed. It has been pointed out (C10, C32) that coordination of olefinic monomer is most favored when the metal is in a high oxidation state, since donation of olefinic π electrons to the metal is involved. On the other hand, while metal reduction, which involves progressive filling of the orbitals, decreases the ability of the metal to coordinate olefins, it also leads to increased polarity of the metal–alkyl bond. As a result coordinated monomer is more readily incorporated into the growing chain. Optimum catalytic activity may involve a balance between these two features, the tetravalent sites functioning as transfer agents.

On the other hand, the theory has been advanced that the symmetry requirements of the metal are more important than its valence state (H8). In heterogeneous catalysis the titanium in the $TiCl_3$ crystal is already in an octahedral environment, whereas with soluble systems only the initial complex is octahedral. In the latter case, as reduction proceeds by dealkylation, the titanium becomes tetrahedral and inactive.

Armstrong et al. (A20) support this view in a theoretical treatment of the system $TiCl_4 + CH_3AlCl_2$. The function of the AlR_2 group is to force the titanium to adopt a high coordination number, the symmetry of the site being such that the titanium–alkyl bond can be localized in a labile molecular orbital of considerable d character. Cossee and Arlman proposed that coordination of an olefin monomer into a vacant sixth position on the metal weakened the titanium–alkyl bond and allowed insertion of olefin (A19, C46), but Armstrong et al. deduced that the change in the lability of the methyl group on titanium caused by coordination of ethylene is not sufficient to allow fission of the C–Ti bond. A mechanism was proposed whereby the alkyl group and the ethylene remain bonded to the titanium at all stages. Migration of the alkyl group to the ethylene carbon is initiated by charge distribution (the methyl carbon is negative, while the olefin is weakly positive). There is no significant loss in bonding energy, the driving force for the migration being the extra stability conferred by the formation of a new carbon–carbon σ bond.

The most energetically favorable conformation for a species containing only a growing alkyl chain is trigonal-bipyramidal (A20). Coordination of another ethylene molecule then involves a reorganization to octahedral, the energy released by ethylene coordination being more than sufficient for this change in geometry.

A special system which has given more reliable results than many is methyltitanium trichloride, with or without added alkylaluminum compounds. In the hands of Bestian, Clauss, and Beerman, this system has given definitive evidence on oxidation number and nature of the active sites. Although the conditions of its use (−70° in methylene dichloride) are not typical for Ziegler–Natta polymerizations, the system deserves some attention. Under the conditions of the normal Ziegler polymerization (80°), CH_3TiCl_3 initiates the polymerization of ethylene in hydrocarbon solvents only when a sufficient quantity of titanium trichloride has been formed by partial decomposition

Table V-3

"Soluble" Polymerization Catalyst Systems[a]

System	Monomer polymerized	Remarks	Refs.
$Cp_2TiCl_2 + (C_2H_5)_3Al$	Ethylene	Weakly active in heptane because of rapid reduction; becomes effective with alkyl chlorides or alcohols	F31, F32, K3, N24
	Vinyl monomers	Quite active with vinyl ethers, vinyl chloride, acrylates, etc., apparently by an ionic mechanism; copolymerizes vinyl toluene + epichlorohydrin, also acrylonitrile + methacrylate	A1, A10, B40, K35, N13, N14, S43, S44
$Cp_2TiCl_2 + (C_6H_5)_3Al$	Ethylene	Weakly active or inactive	H5, N24
$Cp_2TiCl_2 + (C_2H_5)_2AlCl$	Ethylene	Active in benzene, but rate drops with time; in presence of alkyl halide rate stays fairly constant; copolymerizes ethylene + butylene in ethyl chloride or toluene	A21, B34–B39, B89, F29, F31, L23, M27, M28, W26, O3, O4, T13, Z3, F11
	Styrene	—	
	Propylene	Dimerizes propylene	
$Cp_2TiCl_2 + (CH_3)_2AlCl$	Ethylene and ethylene-d_4	Electrodialysis indicates active center could be $(Cp_2TiCH_3)^+$; similar to system $Cp_2Ti(CH_3)Cl + CH_3AlCl_2$; solvent effects (G32)	B1, C24, C25, D45, G31–G33, S35, S61
$Cp_2TiCl_2 + C_2H_5AlCl_2$	Ethylene	Rate of polymerization proportional to concentration of Ti(IV); rate increased by adding ether, amine, thio ether	A18, H18
$Cp_2TiCl_2 + CH_3AlCl_2$	2,4-Hexadiene	Polymerizes *trans*-1,4-	M54
	Ethylene	Forms linear polyethylene at a rate proportional to concentration of the initial 1:1 complex	L20

190

Catalyst system	Monomer	Remarks	References
$Cp_2Ti(C_6H_5)_2 + (C_6H_5)_3Al$	Ethylene	Weakly active	H5, N24
$Cp_2Ti(R)Cl + (C_2H_5)AlCl_2$, R = C_2H_5, C_8H_{17}	Ethylene	Active species has Al:Ti ratio = 1	H9, H15
$(CpTiClO)_4 + (iso\text{-}C_4H_9)_3Al$	Ethylene	Weak activity increased by alkyl halides	M20
$(CpTiClO)_4 + (iso\text{-}C_4H_9)_3Al$, $(iso\text{-}C_4H_9)_2AlCl$, or $(C_2H_5)_3Al$	Vinyl monomers	Moderately active in polymerizing vinyl ethers, acrylates, styrene, etc.	K36, K37
$CpTiCl_3 + R_nAlCl_{3-n}$, R = CH_3 or C_2H_5	Ethylene	Poorly active with R_3Al or R_2AlCl compounds; more active with $RAlCl_2$	D37, G23, L21
$Cp_2TiCl_2 + (iso\text{-}C_4H_9)_2AlCl$	Ethylene	Weakly active	K2
$Ti(OC_4H_9)_4 + (C_2H_5)_3Al$	Ethylene, styrene, or conjugated dienes	—	C7, D47, H32, T5
$Ti(OR')_4 + RAlCl_2$, R = CH_3, C_2H_5; R' = C_2H_5, $n\text{-}C_4H_9$	Butadiene, isoprene	Al:Ti ratio of 4 gives best yields of polymer	C76
$CpTi(OCH_3)Cl_2 + (C_2H_5)_3Al$	Ethylene	100°, 1000 psi pressure	G21
$R_4Ti + AlCl_3$, R = C_4H_9, C_2H_5	Propylene, 1-butene, styrene, and vinyl monomers	—	D2, D4
$CH_3TiCl_3 + CH_3AlCl_2$	Ethylene	Very active in CH_2Cl_2 at −80°	B49
$Cp_2TiX_2 + (C_2H_5)_2AlCl$, X = inorganic anion	Ethylene	Halide 15 times more active than rest	T13
$(Allyl)_4Zr + (C_2H_5)_{1.5}AlCl_{1.5}$ or $C_2H_5AlCl_2$	Isoprene, ethylene	Mostly dimers obtained; sesquichloride best cocatalyst	U1, A20a
Aluminum-free catalyst systems			
$Cp_2Ti(CH_3)_2 + TiCl_3$	Propylene	Weakly active; Ti-CH_3 active site	A22, S31
$Cp_2TiX_2 + TiCl_3 + Na + H_2$, $X_2 = Cl_2$ or borate	Propylene	Lower yields without Cp_2TiX_2	A24, H29, H30, M3
$C_{10}H_{10}Ti$ ($+TiCl_3$)	Ethylene	—	A23, S30, Y4
Cp_2TiR_2 + halogen-containing compound, R = CH_3, C_6H_5	Vinyl chloride	Introduction of Cp lowers efficiency	M39, M46, R33

191

(continued)

Table V-3 (continued)

System	Monomer polymerized	Remarks	Refs.
$(C_6H_5)_4Ti + CCl_4$	Vinyl chloride	More efficient than $Cp_2Ti(C_6H_5)_2$	R33
$Cp_2TiCl_2 + CCl_4$	Vinyl monomers	—	B9, B10
Cp_2TiR_2 $R = CH_3, C_6H_5$	1,5-cyclooctadiene	Isomerizes to 1,3-cyclooctadiene at 95°	M22
$CH_3TiCl_3 + TiCl_3$	Ethylene	—	A2, B24
CH_3TiCl_3	Ethylene	Weakly active; becomes active with aluminum alkyls added	B29, D13, K9, K10, K60, L3, M33
CH_3TiCl_2	Ethylene	High polymerization activity even at $-70°$	K60
$RTiCl_2$ $R = C_2H_5, iso\text{-}C_3H_7, C_8H_{17}$	Ethylene or propylene	—	L15, N7, S59
$Ti(\pi\text{-methylallyl})_4$	Acrylates	Weakly active [also $Zr(allyl)_4$]	B5, B6, O1
$(CH_3)_2Ti(OR)_2 + TiCl_3$ or $TiCl_4$ $R = iso\text{-}C_4H_9$ or $[C(CH_3)_2CH_2CHCH_3]$	Ethylene, propylene	—	F5, I1, S39, S68—S70
$(Allyl)_2Ti(OR)_2$ $R = C_4H_9, C_2H_5$	Ethylene	—	K8
R_nTiX_{4-n} $R = CH_3, C_6H_5; X = OR, Cl$	Styrene	Polymerization of styrene with $C_6H_5Ti(OC_3H_7)_3$ is free radical between $15°\text{-}50°$, the compound decomposing to $[Ti(OC_3H_7)_3]_n$ and C_6H_5 radicals	H21–H23, H25, H26, N8, N43, S57
$Ti(CH_2C_6H_5)_nX_{4-n}$ $X = OC_2H_5, Cl; n = 0\text{-}2$	Ethylene, butadiene	Slow; more active in presence of $(C_6H_5CH_2)_3Al$ or other alkylaluminum (L22); copolymerizes ethylene and butadiene to vinylcyclobutane (C7)	C7, D21, G5, G6, G42, G43, L22, P9
$Zr(CH_2C_6H_5)_4$	Vinyl monomers	Photoinduced polymerization	B7, A20a
$C_6H_5Ti(O\text{-}iso\text{-}C_3H_7)_3 + TiCl_4$	Propylene	Polymer yield high with excess $TiCl_4$	R10

192

System	Compound hydrogenated	Remarks	Refs.
$R_nTiR'_{4-n} + TiCl_4$; $R,R' = C_2H_5, C_4H_9$	Terminal olefins	—	D3
$Cp_2ZrCl_2 + NaC_5H_{11}, LiC_4H_9,$ $Mg(C_2H_5)_2,$ or $Al(C_2H_5)_3$	Ethylene	—	B87
$Cp_2ZrCl_2 + ZrCl_3 + Na$	Propylene	—	T39
$TiCl_4 + 4RMgX$; $R = CH_3, C_6H_5, CH_2=CH,$ C_2H_5	Butadiene	Low yields of polybutadienes	T41
$Cp_2Zr[N(CH_3)_2]_2$	Acrylonitrile	—	J8
$M(Allyl)_4 + MCl_4$; $M = Ti$ or Zr	Propylene	40°	F6, J14

System	Compound hydrogenated	Remarks	Refs.
	Hydrogenation catalysts		
$Cp_2TiCl_2 + (C_2H_5)_3Al$	Mono- and disubstituted olefins	Ambient temp., 4.4 atm pressure	K5, K6, S52
$Cp_2TiCl_2 + LiAlH_n(OR)_{4-n}$	As above	20°, 1.15 kg/cm² pressure	S62
$Cp_2Ti(OR)_2 + LiAlH_2(OR')_2$; $R = C_6H_5, R' = tert\text{-}C_4H_9$	As above	As above	S62
$Cp_2Ti(\pi\text{-methylallyl})$	As above	Isomerizes olefins	M11, M12
$Cp_2TiCl_2 + C_4H_9Li, C_2H_5MgCl,$ or C_6H_5MgBr	Cyclohexene	—	S33
$Ti(O\text{-}iso\text{-}C_3H_7)_4 + (iso\text{-}C_4H_9)_3Al$	Mono-, di-, tri-, and tetrasubstituted olefins	25°, 3.5–3.7 atm pressure	S52
$Cp_2ZrCl_2 + (iso\text{-}C_4H_9)_3Al$	Cyclohexene	25°, 3.5–3.7 atm pressure	S52
Cp_2ZrH_2	Olefins and acetylenes	80°, 100 atm pressure	W10

[a] For heterogeneous catalyst systems, see ref. (B55)

(B29). The system is, therefore, CH_3TiCl_3–$TiCl_3$, which gives high molecular weight, largely unbranched polyethylene.

At $-70°$ in methylene dichloride or toluene, ethylene was oligomerized rapidly to chains with an average length of 15 carbon atoms (B49). Addition of CH_3AlCl_2 under these conditions increased the activity of the catalyst by a factor of 300 (1 mole of CH_3TiCl_3–CH_3AlCl_2–CH_2Cl_2 will oligomerize 3 moles of ethylene per minute at $-70°$). There is little doubt that titanium(IV) centers are involved here. At this low temperature it has been shown that ligand exchange between titanium and aluminum is virtually absent and no reduction occurs (B49). By alcoholysis of the polymerizing mixture at low temperatures, methane equivalent to the CH_3–Al groups only was obtained. In this way Bestian and Clauss were able to show that the methyl groups attached to aluminum remained unchanged throughout the oligomerization. The reaction therefore proceeded exclusively at the titanium–carbon bond. The original methyl groups on titanium added to the olefinic double bond and appeared in the higher polymerization products.

An ionic mechanism was proposed for this system (B49), the olefin being coordinated to the solvated titanium cation of the complex, $[CH_3TiCl_2]^+$-$[CH_3AlCl_3]^-$, before reaction with the titanium–carbon bond.

A similar cationic site was proposed by D'yachkovskii (D42, D46; see also Z3) for the system Cp_2TiCl_2–$(CH_3)_2AlCl$ in dichloroethane.

4. Aluminum-free Catalysts

Organotitanium compounds without added alkylaluminum, but sometimes in combination with titanium halides, almost certainly polymerize olefins by a mechanism similar to that discussed above, using titanium–alkyl sites. Examples are $Cp_2Ti(C_6H_5)_2$, $Cp_2Ti(CH_3)_2$, and CH_3TiCl_3, each in combination with $TiCl_3$. More examples are given in Table V-3. Ligand exchange in systems such as these is known to be facile so that the formation of active sites could reasonably be expected. In many cases addition of alkylaluminum derivatives enhances considerably the activity of such catalysts.

Recently, benzyl derivatives of titanium and zirconium have been shown to polymerize ethylene, propylene, 4-methyl-1-pentene, and butadiene, the catalyst remaining apparently soluble throughout the polymerization (C7, G6, G42, G43). Boor has suggested (B55) that some colloidal low-valent catalyst could be present. Without this explanation the stereospecificity of these catalysts is difficult to explain by presently accepted views. The formation of isotactic polymers from soluble metal catalysts is unknown.

In the presence of tribenzylaluminum (G6) or other alkylaluminums (A20a, L22), the activity of the benzyltitanium and zirconium catalysts is considerably increased. Giannini and co-workers point out (G6) that the relatively low activity of the aluminum-free compounds, together with the high molecular weight of the polymers obtained, suggests that the number of active centers is extremely low and that the organometallic compounds are only precursors of the true catalyst.

The data and theories just presented represent only a small sample of the information published on this subject. The apparent inconsistencies are perhaps good evidence in themselves that Ziegler–Natta catalysts show remarkable diversity in physical characteristics and activity, depending on the metal derivatives used and the conditions employed.

In addition, many of these catalysts show activity as hydrogenation catalysts (see Table V-3), while some are active in nitrogen fixation. The versatility of such systems is a good example of the effect of creating vacant coordination sites on a transition metal.

Chapter VI

Organometallic Compounds of Titanium(III) and Zirconium (III)

Many of the early experiments using either titanium(IV) and titanium(III) halides or alkoxides with alkylaluminum derivatives (B2, M6), Grignard reagents (G16, T31), or an alkyl- or aryllithium (G16) undoubtedly led to organotitanium(III) derivatives which decomposed before isolation. The thermal stability of alkyl- and aryltitanium(III) compounds is normally so low that few have been isolated uncomplexed with other ligands. The majority of those reported have been prepared during the search for new polymerization catalysts and as a result the titanium(III) compound is complexed with an aluminum (or other metal) derivative. This, undoubtedly, adds to their stability sufficiently to allow characterization and sometimes isolation. Bimetallic complexes of this type are dealt with in Chapter V.

As with the other valence states of titanium, by far the majority of the titanium(III) compounds are cyclopentadienyl derivatives. The unique stabilizing influence of this ligand is immediately obvious from the tables in this chapter.

A. Alkyl-, Aryl-, and Arenetitanium(III) Compounds (Excluding Cyclopentadienyl)

The simplest compound of this type, trimethyltitanium, has been obtained only as deep green solutions in ether from the reactions between the tetrahydrofuran or dimethoxyethane adduct of $TiCl_3$ and methyllithium (C37) at $-50°$ to $-80°$ or methylmagnesium chloride (S12). Above $-20°$ the solutions decompose. The benzyl derivative, $(C_6H_5CH_2)_3Ti$, was obtained by Thiele and Schäfer (T31) as a brown ethereal solution which decomposed above $0°$.

Two pyrophoric alkyltitanium dichlorides, $RTiCl_2$ (where $R = CH_3$ or C_8H_{17}) were mentioned in a French patent (P3) as products of the reaction between titanium tetrachloride and the appropriate trialkylaluminum in cyclohexane, but were never obtained free of polyolefins. The methyl derivative, CH_3TiCl_2, has been mentioned also by Kühlein and Clauss in a communication (K60). Pyridine adducts of this class of compound, i.e., $RTiCl_2 \cdot 3py$ (where $R = CH_3$ or C_6H_5), were isolated (M29) from the action of ethereal Grignard reagents on titanium trichloride in pyridine at $-30°$. These blue complexes were relatively stable at room temperature under nitrogen. Alkyl derivatives other than methyl were not sufficiently stable to permit isolation.

Green to brown solutions of tris(crotyl)- (O2), tris(phenyl)-, tris(o-tolyl)-, and tris(p-tolyl)titanium (S12) have been obtained, but these also were too thermally unstable for isolation. However, bis(phenyl)titanium chloride was isolated as a red etherate, $(C_6H_5)_2TiCl \cdot 3(C_2H_5)_2O$ (S12).

Several titanium(III) compounds have been prepared by the action of alkylaluminum derivatives on titanium alkoxides. The bimetallic compounds are dealt with in Chapter V. Cyclooctatetraene compounds of titanium have been isolated by Wilke and co-workers from the reaction between titanium tetrabutoxide and triethylaluminum in the presence of cyclooctatetraene. Two interconvertible complexes were formed according to the ratio of reactants (B85, W22). These were yellow $Ti_2(COT)_3$ and violet-red $Ti(COT)_2$.

$$Ti(COT)_2 \; \frac{(C_2H_5)_2AlH,\ 60°,\ 40\ hr}{COT,\ 80°,\ 40\ hr} \; Ti_2(COT)_3$$

It is believed that the COT molecules are bonded, at least partly, as *quasi*-aromatic, 10 π-electron systems, i.e., as dianions (B85). $Ti_2(COT)_3$ has a double sandwich structure (D22), the two outer rings being regular octahedrons whose planes are inclined slightly toward the axis of the molecule.

With anhydrous HCl, $Ti_2(COT)_3$ forms the compound, $[(COT)TiCl]_2$, which with pyridine is converted to the 1:1 complex, $(COT)TiCl \cdot py$ (ref. 3 in L14).

Lehmkul and Mehler have prepared the known cyclooctatetraene complexes of titanium (Table VI-1) by electrochemical synthesis from titanium

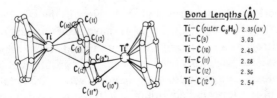

Bond Lengths (Å)

Ti–C (outer C_8H_8)	2.35(av)
Ti–C(9)	3.03
Ti–C(10)	2.43
Ti–C(11)	2.28
Ti–C(12)	2.36
Ti–C(12*)	2.54

Fig. VI-1.   Structure of tris(cyclooctatetraene)dititanium (D22).

Table VI-1

Alkyl-, Aryl-, and Arenetitanium(III) Compounds

Compound	Color	Remarks	Refs.
$(CH_3)_3Ti$	Green	Not isolated	C37, S12
$(C_6H_5CH_2)_3Ti$	Brown	—	T31
CH_3TiCl_2	—	—	K60
$CH_3TiCl_2 \cdot 3py$	Blue	—	M29
$C_6H_6TiCl_2 \cdot 3py$	Blue	—	M29
$C_2H_5TiCl_2$	Black	Impure	P3
$C_8H_{17}TiCl_2$	Black	Impure	P3
$(C_6H_5)_2TiCl \cdot 3(C_2H_5)_2O$	Red	—	S12
$Ti_2(COT)_3$	Yellow	—	B85, L14, W22
$[(COT)TiCl]_2$	—	—	L14
$(COT)TiCl \cdot py$	Brown	—	L14
$(COT)TiCl \cdot THF$	—	—	L14
$(COT)TiCp$	Green	μ_{eff} 1.6 BM	L14
$(COT)Ti(\pi\text{-}C_3H_4CH_3)$	Brown-green	—	H34
$\{Ti_3[C_6(CH_3)_6]_3Cl_6\}Cl^a$	Violet	μ_{eff} 1.84 BM per trimer	F18
$C_6(CH_3)_6 \cdot Ti(OH)Cl_2{}^a$	—	—	W19

^{a} These compounds are described in Chapter V, Section A.

halides or alkoxides (L13, L14) in tetrahydrofuran or pyridine under argon at 20°–40°. Low yields of the known cyclooctatetraenecyclopentadienyltitanium-(III), CpTi(COT), were obtained from Cp_2TiCl_2 or $CpTiCl_3$ in tetrahydrofuran (V4).

With 1-methylallylmagnesium bromide, cyclooctatetraenetitanium chloride dimer was converted to the brownish green methylallyl derivative, $(h^8\text{-COT})$-$Ti(h^3\text{-}C_3H_4CH_3)$, which decomposes at 110°. Evidence for π bonding of the 1-methylallyl groups was found in the infrared absorption band at 1500 cm^{-1}. Fragmentation in the mass spectrometer occurred by successive loss of C_2 fragments (H34).

B. Mono(cyclopentadienyl)titanium(III) Compounds

Considerable work has been carried out in this field in the last 2 years. All the halides, $CpTiX_2$ (where X = Cl, Br, or I), have been characterized, and complexes with nitrogen- and oxygen-containing ligands have been investigated. The carboxylates have also been isolated. This group of compounds lies between the purely inorganic titanium halides, which are octahedral in structure, and the bis(cyclopentadienyl)titanium(III) compounds in which the metal is in an essentially tetrahedral environment.

1. The Halides, $CpTiX_2$

The first member of the halide series, $CpTiCl_2$, was prepared in 1961 by Bartlett and Seidel (B15) by the reaction of diisobutylaluminum chloride with Cp_2TiCl_2 in a toluene–heptane solvent heated to 50° for 3 hr. One of the cyclopentadienyl ligands was lost giving purple $CpTiCl_2$. A more convenient preparation by Coutts and Wailes was by reduction of $CpTiCl_3$ with zinc in benzene containing a stoichiometric amount of tetrahydrofuran (C70). A precipitate of the titanium(III) halide appears almost immediately. The chloride, bromide, or iodide have been prepared as THF complexes by zinc reduction of the corresponding trihalides (C49, C50) at room temperature in

$$2CpTiX_3 + Zn \xrightarrow{\text{THF}} 2CpTiX_2 \cdot THF + ZnX_2 \cdot 2THF$$

THF. The solvent can be removed from these complexes by pyrolysis at 120°–150° under vacuum. Reduction of Cp_2TiCl_2 with aluminum in the presence of a stoichiometric amount of water (W11) also gives $CpTiCl_2$. In the absence of water, a chloride is removed, but in the presence of water, a cyclopentadienyl ligand is lost.

The insolubility of the $CpTiX_2$ halides in nonpolar solvents, together with their involatility, suggests some type of association. Bartlett and Seidel proposed a dimeric structure for the chloride (B15), but offered no suggestion as to how it was bound. Possible structures for the halides were suggested by Coutts et al. (C50).

(16)   (17)   (18)

Structure **16** seems least likely since the tetrahedral environment of the metal would be expected to produce visible spectra similar to those of $CpTiX_2 \cdot$ THF and, in addition, such a structure should produce magnetic interaction between the titanium nuclei similar to the antiferromagnetic behavior (C57) of $(Cp_2TiCl)_2$.

In fact, the spectra of the halides and their THF complexes show significant variations (C45), while the magnetic moments of the halides are close to the spin-only value (1.73 BM) at room temperature and increase only slightly with decreasing temperature, an effect which was more pronounced in the bromide than in the chloride (C8, C50).

The alternative structures for the halide, either the dimeric form (17) or the polymeric form (18), offer a potentially square-pyramidal arrangement.

Monocyclopentadienyltitanium(III) chloride has been shown to form complexes with nitrogen-containing ligands (C51) such as bipyridyl, o-phenylenediamine, pyridine, and 2-picolylamine, of stoichiometry, $CpTiCl_2 \cdot 2N$, where N is one nitrogen of the chelating ligand. Complexes of similar type are formed by both $CpTiCl_2$ and $CpTiBr_2$ with oxygen-containing ligands (C52), e.g., ethanol, methanol, or dimethoxyethane. Many of these solvent ligands are held only weakly and can be displaced readily by warming under vacuum or merely by washing with benzene or diethyl ether. THF, methyl cyanide, and trialkylamines are in this category.

Measurement of physical properties of the complexes does not distinguish between a monomeric five-coordinate and a dimeric six-coordinate species, e.g., for the chloride complexes (where L is the chelating atom of the ligands

mentioned above).

The six-coordinate structure tends to be favored by molecular weight determinations in THF and by the fact that complexes with six groups bound to the metal, e.g., $(CpTiL_5)^{2+}$ (where $L = CH_3CN$ or py), are readily obtained (C47).

The magnetic moments of both oxygen-containing and nitrogen-containing complexes fall in the range 1.5–1.9 BM per Ti atom over a 200° temperature range, with low values of θ. The lack of marked temperature dependency implies that any magnetic interaction via, say halide bridges, must necessarily be very weak.

Although $CpTiCl_2$ and $CpTiBr_2$ form isolable complexes with ethanol and methanol, prolonged exposure to these solvents at room temperature, or heating for 1 hr, brings about alcoholysis of the cyclopentadienyl group giving inorganic complexes of formula, $[Ti(OR)X_2 \cdot 2ROH]_n$ (where n is probably 4) (C56, G15).

The reactions of cyclopentadienyltitanium(III) dihalides with aldehydes and ketones are novel in that different products are formed according to the experimental conditions. In ethereal solvents "redox" reactions take place to yield yellow crystalline titanium(IV) derivatives of pinacols. The nature of the compounds was determined by the shift in the infrared frequency of the carbonyl band from ~ 1700 cm^{-1} to the alkoxide region ~ 1100 cm^{-1}, and also by the isolation of pinacol, $[(CH_3)_2COH]_2$, from hydrolysis of the acetone derivative. The benzophenone analog lacks the stability of the other derivatives since on warming in benzene or toluene to 40° or above, or merely by dissolving the complex in THF, benzophenone was displaced and the cyclopentadienyltitanium(III) halide was reformed (C53). In pure dry acetone instead of in ethereal solvents, the halides, $CpTiX_2$ ($X = Cl$, Br, or I), formed a brown solution from which yellow crystals of the titanoxane polymers, $(CpTiXO)_n$, were precipitated (C55) (see Chapter III).

CpTiCl$_2$ readily splits the S–S bonds of disulfides and thiuram disulfides giving the titanium(IV) derivatives, $CpTiCl_2SR$ and $CpTiCl_2S_2CNR_2$ (C72) (see Chapter III, Section A,3).

2. The Carboxylates, $CpTi(O_2CR)_2$

Monocyclopentadienyltitanium(III) carboxylates have been prepared from the corresponding chloride with the sodium carboxylate in tetrahydrofuran or with the carboxylic acid in the presence of a tertiary amine (C54).

$$CpTiCl_2 \cdot THF + 2NaO_2CR \xrightarrow{\text{THF}} CpTi(O_2CR)_2 + 2NaCl$$

$$CpTiCl_2 \cdot THF + 2HO_2CR + 2base \xrightarrow{\text{THF}} CpTi(O_2CR)_2 + 2base \cdot HCl$$

Russian workers prepared the benzoate in 40–50% yield by the reaction of $Cp_2Ti(C_6H_5)_2$ with benzoyl peroxide (R17, R18).

$$Cp_2Ti(C_6H_5)_2 + (C_6H_5CO_2)_2O \longrightarrow CpTi(O_2CC_6H_5)_2 + C_6H_6 + \text{``polymeric ester''}$$

The low solubility of most of these carboxylates in organic solvents indicated association and prevented the determination of molecular weights in all but the propionate and butyrate which were dimeric in boiling benzene (C54).

The separation of the symmetric and antisymmetric stretching of the carbonyl group is 150–180 cm^{-1}, which is intermediate between the separation observed (< 100 cm^{-1}) in the more symmetrical four-membered bidentate

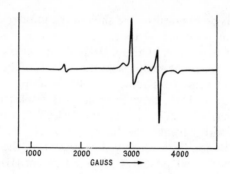

Fig. VI-2.   ESR spectrum of [CpTi(O$_2$CC$_2$H$_5$)$_2$]$_2$ (solid) (C47).

ring system (C64) of Cp$_2$TiO$_2$CR and the unsymmetrical monodentate carboxylates, Cp$_2$Ti(O$_2$CR)$_2$, where the separation is > 300 cm^{-1} (V32).

The carboxylates are virtually diamagnetic over the temperature range 100°–300°K with molar susceptibilities below 150 × 10^{-6} cm^3 mole^{-1} (C59). The presence of pronounced magnetic exchange between pairs of titanium atoms is indicated by X-band electron spin resonance spectra (see Fig. VI-2) (C59) in which $\Delta m = 1$ and $\Delta m = 2$ transitions (W12) are clearly observed.

3. CpTi(COT)

Cyclopentadienylcyclooctatetraenetitanium was reported by van Oven and de Liefde Meijer (V4) in 1969 as a green compound prepared from dipotassium cyclooctatetraene and either CpTiCl$_3$, Cp$_2$TiCl$_2$, or a mixture of TiCl$_4$ and C$_5$H$_5$Na in tetrahydrofuran or toluene. Electrochemical reduction of Cp$_2$TiCl$_2$ or CpTiCl$_3$ in THF containing cyclooctatetraene also gave CpTiC$_8$H$_8$, but in low yield (L14) (25 and 4%, respectively). Both ligands are π-bonded to the metal (K59), the average interatomic distances being as shown in Fig. VI-3. The two rings are virtually parallel, the angle between the planes being only 1.9° ± 0.2°.

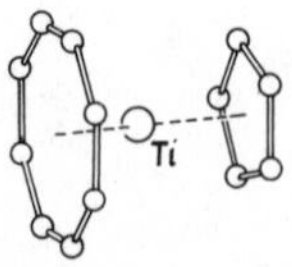

Average Interatomic distances (Å)	
Ti–C (C$_8$H$_8$)	2.323
Ti–C (C$_5$H$_5$)	2.351
C–C (C$_8$H$_8$)	1.395
C–C (C$_5$H$_5$)	1.396

Fig. VI-3.   Structure of cyclopentadienylcyclooctatetraenetitanium (K59).

The magnetic moment of $CpTiC_8H_8$ was found to be 1.60 BM per titanium atom, independent of temperature over the range 90°–300° K. The ESR spectrum of $CpTiC_8H_8$ has been measured by Thomas and Hayes (T34) and indicates that the unpaired electron is in an a_{1g} orbital, which is essentially a metal d_{z^2} orbital.

4. Sulfur and Amido Derivatives

Ungurenasu and Cecal have reported (U2) a mono(cyclopentadieny)-titanium(III) sulfur derivative, $(CpTiS_4)_n$, obtained by ultraviolet irradiation of Cp_2TiS_4 [the existence of which has since been questioned (K45)]. The insoluble orange material was purified by Soxhlet extraction with water, a solvent to which most mono(cyclopentadienyl)titanium compounds are very sensitive. The compound was paramagnetic with a μ_{eff} of 1.76 BM.

Monocyclopentadienyldimethylamido derivatives of titanium have been prepared by Lappert and Sanger (L5) by the reaction

$$\{Ti[N(CH_3)_2]_3\}_2 + 2C_5H_6 \longrightarrow \{CpTi[N(CH_3)_2]_2\}_2 + 2(CH_3)_2NH$$

The compound was formulated with terminal and bridging amido groups (A11).

Isocyanatobenzene was found to insert between the titanium and the dimethylamido group in this compound at room temperature (L6), leading to a phenylureido derivative, $CpTi[N(C_6H_5)CON(CH_3)_2]_2$.

Addition of two equivalents of ethanol to the dimethylamido compound in benzene displaced dimethylamine giving cyclopentadienyldiethoxotitanium-(III), $CpTi(OC_2H_5)_2$, as a red liquid dimer (L6). Use of excess alcohol in this reaction causes alcoholysis of the cyclopentadienyl ligand.

5. Alkyl and Aryl Derivatives

Razuvaev and co-workers (R21a, R21c, R31) have obtained cyclopenta-dienyltitanium(III) phenyl and benzyl derivatives, CpTiRR', by reaction of Cp_2TiR_2 with R'Li (see Table VI-2).

Table VI-2

Monocyclopentadienyltitanium(III) Compounds

Compound[a]	Color	μ_{eff} (20°C)[b]	Other data	Refs.
$(CpTiCl_2)_n$	Purple	1.74^T	ESR (g = 1.975); UV; visible; IR	B15, C8, C49, C50, W11
$(CpTiBr_2)_n$	Dark blue	1.76^T	Visible; IR	C50
$(CpTiI_2)_n$	—	—	Visible; IR	C50
$CpTiCl_2 \cdot$ THF	Sky blue	1.89^T	Visible; IR; MW	C50
$CpTiBr_2 \cdot$ THF	Blue-green	1.74^T	Visible; IR; MW	C50
$CpTiI_2 \cdot$ THF	Blue-green	1.45^T	Visible; IR; MW	C50
$CpTiCl_2 \cdot$ 2py	Deep green	1.75^T	Visible; IR; conductance	C51
$CpTiCl_2 \cdot$ bipy	Gray-blue	1.55^T	Visible; IR; conductance	C51
$CpTiCl_2 \cdot$ phm	Pale green	1.71^T	Visible; IR	C51
$CpTiCl_2 \cdot$ 2pic	Mauve	1.65^T	Visible; IR; conductance	C51
$CpTiCl_2 \cdot$ 1.5en	Blue	1.55^T	Visible; IR	C51
$CpTiCl_2 \cdot 2CH_3OH$	Pale green	1.57^T	Visible; IR; conductance	C52
$CpTiCl_2 \cdot 2C_2H_5OH$	Bright green	1.54^T	Visible; IR; conductance; MW	C52
$CpTiBr_2 \cdot 2CH_3OH$	Bright green	—	Visible; IR; conductance	C52
$CpTiBr_2 \cdot 2C_2H_5OH$	Bright green	1.53^T	Visible; IR; conductance; MW	C52

CpTiCl$_2$·DME	Pale green	1.89^T	Visible; IR	C52
CpTiBr$_2$·DME	Pale green	1.73^T	Visible; IR	C52
[CpTi(O$_2$CCH$_3$)$_2$]$_2$	Green	Diamagnetic	IR; ESR	C54
[CpTi(O$_2$CCF$_3$)$_2$]$_2$	Green	Diamagnetic	IR; ESR	C54
[CpTi(O$_2$CC$_2$H$_5$)$_2$]$_2$	Green	Diamagnetic	IR; MW; ESR	C54
[CpTi(O$_2$CC$_3$H$_7$)$_2$]$_2$	Green	Diamagnetic	IR; MW; ESR	C54
[CpTi(O$_2$CC$_6$H$_5$)$_2$]$_2$	Brown	Diamagnetic	IR; ESR	C54, K7, R17, R18
CpTiC$_8$H$_8$	Green	1.60	IR; MS; X-ray; ESR	K59, L14, T34, V4
(CpTiS$_4$)$_n$	Orange	1.76	IR; UV	U2
{CpTi[N(CH$_3$)$_2$]$_2$}$_2$	Red-brown	Diamagnetic	IR; MW; NMR	A11, L5
CpTi[N(C$_6$H$_5$)CO·N(CH$_3$)$_2$]$_2$	Red-brown	—	IR	L6
CpTiCl$_2$·2AlCl$_3$	—	—	ESR (g = 1.972)	H13, V14
CpTiCl[C(C$_6$H$_5$)$_3$]	—	—	—	R28
CpTi(CH$_2$C$_6$H$_5$)C$_6$H$_5$	Black	—	ESR (g = 1.987)	R31
CpTi(CH$_2$C$_6$H$_5$)$_2$	Black	—	IR (g = 1.984)	R21a, R21c
[CpTi(OC$_2$H$_5$)$_2$]$_2$	Red-brown	—	MW	L6

[a] py, Pyridine; bipy, 2,2'-bipyridyl; phm, o-phenylenediamine; pic, 2-picolylamine; en, ethylenediamine; DME, 1,2-dimethoxyethane.
[b] Superscript T implies that μ_{eff} has been measured at several temperatures.

C. Bis(cyclopentadienyl)titanium(III) Derivatives

The ready availability of Cp_2TiCl, both from reduction of Cp_2TiCl_2 and from $TiCl_3$ with metal cyclopentadienides, has led to the widespread use of this compound as a starting material. Consequently, this group is the largest of the organotitanium(III) compounds. A characteristic of bis(cyclopentadienyl)titanium(III) complexes is the distorted tetrahedral environment around the titanium atom. To achieve this configuration the compounds will polymerize, solvate, or in some way coordinate a ligand into the fourth position. Polymerization generally leads to magnetic exchange between neighboring titanium atoms, resulting in low magnetic moments.

1. Preparation of the Halides

The reduction of Cp_2TiX_2 (where X = halide) has received considerable attention since it provides a convenient way of obtaining lower valent derivatives. Polarographic studies on the halides, Cp_2TiX_2, (X = F, Cl, Br, I), clearly show cathodic waves corresponding to two one-electron steps (C29, D27, G35, G36, S53). The value of the half-wave potential for each step is

$$Cp_2TiX_2 \xrightarrow{\ e^-\ } X^- + Cp_2TiX \xrightarrow{\ e^-\ } X^- + [Cp_2Ti]$$

highly dependent on solvent and supporting electrolyte; in DMF the value for the chloride, bromide, and iodide lies around 0.8–0.6 V or below, while for the fluoride the value is 1.2 V. The second step requires a value of ~ 1.9 V for chloride, bromide, or iodide and ~ 2.1 V for fluoride (G35, S53). It is possible therefore, with the right choice of reductant, to reduce any dihalide selectively to a lower valent form. Among the milder reducing agents which form titanium(III) compounds are zinc and aluminum, while the alkali metals and magnesium will normally reduce further unless used in stoichiometric amounts.

In polar organic solvents such as toluene, methanol, THF, and acetone, Cp_2TiCl_2 will react rapidly with finely divided zinc, giving the green complex

(S2). This same compound was obtained by Birmingham et al. in 1955, although its composition was not recognized at the time (B50). The magnetic moment of the zinc chloride complex shows a spin-only value, independent of temperature, with no magnetic interaction between titanium atoms (S2, W11). Cp_2TiCl can be obtained from this complex in only low yield by pyrolysis (B50), but Green and Lucas have found that the zinc complex may be partially decomposed with diethyl ether to give Cp_2TiCl in almost 50% yield (G30).

If the reduction with zinc is carried out in air-free water, a bright blue solution containing the cation, Cp_2Ti^+, is obtained (presumably hydrated). The aqueous chemistry of Cp_2Ti^+ will be discussed below.

By far the best reducing agent for this preparation is aluminum. A near-quantitative yield of Cp_2TiCl can be obtained by addition of large quantities of Cp_2TiCl_2 to activated aluminum foil in THF (W11). The aluminum chloride remains in solution as the tetrahydrofuranate. An interesting variation of this reduction occurs on addition of water, when one of the cyclopentadienyl groups is removed instead of a chloride (W11), giving $CpTiCl_2$.

Similar reduction procedures with Cp_2TiBr_2 and Cp_2TiI_2 give the corresponding titanium(III) compounds, although the iodide so formed always contains strongly coordinated ether. Even the fluoride can be reduced in this way (C57), although careful activation of aluminum foil is required. In the latter case the product is a blue bimetallic complex containing aluminum and titanium, but this is split by pyrolysis giving a 40% yield of sublimed Cp_2TiF.

The chloride has also been prepared from Cp_2TiCl_2 and a trialkylaluminum or alkylaluminum halide (B89, C38), e.g.,

$$Cp_2TiCl_2 + R_3Al \longrightarrow R\cdot + \begin{array}{c} Cp \\ \diagdown \\ Ti \\ \diagup \\ Cp \end{array} \begin{array}{c} Cl \\ \diagdown \\ \diagup \\ Cl \end{array} \begin{array}{c} R \\ \diagdown \\ Al \\ \diagup \\ R \end{array} \xrightarrow{ether} Cp_2TiCl$$

Reduction with $LiAlH_4$, followed by sublimation, also gave Cp_2TiCl in 70% yield (W1), while all halides with the exception of the fluoride have been formed (N45) by the action of hydrogen halides in ether on Cp_2TiBH_4.

Reaction of Cp_2TiCl with boron trihalides (not fluoride) possibly constitutes the most suitable pathway to the corresponding bromide or iodide (C57).

$$3Cp_2TiCl + BX_3 \longrightarrow 3Cp_2TiX + BCl_3$$
$$X = Br\ or\ I$$

In the case of the iodide the complications caused by abstraction of iodine from BI_3 and subsequent oxidation of the titanium compound to Cp_2TiI_2 can be overcome by addition of aluminum foil to the reaction.

Attempts to form the fluoride, Cp_2TiF, by treatment of the borohydride with $BF_3\cdot(C_2H_5)_2O$, gave the tetrafluoroborate (N45), Cp_2TiBF_4.

Alternative preparations of Cp_2TiCl are from titanium trichloride and sodium cyclopentadienide (N13), or better, magnesium cyclopentadienide (R38) which will place no more than two cyclopentadienyl rings on titanium-(III) halides.

By treatment of "titanocene" with anhydrous HCl various workers have obtained a purple or magenta halide (with liberation of H_2) to which the formula "Cp_2TiCl" has been given (S4, W9). In view of the complex nature of one of the cyclopentadienyl ligands of the starting material (see Chapter VII) there seems no doubt that the rings in magenta "Cp_2TiCl" are not identically bonded. Brintzinger (B43) has formulated the compound as $[(C_5H_5)(C_5H_4)TiCl]_2$. A similarly colored compound of identical composition can be obtained by equilibration of "titanocene" and Cp_2TiCl_2 in boiling toluene (W11). The magnetic moment of this compound varied from 1.07 BM at 360°K to 0.33 BM at 91° K, while absorption in the visible occurred at 18,940 cm^{-1} (ε 448) in benzene.

The bis(indenyl)titanium(III) halides, $(\pi\text{-}C_9H_7)_2TiX$ (X = Cl, Br, or I), have been prepared by reduction of the corresponding titanium(IV) dihalides with triethylaluminum (M8).

2. Properties of Bis(cyclopentadienyl)titanium(III) Halides

All but the iodide can be purified by sublimation under vacuum; they are soluble in air-free water and most organic solvents, although the dissolution of the fluoride is very slow. They are dimeric in benzene.

$$\begin{array}{ccc} Cp & X & Cp \\ \diagdown & \diagup \diagdown & \diagup \\ & Ti \qquad Ti & \\ \diagup & \diagdown \diagup \diagdown & \\ Cp & X & Cp \end{array}$$

X = F, Cl, Br, I

This series offers an opportunity to study the effect of the availability of orbitals of the bridging halide atom on metal–metal interaction in dimeric systems such as these. All are antiferromagnetic. In the case of the chloride, bromide, and iodide the susceptibility passes through a maximum and falls at lower temperatures. The fluoride plot shows no maximum in the temperature range studied. The original magnetic results of Martin and Winter (M19) on Cp_2TiCl have been recalculated using the experimentally observed g value of 1.98, and a T.I.P.* of 62×10^{-6} cm^3 mole^{-1}. The results were in substantial agreement (C8, C47) (see Fig. VI-4). Both investigations showed scattering near 200°K owing to a reproducible change, possibly a phase transition. Minor distortions near the Néel point were also evident in the plot of the iodide.

Although the lighter halides show an increasing interaction of the magnetic centers in the orders F > Cl > Br, the iodide appears to be anomolous. Instead of a Curie temperature above that of the bromide ($T_c > 250$°K), the

* T.I.P. = temperature independent paramagnetism.

Table VI-3
Bis(cyclopentadienyl)titanium(III) Halides

Compound	Color	μ_{eff} (20°C)	Other data	Refs.
Cp_2TiF	Green	1.62 ($J = 62\ \text{cm}^{-1}$)	MW, dimer; IR; visible; ESR ($g = 1.965$)	C47, C57, C58
Cp_2TiCl	Green-brown	1.51 ($J = 159\text{-}186\ \text{cm}^{-1}$)	MW, dimer; IR; visible; ESR ($g = 1.976$)	C47, C57, C58, M19
Cp_2TiBr	Red-brown	1.32 ($J = 276\ \text{cm}^{-1}$)	MW, dimer; IR; visible; ESR ($g = 1.969$)	C47, C57, C58
Cp_2TiI	Black	1.47 ($J = 168\text{-}179\ \text{cm}^{-1}$)	MW, dimer; IR; visible; ESR ($g = 1.973$)	C47, C57, C58

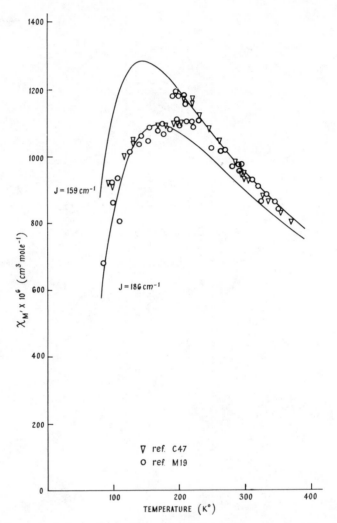

Fig. VI-4. Variation of magnetic susceptibility with temperature for $(Cp_2TiCl)_2$, showing phase change (C47).

iodide showed a T_c of 170°K, possibly due to conformational changes caused by the large ionic radius of iodide (C58).

All the bis(cyclopentadienyl)titanium(III) halides are sensitive to air (the fluoride is slowest to react), oxidizing to yellow-orange oxygen-bridged polymeric solids. With dry oxygen these products are of the type $(Cp_2TiX)_2O$, but in the presence of moisture further reaction occurs leading to loss of cyclopentadienyl ligands (see Chapter III).

In organic solvents Cp_2TiCl will split disulfides readily (C61), giving thio-latotitanium(IV) derivatives, $Cp_2Ti(Cl)SR$ ($R = CH_3$, C_2H_5, $CH_2C_6H_5$, or C_6H_5).

With organic azides, RN_3, nitrogen is liberated and the nitrogen-bridged compounds $(Cp_2TiCl)_2NR$ ($R = C_6H_5$ or p-ClC_6H_4), are obtained (C60). These are of low stability and decompose to Cp_2TiCl_2 and nitrogen-containing polymers. Anhydrous HN_3 reacted similarly with Cp_2TiCl to give (C71) the red-brown product $(Cp_2TiCl)_2NH$.

Several complexes of Cp_2TiCl with nitrogen-containing ligands are known. With bipyridyl, phenanthroline, 2-picolylamine, and o-phenylenediamine complexes have been obtained, which were originally formulated as binuclear (C68), but are now known to be mononuclear 1:1 complexes (C72, G30) of the type

$$\left[\begin{array}{c} Cp \\ \\ Cp \end{array} Ti \left(\begin{array}{c} N \\ \\ N \end{array} \right) \right]^{+} Cl^{-}$$

with the tetrahedral stereochemistry so typical of bis(cyclopentadienyl)-titanium compounds. Similar complexes with $PF_6{}^-$ as the anion were prepared by Green and Lucas (G30). An insoluble terpyridyl adduct is also known (C70).

Monodentate ligands formed monomers of the type, $Cp_2Ti(Cl)N$ (where $N = py$, NH_3, NH_2CH_3, $NH_2C_2H_5$, $NH_2C_3H_5$, or $NH_2C_6H_5$). The pyridine complex, $Cp_2TiCl \cdot py$, has been prepared also by reduction of Cp_2TiCl_2 with 1 mole of sodium pyridyl (W11). A side product in this reaction was the green, less soluble 4,4'-bipyridyl complex $[Cp_2Ti(4,4'\text{-}bipy)]^+Cl^-$. The tetraphenyl-borate derivatives $[Cp_2TiL_2]^+B(C_6H_5)_4{}^-$ (where $L = py$ or CH_3CN), prepared from Cp_2TiCl and $NaB(C_6H_5)_4$ in the presence of the base, have magnetic moments independent of temperature; a Curie–Weiss law is followed with μ_{eff} close to the spin-only value for a d^1 system (C48).

Phosphines such as $(CH_3)_2C_6H_5P$ and $CH_3(C_6H_5)_2P$ formed mononuclear compounds with Cp_2TiCl, while the bidentate phosphine, $(C_6H_5)_2$-$PCH_2CH_2P(C_6H_5)_2$ ("diphos"), gave a binuclear complex with the phosphine bridging two metals (G30).

3. Aqueous Chemistry of Cp₂TiCl

In air-free water Cp_2TiCl dissolves to give a bright blue solution which shows the molar conductance of a 1:1 electrolyte. Wilkinson and Birmingham showed in 1954 that picrate and silicotungstate derivatives could be precipitated from solutions containing the Cp_2Ti^+ ion (W24). Coutts and Wailes

Table VI-4

Complexes of Cp_2TiX with Nitrogen-Containing and Phosphorus-Containing Ligands

Compound[a]	Color	μ_{eff} (20°C)	Other data	Refs.
$(Cp_2Ti \cdot bipy)^+Cl^-$	Blue	1.67^T	Visible; conductance; IR	C68, C72
$(Cp_2Ti \cdot bipy)^+PF_6^-$	Purple	—	Conductance	G30
$(Cp_2Ti \cdot 4,4'\text{-}bipy)^+Cl^-$	Green	1.61^T	Visible	W11
$(Cp_2Ti \cdot phen)^+Cl^-$	Mauve	1.67^T	Visible; conductance; IR	C68, C72
$(Cp_2Ti \cdot pic)^+Cl^-$	Fawn	1.67^T	Visible; conductance; IR	C68, C72
$(Cp_2Ti \cdot phm)^+Cl^-$	Pale blue	1.78^T	Visible; conductance; IR	C68, C72
$(Cp_2Ti \cdot en)^+Cl^-$	Blue	—	—	G30
$(Cp_2Ti \cdot en)^+PF_6^-$	Blue	—	Conductance	G30
$(Cp_2Ti \cdot [(CH_3)_2en])^+PF_6^-$	Blue	—	Conductance	G30
$Cp_2TiCl(NH_3)$	Green	—	—	G30
$Cp_2TiCl(NH_2CH_3)$	Green	—	—	G30
$Cp_2TiCl(NH_2C_2H_5)$	Green	—	ESR	G30
$CpTiCl(NH_2C_3H_5)$	Green	—	—	G30
$Cp_2TiCl(NH_2C_6H_5)$	Green	—	—	G30
$Cp_2TiCl(py)$	Red to brown	1.72^T	Visible; ESR	G30, W11
$[Cp_2Ti(py)_2]^+ B(C_6H_5)_4^-$	Green	1.72^T	Visible	C48
$[Cp_2Ti(NCCH_3)_2]^+ B(C_6H_5)_4^-$	Blue	1.74^T	Visible	C48
$Cp_2TiCl[P(CH_3)_2C_6H_5]$	Green	—	ESR	G30
$Cp_2TiCl[PCH_3(C_6H_5)_2]$	Green	—	—	G30
$(Cp_2TiCl)_2 \cdot diphos$	Green	—	ESR	G30

[a] bipy, 2,2'-Bipyridyl; phen, phenanthroline; pic, 2-picolylamine; phm, o-phenylenediamine; en, ethylenediamine; py, pyridine; diphos, $(C_6H_5)_2PCH_2CH_2P(C_6H_5)_2$.

[b] Superscript T implies that μ_{eff} has been measured at several temperatures.

have prepared a great range of titanium(III) compounds using water as solvent and all (see Table VI-7) are green or blue in color and are readily oxidized in air.

The carboxylato derivatives, Cp_2TiO_2CR, were precipitated from aqueous solution using Cp_2TiCl and the sodium salt of the appropriate acid (C64). The monomeric compounds sublimed readily under reduced pressure and had magnetic moments close to the spin-only value and independent of temperature. In the infrared spectra the separations between symmetrical and unsymmetrical O–C–O stretching frequencies were small (60–100 cm^{-1}) suggesting symmetrically bonded carboxyl groups.

$$Cp_2Ti(O_2CR)$$

The thiocarboxylato derivatives, $Cp_2TiSOCR$ (R = H, CH_3, CH_3CH_2, $CH_3(CH_2)_{16}$, or C_6H_5), have been prepared similarly as dark green, sublimable monomers (C66), structurally similar to the carboxylates. The unusual four-membered ring formed by the ligand and the metal in these compounds was also found in the dithiocarbamates (C74), $Cp_2TiS_2CNR_2$, and the xanthates (C75), Cp_2TiS_2COR. Because of the four-membered ring all these highly colored derivatives were found to be monomeric, volatile, and very soluble in organic solvents.

In a similar class is the sulfate (C67). Although less soluble than the mono-

$$(Cp_2Ti)_2SO_4$$

nuclear derivatives above, this pale green compound was sublimable (200°/ 10^{-4} mm) and magnetically dilute. The infrared-active sulfate bands were moved to higher frequency and split.

The deep blue acetylacetonate, obtained (C69) by vigorously stirring excess acetylacetone with an aqueous solution of Cp_2TiCl, is monomeric and extremely volatile. Its magnetic moment is independent of temperature. The yield of $Cp_2Ti(acac)$ is never more than 50%; the rest of the titanium is found in the red filtrate as the ion $[Cp_2Ti(acac)]^+$, which can be precipitated as the perchlorate or tetrafluoroborate (C69). This titanium(IV) ion was formed probably by oxidation with the HCl liberated during the reaction.

The carbonato complex $(Cp_2Ti)_2CO_3$, prepared from either sodium carbonate or bicarbonate as a pale blue solid, is antiferromagnetic (C67). The magnetic susceptibility could not be measured above 100°C owing to decomposition, when carbon dioxide was liberated leaving a deep blue insoluble residue which is apparently a polymer of titanocene oxide $[(Cp_2Ti)_2O]_n$. Its insolubility, involatility, and low magnetic moment (0.9 BM) suggest a highly polymerized structure, probably bridged through oxygen (C67).

The pale blue, hydrated tetraphenylborato derivative $(Cp_2Ti)_22B(C_6H_5)_4 \cdot 6H_2O$, prepared by addition of $NaB(C_6H_5)_4$ to Cp_2TiCl in water (C48) probably has the structure

$$\left[\begin{array}{c} \overset{\displaystyle H_2}{\underset{\displaystyle}{|}} \\ Cp\diagdown \quad O \quad \diagup Cp \\ Ti \quad\quad Ti \\ Cp \diagup \quad O \quad \diagdown Cp \\ \overset{\displaystyle |}{\underset{\displaystyle H_2}{}} \end{array} \right]^{2+} \quad 2B(C_6H_5)_4{}^- \cdot 4H_2O$$

since the tetraphenylborate anion is noncoordinating. In this way the tetrahedral arrangement about the metal would be preserved, while still explaining the presence of water molecules in two different environments as indicated by the infrared spectrum. There is antiferromagnetic interaction between titanium atoms, either via the bridging water molecules or via metal–metal bonding. The maximum susceptibility occurs at 365°K and at this temperature decomposition to the "titanocene oxide polymer," $[(Cp_2Ti)_2O]_n$, together with triphenylboron, benzene, and water begins, and χ_M increases sharply (C48).

The pseudohalides, Cp_2TiX (X = CN, NCS, NCSe, and NCO), have been prepared by precipitation from aqueous solutions of Cp_2TiCl and the alkali metal pseudohalide (C63). The purple cyanide is almost insoluble in all the common organic solvents, its mass spectrum shows peaks as high as m/e 612 and its magnetic moment is low, around 0.66 BM per titanium atom at room temperature (C72). The compound is, therefore, probably a higher oligomer such as a tetramer, with bridging cyanide ligands, either linear or of the three-center type.

$$Ti—C{\equiv}N—Ti— \quad\quad or \quad\quad \begin{array}{c} —Ti\text{-----}Ti— \\ \diagdown \quad \diagup \\ C \\ {\parallel\parallel} \\ N \end{array}$$

The structures of the isothiocyanate and isoselenocyanate are believed to be similar to the cyanide (C63, C72), but the intermolecular bonding is weaker since these derivatives sublime at a lower temperature than does the cyanide, they are more soluble in polar organic solvents and their mass spectra show no mass peaks above the m/e of the monomer. The pale green isocyanate differs in some respects from the other pseudohalides in that it is monomeric and shows no magnetic exchange between titanium atoms.

4. Alkylbis(cyclopentadienyl)titanium(III) Compounds

The thermal stability of compounds of the type, Cp_2TiR (where R = alkyl or aryl), appears to be so low that few have been isolated until recently. Since the titanium atom in complexes of this type prefers a pseudotetrahedral stereochemistry, the reason for the paucity of such compounds may be the poor bridging properties of most alkyl groups. Nevertheless, it is surprising that the methyl derivative, Cp_2TiCH_3, has not been isolated either as a dimer or as a solvated monomer. However, the dimethyltitanium anion $[Cp_2Ti(CH_3)_2]^-$, has been observed in solution (B91).

Apart from $Cp_2TiCH_2C_6H_5$ and $Cp_2Ti(h^2\text{-}C_5H_5)$ which are discussed elsewhere, the only alkyl derivative which has been claimed is the ethyl compound, $Cp_2TiC_2H_5$, which was mentioned in a paper in 1961 (C27). No details of its preparation or properties were given and its existence in an uncomplexed form must be in doubt. A complex of this ethyl derivative with magnesium chloride and ether has been isolated recently (B43) and is described below.

The reactions of alkylmagnesium halides with Cp_2TiCl_2 or Cp_2TiCl have assumed some importance in recent years in relation to nitrogen fixation and olefin polymerization. These reactions are highly dependent on solvent and temperature.

In methylene dichloride at $-30°$, Long and Breslow (L23) showed in 1960 that it was possible to obtain the mono(ethyl)titanium(IV) compound, $Cp_2Ti(C_2H_5)Cl$, from Cp_2TiCl_2 using 1 mole of C_2H_5MgCl. In the usual ethereal solvents more complex processes occur. On addition of 1 mole of alkylmagnesium halide to Cp_2TiCl_2 in THF at low temperature $(-70°)$, reduction to green-brown $(Cp_2TiCl)_2$ occurs (B90). With excess Grignard reagent (or alkyllithium) at this temperature, the color changes to dark brown as further changes occur. Brintzinger (B91) has detected by ESR spectroscopy the formation of complexes of the type $(Cp_2TiR_2)^-$ [R = CH_3, C_2H_5 or $CH(CH_3)_2$], under these conditions. At room temperature only the methyl derivative is stable, the others decomposing with liberation of olefin probably giving hydrides of formula $(Cp_2TiH_2)^-$. None of these compounds has been isolated. This "reduced solution," which will now absorb nitrogen, will be dealt with further in Chapter VIII.

From the brown solution obtained by addition of a six- to eightfold excess of ethylmagnesium chloride to Cp_2TiCl_2 in diethyl ether at room temperature, a yellow crystalline complex formulated as $[(C_5H_5)_2TiC_2H_5]_2 \cdot 6MgCl_2 \cdot 7(C_2H_5)_2O$ precipitated (B43). It decomposed slowly at room temperature and rapidly above $60°$ giving equal volumes of C_2H_6 and C_2H_4. With diborane, violet Cp_2TiBH_4 was formed in 20% yield, while 1,3-pentadiene converted the complex to the allyl derivative, $Cp_2Ti \cdot \pi\text{-}C_5H_9$, although in low yield (B43). All these reactions are a good indication of the presence of the $Cp_2TiC_2H_5$

Table VI-5

Bis(cyclopentadienyl)titanium(III) Compounds Prepared in Aqueous Media

Compound	Color	μ_{eff} (20°C)[a]	Other data[b]	Refs.
Cp_2TiO_2CR				
R = H	Green	1.69^T	Conductance; IR; MW; MS; visible	C64
R = CH_3	Royal blue	1.71^T	Conductance; IR; MW; MS; visible	C64
R = $CH_3(CH_2)_8$	Royal blue	1.72^T	Conductance; IR; MW; MS; visible	C64
R = $CH_3(CH_2)_{16}$	Royal blue	1.55^T	Conductance; IR; MW; MS; visible	C64
R = C_6H_5	Dark green	1.62^T	Conductance; IR; MW; MS; visible	C64, V32
$(Cp_2TiO_2C)_2R$				
R = $-CH_2CH_2-$	Gray-blue	1.66^T	Conductance; IR; MW; MS; visible	C64
R = $-CH=CH-$ cis	Pale green	1.61^T	Conductance; IR; MW; MS; visible	C64
R = $-CH=CH-$ trans	Mauve	1.68^T	Conductance; IR; MW; MS; visible	C64
$Cp_2TiO(S)CR$				
R = H	Dark green	1.64^T	Conductance; IR; MW; visible	C66
R = CH_3	Dark green	1.74^T	Conductance; IR; MW; visible	C66
R = CH_3CH_2	Dark green	1.64^T	Conductance; IR; MW; visible	C66
R = $CH_3(CH_2)_{16}$	Dark green	1.55^T	Conductance; IR; MW; visible	C66
R = C_6H_5	Teal blue	1.71^T	Conductance; IR; MW; visible	C66
$Cp_2TiS_2CNR_2$				
R = H	Pale green	1.70^T	IR; MW; visible	C73, C74
R = CH_3	Blue green	1.69^T	IR; MW; visible	C73, C74, L6
R = CH_3CH_2	Deep green	1.74^T	IR; MW; visible	C73, C74

Compound	Color	μ_{eff}	Methods	Ref.
R = CH$_3$(CH$_2$)$_2$	Deep green	1.70^T	IR; MW; visible	C73, C74
R = CH$_3$(CH$_2$)$_3$	Deep green	1.53^T	IR; MW; visible	C73, C74
R = CH$_3$(CH$_2$)$_4$	Deep green	—	IR; MW; visible	C73, C74
R$_2$ = (CH$_2$)$_5$	Deep green	1.78^T	IR; MW; visible	C73, C74
Cp$_2$TiS$_2$COR				
R = CH$_3$	Light blue	1.74^T	IR; MW	C75
R = CH$_3$CH$_2$	Bright blue	1.64^T	IR; MW	C75
R = CH$_3$(CH$_2$)$_2$	Bright blue	1.60^T	IR; MW	C75
R = CH$_3$(CH$_2$)$_3$	Dark blue	1.74^T	IR; MW	C75
R = CH$_3$(CH$_2$)$_4$	Dark blue	1.76^T	IR; MW	C75
(Cp$_2$Ti)$_2$SO$_4$	Pale green	1.64^T	Conductance; IR; MW	C67
(Cp$_2$Ti)$_2$CO$_3$	Pale blue	0.8^T	Conductance; IR; MW; visible	C67
[(Cp$_2$Ti)$_2$6H$_2$O]$^{2+}$2B(C$_6$H$_5$)$_4$$^-$	Pale blue	1.4^T	Conductance; IR; MW; visible	C48
Cp$_2$Ti(acac)	Deep blue	1.74^T	IR; MW; visible	C69
Cp$_2$TiCN	Purple	0.6^T	IR; MS; visible	C63
Cp$_2$TiNCS	Red-brown	0.6^T	IR; MS; visible	C63
Cp$_2$TiNCSe	Brown	0.88^T	—	C70
Cp$_2$TiNCO	Green	1.64^T	IR; MS; visible	B100, C63
Cp$_2$Ti(picrate)	Brown	2.3	—	W24

[a] Superscript T implies that μ_{eff} has been measured at several temperatures.
[b] Most of these compounds absorb in the visible at 10–11, 13–14, and 16–17 kK (C70).

217

species, probably in a dimeric form since the complex is diamagnetic. On stirring the complex with nitrogen a titanium nitride was formed and reaction with carbon monoxide led to the known dicarbonyl, $Cp_2Ti(CO)_2$.

5. Allylbis(cyclopentadienyl)titanium(III) Compounds

Allylbis(cyclopentadienyl)titanium(III) has been prepared by reaction of allylmagnesium chloride with Cp_2TiCl_2, or with Cp_2TiCl, e.g.,

$$Cp_2TiCl_2 + 2C_3H_5MgCl \longrightarrow Cp_2TiC_3H_5 + 2MgCl_2 + (C_3H_5) \cdot$$

Methylallyl and phenylallyl derivatives have been prepared in this way (M13–M15). An alternative synthesis of substituted allyl derivatives of Cp_2Ti^{III} involves reaction of Cp_2TiCl_2 with 2 moles of propylmagnesium halide in the presence of an excess of a suitable diene (M15). With *tert*-butylmagnesium

$$Cp_2TiCl_2 + C_nH_{2n-2} + 2C_3H_7MgBr \longrightarrow$$
$$Cp_2Ti(C_nH_{2n-1}) + 2MgClBr + \tfrac{3}{2}C_3H_6 + \tfrac{1}{2}C_3H_8$$

bromide a lower yield of allyl derivative was obtained (M15). Martin and Jellinek proposed that the hydride $(Cp_2TiH_2)^-$, probably coordinated to magnesium, is the active intermediate which forms the allyl derivative by addition of Ti–H across the diene. These views on the hydride intermediate appear to be confirmed by later ESR work (B91, M25). Most of the alkyl-substituted allyl derivatives shown in Table VI-6 have been prepared from the appropriate diene (M15).

On the basis of the band in the infrared spectrum of the allyl derivatives around 1500 cm^{-1}, it was proposed that the allyl ligand was π-bonded to the titanium (M13), a view which was later confirmed by an X-ray crystal structure investigation on the 1,2-dimethylallyl homolog (H6).

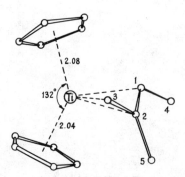

Fig. VI-5. Structure of $Cp_2Ti(1,2\text{-dimethylallyl})$ (H6).

Table VI-6

Allylbis(cyclopentadienyl)titanium(III) Compounds

R in Cp$_2$TiR	Color	μ_{eff} (20°C)	Other data	Refs.
Allyl	Violet	1.65	M.p. 111°–112°(dec); visible	M14, M15
1-Methylallyl	Blue-violet	1.67	M.p. 96.5°–97°(dec); visible	K58, M14, M15
2-Methylallyl	Purple	1.71	M.p. >114°(dec); visible	M14, M15
1,3-Dimethylallyl	Blue	1.70	M.p. 91°–91.5°(dec); visible	M14, M15
1,1-Dimethylallyl	Black	1.48	M.p. >51°(dec);	M14, M15
1,2-Dimethylallyl	Violet	—	M.p. 70.5°–71°(dec); visible	M14, M15
1,2,3-Trimethylallyl	Violet	—	M.p. 73.5°; visible	M15
1-Methyl-3-ethylallyl	Blue	—	M.p. 72°; visible	M15
1-Phenylallyl	—	—	Visible	M15
2-Phenylallyl	—	—	Visible	M15

The distances from titanium to the center of gravity of the allyl plane is 2.14 Å. This plane is not perpendicular to the line joining its center with the titanium atom; the observed angle is 55.9°. The three carbon atoms of the allyl group are symmetrically disposed around the titanium atom, the bond lengths C-1–C-2 and C-2–C-3 being very similar and close to the values observed in π-allylpalladium chloride (S54) (1.36 and 1.40 Å).

These allyl derivatives are active homogeneous catalysts for the hydrogenation of olefins and acetylenes (M12). At room temperature and atmospheric pressure the 1-methylallyl derivative absorbs hydrogen smoothly giving a gray-green low-valent titanium compound believed to be a hydride of the type $(Cp_2TiH)_n$ (M11, M12), which is probably the active catalyst (see Chapter V, Section C). Unsaturated compounds containing chlorine, carbonyl, or cyanide groups deactivate the catalyst (M12).

Shilov and co-workers (K58) have found that solutions of Cp_2TiCl_2 + C_2H_5MgBr in ether in a sealed system form the violet methylallyl derivative, π-crotylbis(cyclopentadienyl)titanium(III), apparently by combination of 2 moles of ethylene to give the crotyl ligand. The compound is identical with that prepared by Martin and Jellinek (M14, M15).

6. Bis(cyclopentadienyl)ethynyltitanium(III) Compounds

Bis(cyclopentadienyl)titanium(III) chloride reacts readily with acetylenic Grignard reagents and alkali metal salts. The phenylethynyl derivative is a deep green crystalline compound (T19), only slightly soluble in even the most polar solvents. Teuben and de Liefde Meijer suggest that the compound is dimeric from mass spectral evidence and has one of the structures shown below.

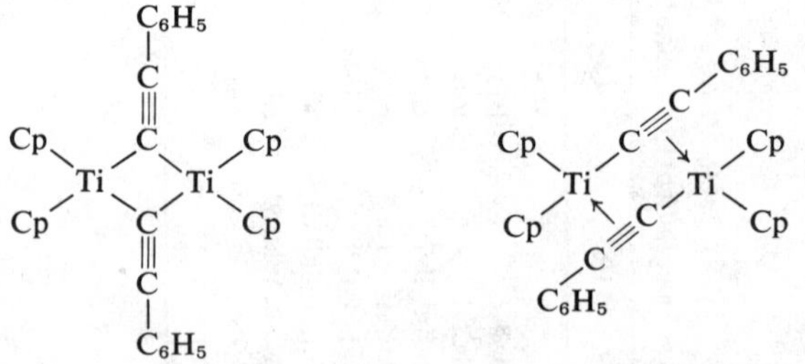

Both these structures should show the C≡C stretching band in the infrared spectrum around 2000 cm^{-1}. The absence of such a band, coupled with the extreme insolubility of the compound, suggests that the triple bond is associated in a more complex manner than a bridging dimer would require.

The related complexes, black $(Cp_2TiC≡CH)_n$ and blue $(Cp_2TiC≡CTiCp_2)_n$, have been isolated from reaction mixtures of the mono- and dimagnesium bromides of acetylene with Cp_2TiCl (W1). Both are air-sensitive and, undoubtedly, polymeric. Like the phenylethynyl derivative they show no acetylenic bands in their infrared spectra.

7. Arylbis(cyclopentadienyl)titanium(III) Compounds

Bis(cyclopentadienyl)titanium phenyl and substituted-phenyl derivatives have been prepared by reaction of the arylmagnesium bromide with Cp_2TiCl in ether at 0° to $-25°$ (T20, T21). The $3\text{-}CH_3C_6H_4$ and $4\text{-}CH_3C_6H_4$ analogs decomposed below 20° and could not be isolated. All were monomeric, so that the titanium has been forced to depart from its preferred tetrahedral stereochemistry. As a result the thermal stability of the aryl compounds is low, although increasing markedly with the degree of substitution of the aryl groups. In fact, the 2,6-dimethylphenyl derivative could be sublimed at 160°/0.1 mm and the 2,4,6-trimethylphenyl compound could be distilled without decomposition (see Table VI-7).

The aryl ligands underwent the usual reactions of σ-bonded groups, in that with HCl at $-78°$ the ligand was removed as the aromatic hydrocarbon, leaving Cp_2TiCl, which was converted by excess HCl at higher temperatures to Cp_2TiCl_2 with liberation of hydrogen. Similarly bromine at $-78°$ gave the aryl bromide together with Cp_2TiBr_2.

At low temperatures ($-80°$ to $-100°$) all but the $2,6\text{-}(CH_3)_2C_6H_3$ and $2,4,6\text{-}(CH_3)_3C_6H_2$ derivatives absorb gaseous nitrogen (T18), giving the binuclear complexes $(Cp_2TiR)_2N_2$, which are deep blue in color and absorb in the visible region around 17 kK (see Table VIII-3). The nitrogen could be reduced by a strong reducing agent and recovered as hydrazine or ammonia after hydrolysis (see Chapter VIII).

Table VI-7
Aryl and Benzyl Derivatives of Bis(cyclopentadienyl)titanium(III)[a]

Compound	Color	Decomposition temp. (°C)	μ_{eff} (20°C)	Other data
$Cp_2TiC_6H_5$	Green	29	1.58	Visible; ESR; MW
$Cp_2Ti(2\text{-}CH_3C_6H_4)$	Green	69[b]	1.81	Visible; ESR; MW
$Cp_2Ti(3\text{-}CH_3C_6H_4)$	Green	<20	—	Visible; ESR
$Cp_2Ti(4\text{-}CH_3C_6H_4)$	Green	<20	—	Visible; ESR
$Cp_2Ti[2,6\text{-}(CH_3)_2C_6H_3]$	Green	180[b]	1.66	ESR; MW
$Cp_2Ti[2,4,6\text{-}(CH_3)_3C_6H_2]$	Green	78[b]	1.72	ESR; MW
$Cp_2TiC_6F_5$	Brown-purple	106	1.92	Visible; ESR; MW
$Cp_2TiCH_2C_6H_5$	Brown	43	—	Visible; ESR; MW

[a] Data from ref. (T21).
[b] Melting point.

8. Alkoxides and Thioalkoxides of
Bis(cyclopentadienyl)titanium(III)

The preparation of the alkoxides $(Cp_2TiOR)_2$ ($R = CH_3$, C_2H_5, and *tert*-C_4H_9), was reported by Coutts and Wailes (C70) and Lappert and Sanger (L6) (where $R = C_2H_5$ and C_6H_5), using somewhat different methods. Elimination of alkali metal chloride from a mixture of Cp_2TiCl with sodium methoxide, ethoxide, or *tert*-butoxide (C70) or lithium ethoxide (L6) or, alternatively, elimination of dimethylamine from $[Cp_2TiN(CH_3)_2]_2$ and ethanol or phenol (L6) gave the alkoxides and phenoxide as brown solids. Care must be taken in the amine elimination method not to use excess alcohol or solvolysis of cyclopentadienyl groups occurs.

The variation with temperature of the magnetic susceptibilities of the alkoxides clearly indicates some form of association, probably through bridging groups (C70, C72). The magnetic moment of the methoxide, which has been shown to be dimeric, varied from 1.38 BM at 333°K to 0.70 BM at 93°K.

Only one example of a thio alcohol derivative has been prepared to date (L6), namely $(Cp_2TiSC_6H_5)_2$, isolated as a magenta solid after amine elimination from $[Cp_2TiN(CH_3)_2]_2$ with HSC_6H_5. Magnetic interaction between the titanium atoms is strong since the compound is diamagnetic.

9. Amido, Phosphido, and Arsenido Derivatives of
Bis(cyclopentadienyl)titanium(III)

Several such compounds of general formula, Cp_2TiMR_2, have been prepared by treatment of Cp_2TiCl with the alkali metal salt of the appropriate disubstituted metalloid. These complexes are usually dimeric, bridging through the metalloid atoms. In the presence of excess $NaMR_2$ reagent, there is reasonable evidence from ESR that monomeric complex anions of the type $[Cp_2Ti(MR_2)_2]^-$, exist (K21).

$[Cp_2TiN(CH_3)_2]_2$ is a useful intermediate in the synthesis of other titanium derivatives (L5, L6). It reacts with metal hydrides to give titanium–metal-bonded complexes, e.g., $Cp_2TiMo(CO)_3Cp$; with thiophenol to give $(Cp_2TiSC_6H_5)_2$; with alcohols to give the alkoxides $(Cp_2TiOR)_2$ ($R = C_2H_5$ or C_6H_5). Carbon disulfide inserts into the dimethylamino–titanium bond (L5) to give a partially solvated form of the N,N-dimethyldithiocarbamate, $Cp_2TiS_2CN(CH_3)_2$. Although this compound is best prepared by other means (C74), this is an interesting example of an insertion reaction into a Ti–N bond. Isocyanatobenzene inserts in like manner giving a ureido derivative, $Cp_2TiN(C_6H_5)CON(CH_3)_2$, albeit in impure form (L6).

Although diphenylamido and diethylamido complexes could not be obtained, potassium pyrrolyl reacted readily with Cp_2TiCl_2 to give a red-brown solution with a 1:2:3:2:1 quintet ESR spectrum (K21) corresponding to coupling to two equivalent nitrogen nuclei in a complex of the type

$$\left[Cp_2Ti \left(N \overline{\underset{}{\bigcirc}} \right)_2 \right]^-$$

In 1966 Issleib and Häckert reported the formation of the titanium(III) phosphides believed to have the structure

$$\begin{array}{ccc} & R \quad R & \\ Cp & \diagdown P \diagup & Cp \\ \diagdown & Ti \diagdown \quad \diagup Ti & \diagup \\ Cp \diagup & \diagup P \diagdown & \diagdown Cp \\ & R \quad\quad R & \end{array}$$

$$R = C_2H_5, \; n\text{-}C_4H_9$$

by reduction of Cp_2TiCl_2 with 2 moles of lithium dialkylphosphide in tetrahydrofuran (I4). The phenyl and cyclohexyl derivatives could not be prepared in this way. The violet compounds are virtually diamagnetic owing to spin-spin interaction. With methyl iodide, Cp_2TiI (not isolated) was formed, together with dimethyldialkylphosphonium iodide.

An ESR study of similar systems was later carried out by Kenworthy, Myatt, and Todd (K20, K21). After addition of an excess of the alkali metal phosphides, $NaPR_2$ (R = CH_3 or C_6H_5), to Cp_2TiCl_2 in THF or dimethoxyethane, 1:2:1 triplets were observed, indicating interaction of the unpaired titanium(III) electron with two phosphorus nuclei. The complexes were formulated as the monomeric anions $[Cp_2Ti(PR_2)_2]^-$. Similar effects were

Table VI-8
Phosphido, Arsenido, and Amido Derivatives of Bis(cyclopentadienyl)titanium(III)

Compound	Color	Isolated	Other data	Refs.
$[Cp_2TiN(CH_3)_2]_2$	Brown	Yes	ESR; Diamagnetic	L5, L6
$[Cp_2TiPR_2]_2$	Violet	Yes	Diamagnetic	I5
R = C_2H_5, n-C_4H_9				
$[Cp_2Ti(PR_2)_2]^-$	Brown	No	ESR; Paramagnetic	K20, K21
R = CH_3, C_6H_5				
$[Cp_2Ti(AsR_2)_2]^-$	Brown	No	ESR; Paramagnetic	K21
R = C_6H_5				
$[Cp_2Ti(pyrrolyl)_2]^-$	Red-brown	No	ESR; Paramagnetic	K21

noticed with the diphenylarsenide complex (K21). A septet of lines was obtained corresponding to coupling to two equivalent arsenic nuclei in the anion.

10. Metal–Metal-Bonded Compounds Containing Bis(cyclopentadienyl)titanium(III)

Compounds containing metal–metal bonds between elements of Groups IVA and IVB have been prepared. Titanium(III)–tin and titanium(III)–germanium compounds were isolated as green paramagnetic solvated complexes, $Cp_2TiM(C_6H_5)_3 \cdot THF$, from the reaction between Cp_2TiCl and sodium tin triphenyl or lithium germanium triphenyl in THF (C65). The corresponding lead derivative was formed as a bright green solution at low temperature, but deposition of lead occurred rapidly at room temperature and the compound could not be isolated. The tin and germanium compounds were magnetically dilute. A Curie–Weiss law was followed, there being one unpaired electron per titanium atom. In chloroform solution these compounds eventually showed the PMR spectrum of the titanium(IV) compound, $Cp_2Ti(Cl)M(C_6H_5)_3$, formed apparently by abstraction of chloride from solvent (C65).

An ESR study in 1,2-dimethoxyethane of the system Cp_2TiCl_2 + (excess) $KM(C_6H_5)_3$ (M = Sn or Ge) (K17, K19) has shown the presence of the anionic complexes $\{Cp_2Ti[M(C_6H_5)_3]_2\}^-$. With M = Pb the complex in solution was $Cp_2TiPb(C_6H_5)_3$.

Titanium–tin- and titanium–molybdenum-bonded complexes have been obtained by an amine elimination procedure using $[Cp_2TiN(CH_3)_2]_2$ and metal hydrides (L6). The titanium–tin compound decomposed rapidly at

$$[Cp_2TiN(CH_3)_2]_2 + 2HSn(C_6H_5)_3 \longrightarrow 2Cp_2TiSn(C_6H_5)_3 + 2(CH_3)_2NH$$

$$[Cp_2TiN(CH_3)_2]_2 + 2CpMo(CO)_3H \longrightarrow 2Cp_2TiMo(CO)_3Cp + 2(CH_3)_2NH$$

Table VI-9
Metal–Metal-bonded Compounds of Bis(cyclopentadienyl)titanium(III)

Compound	Color	Other data	Refs.
$Cp_2TiSn(C_6H_5)_3 \cdot THF$	Green	One unpaired electron	C65
$Cp_2TiSn(C_6H_5)_3$	Green	Unstable	L6
$Cp_2TiGe(C_6H_5)_3 \cdot THF$	Green	One unpaired electron	C65, R21d
$Cp_2TiPb(C_6H_5)_3$	Green	Unstable, not isolated	C65, L6
$\{Cp_2Ti[Sn(C_6H_5)_3]_2\}^-$	—	ESR; not isolated	L6
$\{Cp_2Ti[Ge(C_6H_5)_3]_2\}^-$	—	ESR; not isolated	L6
$Cp_2TiMo(CO)_3Cp \cdot \frac{1}{2}THF$	Brown	μ_{eff} 1.74 BM/Ti	L6

room temperature and could not be studied further. The titanium–molybdenum complex, which could also be prepared from Cp_2TiCl and $NaMo(CO)_3Cp$, on the other hand, was more stable. It retained THF strongly (0.5 mole of THF per mole of titanium compound). The magnetic moment of the solvated form was 1.74 BM per titanium atom at 22°C. In the infrared spectrum four bands were observed in the metal carbonyl region, at frequencies very close to those observed in $[Mo(CO)_3Cp]_2$.

D. Tris(cyclopentadienyl)titanium(III)

Tris(cyclopentadienyl)titanium(III) was first prepared in 1960 by Fischer and Löchner (F17) in "low yield" from sodium cyclopentadienide and Cp_2-$TiCl_2$ [7.5% yield was obtained in a repeat of this method (S40)]. A much better yield (70%) was obtained (C8) from $(\pi\text{-}C_5H_5)_2TiCl$ and sodium cyclopentadienide. A recent preparation (S40) involved pyrolysis of tetra(cyclopentadienyl)titanium at 125°/10^{-3} mm to give a 41% yield of the green $(C_5H_5)_3Ti$ as a sublimate.

Over the range 100°–300°K the magnetic moment of this compound was found to be close to the spin-only value for a d^1 system, so there is no interaction between titanium atoms. In boiling benzene the compound is monomeric, in line with its high volatility and solubility.

Siegert and de Liefde Meijer proposed that one of the cyclopentadienyl ligands was σ-bonded on the basis of infrared data and the formation of 0.85 mole of C_5H_6 per mole of $(C_5H_5)_3Ti$ on treatment with HCl in toluene (S40). It is difficult to correlate the physical properties of tris(cyclopentadienyl)titanium (in particular its thermal stability) with such a structure. A recent crystal structure determination has shown (L23a) that the third cyclopentadienyl ligand is bonded through two carbon atoms forming a three-center, four-electron bond (Fig. VI-6).

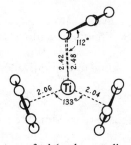

Fig. VI-6.   Structure of tris(cyclopentadienyl)titanium (L23a).

Under 150 atm pressure of carbon monoxide at 80°, tris(cyclopentadienyl)-titanium is converted to the dicarbonyl $Cp_2Ti(CO)_2$ (F17). It will also react immediately with dialkyl and diaryl disulfides, RSSR, the color changing from green to red, giving an equilibrium mixture of thiolatotitanium(IV) compounds of type $(C_5H_5)_nTi(SR)_{4-n}$ (C72).

E. Compounds of Zirconium(III)

It is surprising, at first sight, that there are so few organometallic zirconium(III) compounds known. In fact, there are no alkyl- or arylzirconium-(III) compounds known and no hafnium(III) species of any kind. Bis-(cyclopentadienyl)titanium(III) compounds are readily obtained (*vide supra*) by reduction of Cp_2TiCl_2 with zinc, Grignard reagents, aluminum, and a host of other reductants. On the other hand, Cp_2ZrCl_2 is reduced only by the more electropositive elements such as the alkali metals. It is certainly unaffected by aluminum. With magnesium in THF a reduction may possibly occur since a crimson by-product is formed, in addition to $Cp_2Zr(H)Cl$.

Attempts to prepare Cp_2ZrCl by reduction of Cp_2ZrCl_2 with stoichiometric amounts of sodium, sodium amalgam, or sodium/naphthalene (W5) were not successful in that the red-brown polymeric products isolated were not discrete compounds. Cp_2ZrCl could not be made from $ZrCl_3$.

Both the first and best established zirconium(III) species were made by Issleib and Häckert (I4) by the reaction

$$2Cp_2ZrBr_2 + 4LiPR_2 \xrightarrow[\text{room temp.}]{\text{THF}} Cp_2Zr\underset{\underset{R_2}{\overset{|}{P}}}{\overset{\overset{R_2}{\overset{|}{P}}}{\diagdown\diagup}}ZrCp_2 + 4LiBr + R_2PPR_2$$

$$R = C_2H_5,\ n\text{-}C_4H_9$$

The molecular weight and diamagnetism of these red-brown compounds are in keeping with the proposed dimeric structure. They are not readily soluble, but they do sublime under high vacuum. While thermally quite stable, they are oxidized readily by air, iodine (quantitatively), or methyl iodide, e.g.,

$$(Cp_2ZrPR_2)_2 + 5I_2 \longrightarrow 2Cp_2ZrI_2 + 2R_2PI_3$$

Equilibration of zirconocene [prepared by the method of Watt and Drummond (W17)] and Cp_2ZrCl_2 in refluxing toluene (W5) gave both a red and a green compound. The red, benzene-soluble product was isolated as

$(Cp_2ZrCl)_2C_6H_6$, while the major green product, which was diamagnetic, pyrophoric, and insoluble in all solvents, was assigned the structure $[Cp(C_5H_5)ZrCl]_n$, mainly from its infrared spectrum. The presence of σ-bonded cyclopentadienyl groups is to be expected considering the complexity of the structure of zirconocene.

Reduction of the μ-oxo compound $(Cp_2ZrCl)_2O$, with sodium/naphthalene gave a μ-oxozirconium(III) species as a dark brown solid (W5). The naph-

$$2(Cp_2ZrCl)_2O + 4Na/C_{10}H_8 \xrightarrow[60°, \ 1 \ hr]{THF} [(Cp_2Zr)_2O]_2 \cdot C_{10}H_8 + 4NaCl$$

thalene could not be expelled from the complex by heating at 110° *in vacuo* for 5 hr. The compound is weakly paramagnetic, the cyclopentadienyl protons are found as a broad band at $\delta 5.88$ in the PMR spectrum.

The reaction of $Cp_2Ti(CH_3)_2$ with hydrogen is known to give bis(cyclopentadienyl)titanium(III) hydrides (B42) or titanocene (C38). With $Cp_2Zr(CH_3)_2$ and hydrogen (W8) in light petroleum at 110°–120°, the crimson product $[Cp_2Zr(CH_3)]_2$, formed. It was suggested that the methyls may be the bridging groups.

Chapter VII

Organometallic Compounds of Titanium(II), Zirconium(II), and Hafnium(II)

A lowering in the oxidation state in compounds of these metals can be achieved by numerous means. However, for reduction beyond the trivalent state, alkali metals, their amalgams, naphthalene complexes, pyridyls, and the like are most effective. Magnesium has been reported to reduce titanium(IV) to titanium(II) (V11), but with Cp_2ZrCl_2 in THF, the hydride, $Cp_2Zr(H)Cl$, is the major product (W2). Thermal degradation of alkyl or aryl derivatives or, in some cases, their reaction with hydrogen or carbon monoxide will often give metal(II) compounds.

Smirnova and Gubin (S53) have shown by polarography that the fluoride complex is the most difficult of the bis(cyclopentadienyl)titanium dihalides to reduce to the divalent state. Substitution of alkyl groups on the cyclopentadienyl rings also impeded reduction. Replacement of a cyclopentadienyl with a third chloride has little effect on the polarographic behavior of the compound (compared to the dichloride) except to permit the reduction of the metal to the monovalent state.

It is a general trend in the chemistry of this group of elements that the number of compounds known diminishes with decrease in oxidation number and increase in atomic weight. This chapter therefore deals predominantly with compounds of titanium and, in particular, the problem of titanocene.

The "titanocene problem" arises from the ease of hydrogen abstraction from a cyclopentadienyl ring on lowering the oxidation state of the metal to two. Thus, when $CpTiCl_3$ is reduced with magnesium (V18), a hydride species is formed, the hydrogen having originated from the C_5H_5 ring. The corresponding pentamethylcyclopentadienyl compound fails to form a hydride under similar conditions (V18). Ready exchange of the C_5H_5 ring hydrogens with deuterium (B43, M17) is also attributed to the facile abstraction of the hydrogens from the ring.

A. Titanocene

Since the initial report on the preparation of titanocene by Fischer and Wilkinson (F13) in 1955, much work has been done on this and the corresponding zirconium and hafnium compounds to elucidate their structures. The considerable controversy that still persists concerning their molecular configuration has arisen from the unusual behavior of the cyclopentadienyl groups on these metals in a low-oxidation state. Hydrogen abstraction from the cyclopentadienyls and dimerization through these ligands are the basic complications which occur in these compounds. Failure to form suitable crystals has excluded the application of X-ray analysis to this problem. Further, the absorptions in the PMR spectrum of titanocene, the most soluble compound, have been broad (S4) and have provided little useful information other than to negate the presence of two symmetrically π-bonded cyclopentadienyl groups in the sample studied.

The literature on titanocene implies that the nature of the compound isolated varies with the method of preparation; thus, the parent titanocene $(Cp_2Ti)_2$, its rearrangement products $[(C_5H_5)(C_5H_4)TiH]_x$, and $[(C_5H_5)-(C_5H_4)TiH]_2$ can all be formed. On this account an attempt has been made to describe the various preparative procedures employed together with the reactions undergone by that particular form of "titanocene." A difficulty encountered here has been the failure of some authors to indicate the method of syntheses of the "titanocene" used in their studies.

Several authors (B17, B53, M33, M41, R42, S21, S38) have assumed a ferrocene-type structure for titanocene and have attempted to calculate the electronic energy states for titanocene (and zirconocene) in order to account for their diamagnetism. It has been contended that the high reactivity and, hence, instability of the monomeric (Cp_2Ti) species lies in its carbenoid-type behavior (B92, V19).

1. Methods of Preparation

a. FROM TiCl₂ [*Method of Fischer and Wilkinson (F13)*]

The product obtained from the reaction between titanium(II) dichloride and sodium cyclopentadienide was found to be dependent on the purification procedure.

The titanocene species $[(C_5H_5)_2Ti]$ was diamagnetic in solution and also in the solid state between 100° and liquid nitrogen temperature. Unlike other forms of titanocene, this was very soluble in light petroleum (0.1 M solution) and other hydrocarbon solvents. It failed to form a solvate in liquid ammonia.

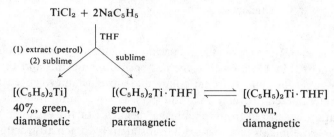

Oxidation with concentrated hydrochloric acid yielded Cp_2TiCl_2, while with ferrous chloride in THF an indigo-colored titanium(III) complex was formed which could be converted in the presence of water and a little nitric acid to the yellow $[Cp_2Ti(OH)]^+$ species.

The green, crystalline $[(C_5H_5)_2Ti \cdot THF]$, although stable at $-80°$, was transformed spontaneously and completely to a brown form in 30 min even at $0°$. The brown isomer was stable up to its melting point of $81° \pm 0.5°$, when it reconverted to the green form. Bercaw et al. (B43) noted that their titanocene dimer $(Cp_2Ti)_2$ gave an initially bright green THF adduct (dark green in the solid state), which degraded in about 30 min to a brown product considered to be an alkoxide formed by ring opening of the THF.

Attempts by Watt and Baye to repeat this synthesis from $TiCl_2$ were not successful (B19, W15). However, Brantley (B76) in 1961 claimed in the patent literature to have made titanocene by this method and found it to react rapidly with HBr in benzene to form Cp_2TiBr_2.

b. FROM Cp_2TiCl_2 AND SODIUM/NAPHTHALENE (*Method of Watt, Baye, and Drummond*)

In 1965, Watt, Baye, and Drummond (W16) reported the synthesis of titanocene by the reaction

$$Cp_2TiCl_2 + 2Na/C_{10}H_8 \xrightarrow[25°]{THF} \text{"titanocene"} + 2C_{10}H_8 + 2NaCl$$

After an initial extraction with benzene to remove the NaCl, the naphthalene, which need only be used in catalytic amounts (W17), was sublimed at about $100°$ from the green titanocene product. Watt and Drummond (D38, W17, W18) have extended this reaction to include the preparation of zirconocene and hafnocene.

This titanocene is dimeric in benzene. Its infrared spectrum suggests that the cyclopentadienyl groups are not both symmetrically π-bonded. Further, there are no bands attributable to Ti–H stretching vibrations, although these are found in several other isomers of titanocene. The absence of an hydridic group was confirmed by its reaction with methanol and anhydrous HCl

(W11). Methanol, which oxidizes the metal only to the trivalent state, reacts further with the reaction product to form $CpTi(OCH_3)_2$. With HCl, the reaction is

$$(C_{10}H_{10}Ti)_2 + 2HCl \xrightarrow{\text{fast}} 2/n(C_{10}H_{10}Ti^{III}Cl)_n + H_2$$
violet, soluble, paramagnetic

$$+ 2HCl \xrightarrow{\text{slow}} 2/m(C_{10}H_{10}Ti^{IV}Cl_2)_m + H_2$$
green, insoluble

The products of this reaction with HCl appear to be analogous to those obtained by Salzmann and Mosimann (S4) with titanocene obtained using sodium sand as reductant. Although this titanocene is diamagnetic (W16), the PMR spectral bands in toluene-d_8 are quite broad and lie between $\delta 5.19$ and 6.79 at 30° and move downfield to between $\delta 5.66$ and 7.56 at 90° (W11). It is soluble to some extent in aromatic solvents and THF, but nearly insoluble in saturated hydrocarbons. It is thermally stable up to 200°, but reacts with water, oxygen, halogenated solvents, carbon disulfide, sulfur dioxide, dimethyl sulfide, and nitromethane (M32b, W16).

Reduction of Cp_2TiCl_2 with sodium/naphthalene in the presence of diphenylacetylene (W18) gave 1,1-bis(π-cyclopentadienyl)-2,3,4,5-tetraphenyl-titanacyclopentadiene. This has also been prepared by Vol'pin et al. (V19, V23) by treatment of $TiCl_4$ with 4 moles of sodium cyclopentadienide in the presence of diphenylacetylene.

The oxidation of titanocene with boron halides has been investigated by Nöth and co-workers (S14). The black-violet products (dec. 300°) were ob-

$$(C_{10}H_{10}Ti)_2 + 2XBY_2 \xrightarrow[\text{room temp.}]{\text{ether}} 2/n[C_{10}H_{10}Ti(X)BY_2]_n$$

$Y = C_6H_5$, $X = Cl$; $Y = X = Cl$ or Br

tained in 75–80% yield. With Cp_2TiCl_2, this titanocene reacts to give a red, antiferromagnetic titanium(III) dimer, presumably isomeric with the green $(Cp_2TiCl)_2$ (W11). An oxidative addition reaction apparently proceeds also with hydrogen, as this form of titanocene, and for that matter zirconocene, is an alkene hydrogenation catalyst (W10).

This titanocene reacts with 2,2'-bipyridyl to give a dark blue, antiferromagnetic species of composition $[(C_{10}H_{10}Ti)bipyridyl]_x$ (see Section E on bipyridyl adducts). With $[CpCr(CO)_3]_2$, a compound very soluble in benzene and analyzed as $[C_{10}H_{10}Ti][CpCr(CO)_3]$ was obtained (M40).

With triethylaluminum titanocene reacts smoothly to give the dimer $[(C_5H_5)(C_5H_4)TiAl(C_2H_5)_2]_2$ and ethane as the only gaseous side product (W4). This titanium–aluminum complex was considered to be identical with that prepared by Natta and co-workers (M44, M46, N14, N15, N18) from

Cp_2TiCl and excess triethylaluminum and whose crystal structure was measured by Corradini and Sirigu (C45) (see Chapter V, Section C). Tebbe and
Guggenberger believe that this compound contains Ti–H–Al bridges (T14).

c. FROM Cp_2TiCl_2 AND SODIUM SAND (*Method of Salzmann and Mosimann*) OR SODIUM AMALGAM (*Method of Shikata and co-workers*)

Salzmann and Mosimann (S4) modified the procedure of Watt *et al.* for
making titanocene by utilizing sodium sand (10% excess) instead of its
naphthalene complex, as this eliminated the need to sublime the naphthalene
from the product. A consequence of this has been the preparation of other
titanocene species.

$$2Cp_2TiCl_2 + 4Na \xrightarrow[\text{room temp., 4–8 days}]{\text{toluene or THF}} \text{“titanocene”} + 4NaCl$$

The PMR spectrum of the dimeric product shows at least four absorptions
with chemical shifts in the $\delta 5.9$ to 6.7 region (S4). Further, the infrared spectrum is considerably more complicated than expected for a sandwich bis(π-
cyclopentadienyl)titanium structure; in fact, it resembles that of compounds
known to contain both π- and σ-bonded cyclopentadienyl groups.

In THF with anhydrous HCl this titanocene first forms a violet, soluble,
paramagnetic compound of composition $(C_{10}H_{10}TiCl)_{10}$ (S4) and then in a
slower reaction with more HCl, the corresponding green, insoluble titanium-
(IV) dichloride. The infrared spectra of both of these compounds are strongly
indicative of the presence of more than one type of cyclopentadienyl ring.

Brintzinger and Bercaw (B93) maintain on two counts that the compound
is, in fact, the hydride $[Cp(C_5H_4)TiH]_2$. First, on reaction with DCl nearly
pure DH is formed in quantities required by the stoichiometry of the reaction.

$$[Cp(C_5H_4)TiH]_2 + 2DCl \longrightarrow 2/n[Cp(C_5H_4)TiCl]_n + 2DH$$

Second, the infrared band at 1228–1230 cm^{-1} can best be assigned to bridging
hydrides.

The peak of highest m/e in the mass spectrum is due to the parent dimer
$[Cp(C_5H_4)TiH]_2$, which fragments by elimination of hydrogen to the species
(parent-2H) and (parent-4H). No other metal-containing fragments giving
rise to peaks of relative intensity greater than five were reported (B93). It is
of interest to compare this mass spectrum with that reported by Takegami
and co-workers (T8). In a high-resolution spectral analysis these workers
found that the titanocene species prepared by the method of Shikata *et al.*
(S34) (using 0.1–0.2% sodium amalgam for 1.5 hr at 50°), did not vaporize
in the mass spectrometer as a dimer, although Shikata and co-workers had
established that their product was a dimer in solution. The monomer peak in

the mass spectrum was very intense as was that due to the $TiC_5H_5^+$ ion. Species of mass number higher than that of the monomer were attributed to the oxygen-containing ions $(C_5H_5)_2TiO^+$, $(C_5H_5)_2TiOH^+$, and $(C_5H_5)_4Ti_2O^+$. It is worthy of note that the infrared spectrum given by Takegami et al. (T8) differs from that published by Salzmann and Mosimann (S4) or Brintzinger and Bercaw (B93).

In a recent communication, van Tamelen and co-workers made the observation (V8) that several species of titanocene are formed in sequence on reaction of Cp_2TiCl_2 with sodium in toluene under argon at room temperature. The reaction pathway proposed involved the initial relatively rapid reduction of Cp_2TiCl_2 to $(Cp_2TiCl)_2$ and then the slow (6–10 days) reduction to the metastable compound $(Cp_2Ti)_n$ ($n = 1$–2) in which both cyclopentadienyls are symmetrically π-bonded. The species $(Cp_2Ti)_n$ ($n = 1$–2) was first prepared and identified by Brintzinger's group (B43, M21) and its chemistry is described in Section A,1,d on the reaction of $Cp_2Ti(CH_3)_2$ with hydrogen. Surprisingly $(Cp_2Ti)_n$ ($n = 1$–2), although requiring 6–10 days for preparation by the above procedure, is said to be highly unstable at room temperature under argon and decomposes to the isomeric hydride $[Cp(C_5H_4)TiH]_x$, in which the Ti–H stretching vibrations occur at 1815 and 1960 cm^{-1} [ν(Ti–D) = 1305 and 1355 cm^{-1}]. On stirring this hydride in toluene over sodium for extended periods or on heating it to 100° for several hours, the final stable isomer $[Cp(C_5H_4)TiH]_2$, was obtained.

The number of isomers of titanocene formed in the reaction of Cp_2TiCl_2 with sodium provides a ready explanation for the inconsistent mass and infrared spectral results reported. Furthermore, the variable yields of Cp_2Ti^{III} or Cp_2Ti^{IV} compounds obtained from oxidative addition reactions may well be due to the same reason. Several such reactions are (K44, R17)

$$\text{Titanocene} + X—X \longrightarrow Cp_2TiX_2$$
$$18\text{–}37\% \text{ yield}$$
$$X = SCH_3, SC_6H_5, SeC_6H_5, \text{ and } I$$

$$\text{Titanocene} + Cp_2Ti(O_2CC_6H_5)_2 \longrightarrow Cp_2Ti(O_2CC_6H_5)$$
$$42\%$$

A 70% yield of an insoluble, yellow-brown, diamagnetic titanocene peroxide was obtained on oxidation of titanocene in toluene or THF with molecular oxygen (S3). Nitric oxide did not form a nitrosyl compound with titanocene, but led to a brown, weakly paramagnetic, compound containing the cis-hyponitrite group (S3). If the "titanocene" used in these reactions was a hydride, then an explanation of the results is not obvious. One may speculate that the nitric oxide may have inserted into a Ti–C σ bond by analogy with its behavior with $Cp_2Zr(CH_3)_2$, when $Cp_2Zr(CH_3)ON(NO)CH_3$ is formed (W8).

Takegami and co-workers (T8) have treated this titanocene with methyl-acetylene to form a 1,1-bis(π-cyclopentadienyl)dimethyltitanacyclopentadi-ene complex.

Shikata *et al.* (S32, S34) have shown that titanocene prepared using sodium sand or amalgam behaved as an hydrogenation catalyst for olefins. It can also initiate polymerization of styrene (S32).

The ultraviolet absorption spectrum has been recorded by Salzmann and Mosimann (S4), who made the tentative suggestion that the band at 12,200 cm^{-1} may be associated with a Ti–Ti bond.

d. FROM $Cp_2Ti(CH_3)_2$ AND HYDROGEN (*Method of Clauss and Bestian and Brintzinger et al.*)

In 1962 Clauss and Bestian (C38) observed that the facile reaction between $Cp_2Ti(CH_3)_2$ and hydrogen led to a titanocene species. The reaction is pre-

$$Cp_2Ti(CH_3)_2 + H_2 \xrightarrow[20°]{hexane} 1/n(C_{10}H_{10}Ti)_n + 2CH_4$$

ceded by an induction period, short in aliphatic solvents and longer (several hours) in aromatic hydrocarbon solvents,* and proceeds via a short-lived, violet intermediate shown by Bercaw and Brintzinger (B41) to be the hydride $(Cp_2TiH)_2$. This violet hydride, which can be isolated only in the absence of solvent (B41), decomposes on warming to give the titanocene species. The hydride is unstable even below room temperature and interconverts into a more stable gray-green hydride of identical composition, but apparently of significantly increased molecular weight (B43, M21).

On stirring the polymeric gray-green hydride $(Cp_2TiH)_x$, at room temperature in toluene or diethyl ether, it dissolves to give a dark solution of the relatively unstable dimer $(Cp_2Ti)_2$. On heating in toluene, $(Cp_2Ti)_2$ is con-

$$2/x(Cp_2TiH)_x \xrightarrow[room\ temp.]{toluene} (Cp_2Ti)_2 + H_2$$

verted into an isomeric hydride. The violet transient hydride dimer undergoes a similar transformation in toluene.

$$(Cp_2Ti)_2 \xrightarrow[2\ hr]{100°} [(C_5H_5)(C_5H_4)TiH]_2$$

$(Cp_2Ti)_2$ has been isolated, but not as a crystalline solid. It has been inferred from the absence of splitting of the cyclopentadienyl bands in the infrared spectrum that the cyclopentadienyl ligands are symmetrically π-bonded. Solutions of the compound show some paramagnetism. In toluene at room temperature it is stable for several days, but in THF it is unstable, the bright green adduct decomposing in about 30 min.

* The reaction of $Cp_2Ti(C_6H_5)_2$ with hydrogen (C38) proceeds at a noticeable reaction rate only on warming to 60°.

With anhydrous HCl at $-80°$, $(Cp_2Ti)_2$ forms the red Cp_2TiCl_2 and not the green insoluble isomer obtained from other titanocene species. The pentahapto nature of the cyclopentadienyl groups in $(Cp_2Ti)_2$ is further exemplified by its reaction with carbon monoxide (B43) (see Section C).

Reaction of $(Cp_2Ti)_2$ with hydrogen at atmospheric pressure is not stoichiometric (0.2–0.4 mole H_2/Ti), and the suggestion has been made that the dimer is oxidatively cleaved by the hydrogen to form a monohydride and also a dihydride, analogous to that formed when bis(π-pentamethylcyclopentadienyl)titanium reacts with hydrogen. In toluene, all the cyclopentadienyl protons are readily replaced by deuterium.

$(Cp_2Ti)_2$ reacts reversibly with nitrogen (B43) to form an intense blue dinitrogen complex, $(Cp_2Ti)_2N_2$, which is discussed in Chapter VIII.

A dimeric triphenylphosphine adduct of titanocene has also been isolated on reaction of the violet dimeric or gray-green polymeric hydride or the metastable titanocene, $(Cp_2Ti)_2$, with triphenylphosphine (B43). It is also

$$(Cp_2TiH)_2 + 2P(C_6H_5)_3 \xrightarrow{-20°} 2Cp_2Ti(H)[P(C_6H_5)_3]$$

$$\updownarrow \text{room temp.}$$

$$\{Cp_2Ti[P(C_6H_5)_3]\}_2 + H_2$$
$$\text{purple}$$

formed as a precipitate on reaction of $Cp_2Ti(CH_3)_2$ with hydrogen in hexane containing triphenylphosphine. This phosphine adduct has a limited stability

$$Cp_2Ti(CH_3)_2 + P(C_6H_5)_3 + H_2 \xrightarrow[0°]{\text{hexane}} 1/2\{Cp_2Ti[P(C_6H_5)_3]\}_2 + 2CH_4$$

in solution in the absence of excess phosphine. It has a low solubility in toluene (ca. $10^{-2}\ M$), THF, and diethyl ether, and is insoluble in aliphatic hydrocarbons. On oxidation with anhydrous HCl it forms the crimson titanium(III) chloride, $[(C_5H_5)(C_5H_4)TiCl]_x$, and on heating to 110°, the corresponding dark green hydride, $[(C_5H_5)(C_5H_4)TiH]_2$. This latter reaction is not reversible. The pentahapto character of the cyclopentadienyl rings in this phosphine adduct follows from the products of its reaction with carbon monoxide under ambient conditions of temperature and pressure. It absorbs

$$\{Cp_2Ti[P(C_6H_5)_3]\}_2 + 4CO \longrightarrow 2Cp_2Ti(CO)_2 + 2P(C_6H_5)_3$$

nitrogen to give a complex and reacts reversibly with hydrogen to form $Cp_2Ti(H)[P(C_6H_5)_3]$. A complete exchange of all the cyclopentadienyl hydrogens can be effected with deuterium (B43).

e. PYROLYSIS OF ALKYL- AND
ARYLBIS(CYCLOPENTADIENYL)TITANIUM COMPOUNDS

i. $Cp_2Ti(C_6H_5)_2$.   Razuvaev and co-workers (R22, R23, R28) have contended in a number of publications that the thermal decomposition of Cp_2Ti-$(C_6H_5)_2$ led to a green form of titanocene. However, in the light of the results of O'Brien et al. (D40), it would appear that the decomposition product is rather a phenylene derivative having the formula $Cp_2Ti(C_6H_4)$. The oxidation state of the titanium in this compound would appear to be tetravalent, judging by its inertness toward alcohol. An o-phenylene-type structure such as

is in keeping with the chemical behavior of this compound. Its reactions with carbon dioxide (A4, K38) and nitrogen (S37) are of interest and are discussed in greater detail in Chapter III, Section B,5,a. The high yields of Cp_2TiCl_2 obtained on reaction of the pyrolysis product with chlorine-containing compounds such as HCl, CCl_4, $HgCl_2$, and $CHCl_3$ can also be rationalized by assuming the presence of a phenylene compound. A bis(cyclopentadienyl)-titanium(II) species can be made from $Cp_2Ti(C_6H_5)_2$ on reaction with $(C_6H_5)_2Hg$ (R28).

ii. $Cp_2Ti(CH_3)_2$.   According to Razuvaev et al. (R27, R30), the thermal decomposition of $Cp_2Ti(CH_3)_2$ proceeds via the elimination of methyl radicals. In hexane a green product thought to be titanocene was obtained,

$$Cp_2Ti(CH_3)_2 \xrightarrow[\text{5 hr}]{90°} Cp_2TiCH_3 + CH_3 \cdot$$
$$\downarrow$$
$$Cp_2Ti + CH_3 \cdot$$

while in THF a brown compound was formed which was claimed to be the tetrahydrofuranate of titanocene. These two products were not interconvertible.

The only gaseous product evolved in the pyrolysis was methane, liberated in 80–85% yield. If this is, indeed, a radical reaction as suggested (R30), it is most surprising that no ethane was formed. Wailes, Weigold, and Bell (W11) have confirmed that only methane is evolved and, further, that in both cyclohexane and THF only 75% of the methyl groups were converted to methane. Based on the amount of methane evolved, the reactions are complete in 1.5 hr. The brown THF product is a dimer, which must contain one methyl or methylene group. The pyrolysis products of $Cp_2Ti(CH_3)_2$ in both cyclohexane and THF react with methanol and anhydrous HCl to give, respectively, 0.5 and 1.5 moles of hydrogen per dimer. Since methanol oxidizes titanium only to the trivalent state whereas HCl takes it to the tetravalent state, the complex behaves as a Ti(III)–Ti(II) dimer ($g = 1.97$). The absence of any gaseous hydrocarbon on acid hydrolysis of this complex suggests that the remaining methyl or methylene group is attached to a cyclopentadienyl ring. The product formed by oxidation with HCl is not the normal red Cp_2TiCl_2, but is an insoluble green complex of similar composition. It would appear that the methane evolved is formed, in part at least, by hydrogen abstraction from cyclopentadienyl groups.

ii. $Cp_2Ti(CH_2C_6H_5)_2$. This compound, unlike its zirconium analog, decomposes readily in benzene at 30° to a sparingly soluble and extremely oxygen-sensitive titanocene-type compound (F1). This mode of decomposition

$$Cp_2Ti(CH_2C_6H_5)_2 \longrightarrow 1/n[(C_5H_4)_2Ti]_n + 2C_6H_5CH_3$$

was established from the relative intensities of the CH_2 and CH_3 PMR signals before and after reaction.

iv. $Cp_2Ti(C_2H_5)$. The thermal degradation of a $Cp_2Ti(C_2H_5)$ complex (B43) also yields a titanocene species which can be converted into $Cp_2Ti(CO)_2$ or a titanocene nitride derivative by reaction with either carbon monoxide or nitrogen. In the nitride species, $[(Cp_2Ti)^{2+}]_3[N^{3-}]_2$, the cyclopentadienyl rings remain π-bonded.

B. Bis(pentamethylcyclopentadienyl)titanium (or Decamethyltitanocene)

Unlike the unsubstituted cyclopentadienyl forms of titanocene, decamethyltitanocene does not form on direct reduction of the corresponding titanium-(IV) dihalide with sodium amalgam or sodium/naphthalene (B43). It was synthesized by Brintzinger and Bercaw (B42, B43) using the procedure shown below. A methylene group bridging between a titanium and a cyclopentadienyl group such as in the turquoise intermediate has also been proposed (W11) in the pyrolysis product of $Cp_2Ti(CH_3)_2$.

A molecular weight determination on decamethyltitanocene in benzene at one concentration of solute led to a value of 520 ± 20, somewhat below the dimer weight of 636. However, as the compound was contaminated by some monomeric decamethyltitanocene dihydride, assignment of a dimeric structure to decamethyltitanocene would appear to be quite compatible with the experimental result. No analytical data on this compound are available.

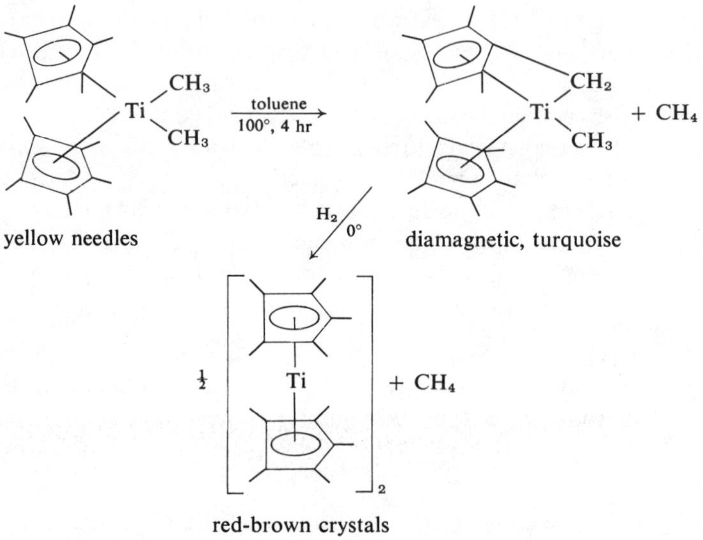

yellow needles   diamagnetic, turquoise

red-brown crystals

The highest mass peak in the mass spectrum of decamethyltitanocene was the monomer parent (relative intensity 50). The most intense peaks correspond to the fragments (parent-H)$^+$, $C_5(CH_3)_4CH_2{}^+$ (intensity 100) and $C_5(CH_3)_5H^+$ (intensity 91). The pentahapto nature of the cyclopentadienyl rings on the compound has been confirmed by infrared spectroscopy (B43). Owing to some paramagnetism, the methyl resonance in the PMR spectrum in toluene-d_8 was a broad singlet at $\delta 2.26$.

Decamethyltitanocene decomposes on heating to 35° giving a diamagnetic hydride, which is presumably a titanium(IV) derivative. A number of absorp-

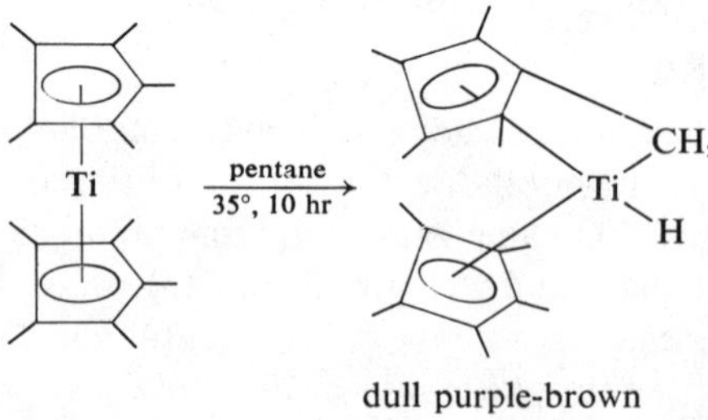

dull purple-brown

tions in the region $\delta 2.2$ to 0.25 were observed in the PMR spectrum of fresh benzene solutions of this hydride. The solutions are not stable (B43).

The reactions of decamethyltitanocene with hydrogen chloride, carbon monozide, nitrogen, and hydrogen have been studied (B42, B43). Orange-red

$$[\pi\text{-}C_5(CH_3)_5]_2Ti + 2HCl \xrightarrow[-80°]{toluene} [\pi\text{-}C_5(CH_3)_5]_2TiCl_2 + H_2$$

purple-brown needles

$[\pi\text{-}C_5(CH_3)_5]TiCl_3$ was also produced during this reaction. With nitrogen

$$[\pi\text{-}C_5(CH_3)_5]_2Ti + 2CO \xrightarrow[-80°, \ 1 \ atm]{\text{pentane}} [\pi\text{-}C_5(CH_3)_5]_2Ti(CO)_2$$
$$\text{yellow-brown}$$

at atmospheric pressure, decamethyltitanocene reacts at $-100°$ in pentane to give an olive green solution containing a compound thought to be the dinitrogen analog of the carbonyl compound. Removal of the nitrogen atmosphere above this solution, even at $-100°$, converts the olive green compound to a purple-blue species containing one dinitrogen molecule per two titanium atoms. Reaction of decamethyltitanocene at $-80°$ in hexane or toluene with nitrogen leads directly to this species

$$\{[\pi\text{-}C_5(CH_3)_5]_2Ti\}_2 + N_2 \longrightarrow [\pi\text{-}C_5(CH_3)_5]_2Ti\text{---}N_2\text{---}Ti[\pi\text{-}C_5(CH_3)_5]_2$$
$$\text{purple-blue}$$

Hydrogen adds reversibly to decamethyltitanocene, or to its turquoise precursor, yielding a monomeric, diamagnetic dihydride.

orange crystals

C. Carbonyl Compounds

To date only dicarbonyl derivatives of cyclopentadienyltitanium(II) and -zirconium(II) are known. $Cp_2Ti(CO)_2$ was first obtained by Murray (M55, M56) in 1958 as dark reddish brown crystals in 18% yield by the reactions

$$Cp_2TiCl_2 + RM(\text{excess}) + 2CO \xrightarrow[100°, \ 8 \ hr]{110 \ atm} Cp_2Ti(CO)_2$$

$$R = C_5H_5, \ M = Na$$
$$R = C_4H_9, \ M = Li$$

The reactions probably proceeded via the precursors $Cp_2Ti^{IV}R_2$ or $Cp_2Ti^{III}R$. The reactions below illustrate the utility of such alkyl species in preparing

$$Cp_2Ti(CH_2C_6H_5)_2 + CO \xrightarrow[25°-30°, 1\ atm]{heptane} Cp_2Ti\overset{\overset{\textstyle O}{\|}}{-}C-CH_2C_6H_5$$

$$Cp_2Ti(CO)_2 + (C_6H_5CH_2)_2CO$$

$Cp_2Ti(CO)_2$. This is a convenient synthesis leading to 80% yield of the dicarbonyl (F1). In benzene a thermal decomposition to give the titanium(II) complex, $[Ti(C_5H_4)_2]_n$, accompanies this carbonylation reaction. $Cp_2Ti(CH_3)_2$, unlike its zirconium analog, provides another convenient route to the di-

$$Cp_2Ti(CH_3)_2 + CO \xrightarrow[25°,\ \frac{1}{2}\ hr]{petrol,\ 120\ atm} Cp_2Ti(CO)_2$$

carbonyl (W11). The diphenyl compound, $Cp_2Ti(C_6H_5)_2$, has also been used in this way (M22). Tris(cyclopentadienyl)titanium(III), $Cp_2Ti(h^2\text{-}C_5H_5)$, formed by heating $Cp_2Ti(h^1\text{-}C_5H_5)_2$ to 125° (L23a, S40) reacts similarly with carbon monoxide (F17) to give $Cp_2Ti(CO)_2$. In another preparative

$$Cp_2Ti(h^2\text{-}C_5H_5) + 2CO \xrightarrow[80°]{125\ atm} Cp_2Ti(CO)_2$$

procedure the reaction of titanocene with carbon monoxide (B43, C2, M21) was utilized. Calderazzo, Salzmann, and Mosimann (C2) found that 10% of the required amount of carbon monoxide was absorbed by the titanocene obtained by the method of Watt and co-workers (W16). Marvich and Brintzinger (M21) were able to convert their titanocene quantitatively to the dicarbonyl. The uptake of carbon monoxide by this titanocene is irreversible.

$$Cp_2Ti(CH_3)_2 + \tfrac{3}{2}H_2 \longrightarrow 2CH_4 + \tfrac{1}{2}(Cp_2TiH)_2 \longrightarrow 1/n(Cp_2TiH)_n$$
$$\qquad\qquad\qquad\qquad\quad \text{violet} \qquad\qquad\qquad \text{gray-green}$$

$$Cp_2Ti(CO)_2 \xleftarrow[CO,\ room\ temp.]{ether} (Cp_2Ti) + \tfrac{1}{2}H_2$$

A purple phosphine adduct of titanocene (B43) reacts similarly with carbon monoxide. When the reduction of Cp_2TiCl_2 was performed electrolytically under an atmosphere of carbon monoxide, the dicarbonyl was again formed (B43).

Although $Cp_2Ti(CO)_2$ is thermally unstable above 90° (M56), it is volatile and can be purified readily by sublimation at $40°$–$80°/10^{-3}$ mm (C2, F17). It is quite soluble in organic solvents, and in benzene it is monomeric (C2).

The positions of the two carbonyl stretching bands are somewhat solvent dependent (C2). In hexane the bands are found at 1975 and 1897 cm^{-1}, corresponding, respectively, to the A_1 and B_1 bands for a molecule of C_{2v} symmetry (C2, F28). Bands thought to be metal–carbonyl carbon stretches (F28) appear at 485 (A_1) and 551 cm^{-1} (B_1). Infrared bands pertaining to the cyclopentadienyl ligands of the molecule have been tabulated (F17, F27, S4), as has the ultraviolet absorption spectrum in hexane (F14). Diamagnetism down to $-50°$ has been confirmed by the sharp singlet PMR line at $\delta 5.06$ in acetone-d_6 and at $\delta 4.58$ in benzene-d_6.

Cp$_2$Ti(CO)$_2$ is readily oxidized by oxygen (S3) and other oxidants such as iodine (F21), HCl (M56), phenylacetylene (S60), and tolan (S60). For example,

$$\text{Cp}_2\text{Ti(CO)}_2 + \text{O}_2 \xrightarrow[\text{3-5 days}]{\text{THF or toluene}} (\text{Cp}_2\text{TiO}_2)_n$$
diamagnetic, insoluble, yellow-brown

$$\text{Cp}_2\text{Ti(CO)}_2 + \text{C}_6\text{H}_5\text{C}\equiv\text{CH} \xrightarrow[\text{60°, 20 min}]{n\text{-hexane}} \text{Cp}_2\text{Ti}\begin{matrix} \text{CH}=\text{C}-\text{C}_6\text{H}_5 \\ | \\ \text{CH}=\text{C}-\text{C}_6\text{H}_5 \end{matrix}$$
air-stable, green needles

The carbonyl groups are also replaced by diphenylketene (H36). The

$$\text{Cp}_2\text{Ti(CO)}_2 + (\text{C}_6\text{H}_5)_2\text{C}=\text{C}=\text{O} \xrightarrow[\text{60°, 1 hr}]{\text{hexane}} \text{Cp}_2\text{Ti}[(\text{C}_6\text{H}_5)_2\text{C}=\text{C}=\text{O}].$$

structure of this monomeric orange-yellow diamagnetic complex is not known. Apart from two infrared absorption bands at 1555 and 1575 cm^{-1}, no others were observed up to 2800 cm^{-1}. It is unstable toward CH$_2$Cl$_2$ and CHCl$_3$, and gives tetraphenylethane with protic reagents such as water or alcohol. Other ketones such as 9,10-phenanthroquinone add oxidatively to the dicarbonyl (F21).

Bases such as bipyridyl (F14), terpyridyl (B33), and o-phenanthroline (B33) react with Cp$_2$Ti(CO)$_2$ displacing carbon monoxide. For example,

$$Cp_2Ti(CO)_2 + 2 \text{ terpyridyl} \xrightarrow[160°]{\text{cyclohexane}} Ti(\text{terpyridyl})_2 + 2CO + 2(C_5H_5\cdot)$$
black

$$Cp_2Ti(CO)_2 + 3o\text{-phenanthroline} \xrightarrow[160°]{\text{cyclohexane}} Ti(o\text{-phenanthroline})_3 +$$
$$2CO + 2(C_5H_5\cdot)$$
violet-black

$Cp_2Ti(CO)_2$ can behave as a homogeneous hydrogenation catalyst (S60), reducing 1-pentyne to 1-pentene, tolan to bibenzyl, and phenylacetylene to ethylbenzene and styrene.

The only other titanium carbonyl reported was made in good yields from bis(π-pentamethylcyclopentadienyl)titanium (B42, B43). This compound is

$$[\pi\text{-}C_5(CH_3)_5]_2Ti + 2CO \xrightarrow[\text{room temp.}]{\text{toluene or pentane}} [\pi\text{-}C_5(CH_3)_5]_2Ti(CO)_2$$
yellow-brown in solution

also volatile, subliming at $80°/10^{-3}$ mm. The methyl hydrogens give rise to a singlet at $\delta 1.69$ in the PMR spectrum. Both carbonyl stretches (at 1930 and 1850 cm^{-1}) are at lower energy than the corresponding bands of $Cp_2Ti(CO)_2$.

Although $Cp_2Ti(CH_3)_2$ gives the dicarbonyl compound with carbon monoxide, only the acyl complex is obtained in the case of the corresponding zirconium compound (W11). A carbonyl of zirconium can be formed from

$$Cp_2Zr(CH_3)_2 + CO \rightleftharpoons Cp_2Zr(CH_3)COCH_3$$

Cp_2ZrH_2 and carbon monoxide under pressure (W11). It is at least a trimer

$$Cp_2ZrH_2 + CO \xrightarrow[130 \text{ atm}]{\text{benzene, } 50°} [Cp_2Zr(CO)_2]_n$$
brown-black

in boiling benzene and not all of the cyclopentadienyl protons are PMR equivalent. The carbonyl stretches occur at 1849 and 1945 cm^{-1} in the infrared spectrum.

Acyl complexes of titanium have been synthesized by treatment of the dicarbonyl with acetyl chloride or benzoyl chloride (F21). An alternative

$$Cp_2Ti(CO)_2 + RCOCl \xrightarrow{1 \text{ day}} Cp_2Ti(Cl)COR + 2CO$$
80%

approach is to oxidize the dicarbonyl with either methyl or ethyl iodide (F21).

$$Cp_2Ti(CO)_2 + RI \longrightarrow Cp_2Ti(I)COR + CO$$

D. Zirconocene and Hafnocene

Zirconocene and hafnocene have been obtained as pyrophoric dark purple compounds by the sodium/naphthalene reduction method of Watt and Drummond (D38, W17, W18). These compounds have X-ray powder patterns

$$Cp_2MCl_2 + 2Na/naphthalene \xrightarrow{\text{THF}} 1/n(C_{10}H_{10}M)_n + 2NaCl$$

$$M = Zr \text{ or } Hf$$

similar to the titanocene analog (D38, W18). When diphenylacetylene was included in the reaction mixture and M = Zr, the compound 1,1-bis-(π-cyclopentadienyl)-2,3,4,5-tetraphenylzirconacyclopentadiene was formed (D38, W18).

The solubility and volatility of zirconocene is higher than that of hafnocene, although both are less soluble than titanocene. Because of this lack of solubility the molecular weights of zirconocene and hafnocene have not been established. While zirconocene sublimes with difficulty at $300°/10^{-3}$ mm, hafnocene does not sublime and decomposes to a black solid at $345°/10^{-3}$ mm. Oxidation of the compounds with oxygen yields yellow to orange insoluble products (W17). The zirconocene oxidation product analyses for Cp_2ZrO_2 (W11), a result analogous to that found by Salzmann (S3) for titanocene.

Zirconocene also forms on decomposition of $Cp_2Zr(alkyl)_2$ compounds (W10). It has been shown to be an hydrogenation catalyst for alkenes (W10).

E. 2,2'-Bipyridyl Adducts of Titanocene and Zirconocene

Mono(2,2'-bipyridyl) complexes have been prepared by treating titanocene (C2) or zirconocene (W5) (prepared by the sodium/naphthalene method) with 2,2'-bipyridyl. The titanium compound formed as dark blue crystals from hexane and was antiferromagnetic, μ_{eff} dropping from 1.04 BM at 296°K ($g = 1.989$) to 0.85 BM at 90°K. A purple zirconium complex, $(C_{10}H_{10}Zr)\cdot$ 2,2'-bipyridyl$\cdot C_6H_6$ was obtained on heating zirconocene with 2,2'-bipyridyl at 150° in toluene for 25 hr and subsequent extraction of the product with benzene. The benzene can be expelled from the compound on heating to 100°–120° under high vacuum, but is not evolved in boiling THF. This monomeric benzene complex is antiferromagnetic, the bipyridyl apparently again behaving as a "noninnocent" ligand. The magnetic moment decreases from 1.24 BM at 352°K to 0.73 BM at 94°K.

Analogous bipyridyl complexes containing only symmetrically π-bonded cyclopentadienyl groups can be prepared either from $Cp_2Ti(CO)_2$ with

2,2'-bipyridyl (F14) or by reducing Cp_2TiCl_2 or Cp_2ZrCl_2 with disodium or dilithium bipyridyl (C2, W5). Both the titanium and zirconium compounds

$$Cp_2Ti(CO)_2 + 2,2'\text{-bipyridyl} \xrightarrow[80°, 24 \text{ hr}]{\text{hexane}} Cp_2Ti(2,2'\text{-bipyridyl}) + 2CO$$
$$\text{blue-black, } 72\%$$

$$Cp_2TiCl_2 + Na_2(2,2'\text{-bipyridyl}) \xrightarrow[\text{room temp.}]{\text{THF}} Cp_2Ti(2,2'\text{-bipyridyl}) + 2NaCl$$
$$81\%$$

exhibit antiferromagnetic behavior even though they are monomeric. The titanium complex has a moment of 0.6 to 0.8 BM at 295°K (C2, F14) and has two g values, 1.983 and 1.989 (C2). The infrared spectrum in KBr and the visible ultraviolet spectrum in toluene have been measured (F14). μ_{eff} for the zirconium compound drops from 0.44 BM at 356°K to 0.34 BM at 99°K (W5).

Cp$_2$Ti(2,2'-bipyridyl) reacts with oxygen at room temperature to give a mixture containing between 1.5 and 2 molecules of oxygen per titanium, some oxygen presumably reacting with the bipyridyl radical (S3).

F. Cyclopentadienylphenyltitanium

Razuvaev and co-workers (R30, R32) found that at room temperature in ether $CpTi(C_6H_5)_3$ degraded to $CpTi(C_6H_5)$, which was isolated as the bis(etherate). The coordinated ether was readily replaced by an equal number

$$CpTi(C_6H_5)_3 \xrightarrow[20°]{\text{ether}} CpTi(C_6H_5)\cdot 2(C_2H_5)_2O + C_6H_6 + \tfrac{1}{2}C_6H_5\cdot C_6H_5$$

of ammonia molecules. Although the reactions of $CpTi(C_6H_5)$ tend to confirm its general composition, they do not eliminate the possibility of a phenylene species, $CpTi^{III}(C_6H_4)$. A similar phenylene compound, $Cp_2Ti(C_6H_4)$, is produced by the thermal decomposition of $Cp_2Ti(C_6H_5)_2$ (D40).

CpTi(C_6H_5) possesses a high thermal stability degrading to titanium metal, benzene, and an "unsaturated tar" only on heating at 170° for 100 hr. With halogenated compound such as chloroform, carbon tetrachloride, or mercuric chloride it is converted to $CpTiCl_3$. It is also oxidized by alcohols.

G. Cyclopentadienylcycloheptatrienyltitanium

Van Oven and de Liefde Meijer (V5) prepared this sky blue complex by treatment of $CpTiCl_3$ with isopropylmagnesium bromide in the presence of excess cycloheptatriene at $-78°$. It is volatile, subliming at 125°/1 mm. Like other Ti(II) compounds, it is thermally stable, no degradation occurring on heating to 320°, at which temperature the compound sublimed. Reaction solutions absorb nitrogen between $-60°$ and $0°$, but above $10°$ the nitrogen is evolved.

Cyclopentadienylcycloheptatrienyltitanium is a diamagnetic sandwich compound. The C_7H_7 and C_5H_5 hydrogens give rise to singlets in the PMR spectrum in C_6D_6 at δ5.47 and 4.97, respectively. The infrared spectrum reinforced the assignment of planarity to the C_7H_7 ring in this titanium complex. The "sandwich" arrangement of the ligand rings has been confirmed by X-ray diffraction (Z5). The angle between the ring planes is $2.2 \pm 0.3°$ (see Fig. VII-1). The metal-to-carbon distances are 2.32 Å (Ti–C_5H_5) and 2.19 Å

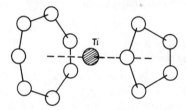

Fig. VII-1.   Structure of cyclopentadienylcycloheptatrienyltitanium (Z5).

(Ti–C_7H_7), this latter value being inexplicably low. Little variation from the mean values of 1.396 Å (C_5H_5 ring) and 1.397 Å (C_7H_7 ring) was found in the C–C distances.

In the mass spectrum the compound decomposes through an arene species.

Methyl-Substituted Cyclopentadienylcycloheptatrienyltitanium Compounds

A series of these compounds has recently been prepared following the procedure used for the parent compound (V13). These compounds too are thermally stable to over 300° and they can be purified by sublimation at 120°/0.1 mm. PMR data are shown in Table VII-1. Their mass spectra are interesting in that the parent ions tend to undergo an initial rearrangement to a bis(arene) species before degrading by loss of benzene or toluene. For example,

$$CpTi(\pi\text{-}C_7H_7) \longrightarrow (C_6H_6)_2Ti^+ \longrightarrow (C_6H_6)Ti^+$$

$$(\pi\text{-}CH_3C_5H_4)Ti(\pi\text{-}C_7H_7) \qquad\qquad\qquad (CH_3C_6H_5)Ti^+$$
$$\text{or} \qquad\longrightarrow (CH_3C_6H_5)Ti^+(C_6H_6) \longrightarrow \qquad \text{and}$$
$$CpTi(\pi\text{-}CH_3C_7H_6) \qquad\qquad\qquad\qquad (C_6H_6)Ti^+$$

Table VII-1

PMR Chemical Shifts in Perdeuterobenzene for the Compounds $(\pi\text{-}C_5H_4R)Ti(\pi\text{-}C_7H_6R')^a$

R	R'	C_5 ring[b]	C_7 ring	$CH_3(C_5)$	$CH_3(C_7)$
H	H	4.97s	5.47s	—	—
CH$_3$	H	4.90m	5.41s	1.81s	—
H	CH$_3$	4.96s	5.43m	—	2.52s
CH$_3$	CH$_3$	4.88m	5.38m	1.79s	2.52s

[a] From ref. (V13).
[b] s, Singlet; m, multiplet.

H. Cycloheptatrienylcycloheptadienyltitanium

This compound was obtained from the reaction between titanium trichloride, isopropylmagnesium bromide, and excess cycloheptatriene (V6) as green crystals from pentane. It is a diamagnetic, thermally stable compound which is reactive toward oxygen and water. The mass spectrum confirmed the composition of the complex and the infrared and PMR spectral measurements indicated the presence of a planar symmetrical $h^7\text{-}C_7H_7$ ring ($\delta 5.37$). The C_7H_9 ring gives rise to five sets of multiplets with chemical shifts in the region $\delta 5.69$ to 1.1. To obtain a 16-electron configuration, van Oven and de Liefde Meijer (V6) maintain that the compound is best viewed as (h^7-cycloheptatrienyl)(h^5-cycloheptadienyl)titanium.

Müller and Mertschenk (M52) have shown that decomposition in the mass spectrometer proceeds in part through ring expansion and ring contraction processes.

I. Cyclooctatetraene(tetraphenylcyclobutadiene)titanium

The preparative method above (i.e., TiCl$_3$ + iso-C$_3$H$_7$MgBr + unsaturated organic compound) has been used also to prepare other π-bonded organotitanium compounds. By addition of TiCl$_3$ and isopropylmagnesium bromide to a solution of cyclooctatetraene and diphenylacetylene, the green sandwich compound octahapto-cyclooctatetraene-tetrahapto-tetraphenylcylobutadiene-titanium was prepared (V3). It is diamagnetic and thermally stable to 300°.

In the solid state at room temperature it is stable in air. In toluene and CDCl₃ the h^8-C₈H₈ protons have a PMR chemical shift of δ6.6 (singlet). The phenyl protons give rise to an unresolved multiplet in the δ7.10–7.35 region. The structure of the compound was deduced from the above PMR information and from infrared and mass spectral data.

J. Bis(phenyl)titanium

This compound is formed as a black benzene- and ether-soluble powder on thermal decomposition of tetraphenyltitanium at room temperature in ether (L9, R30).* On heating at 200°–250° for 6 hr it decomposes to titanium metal, biphenyl, and benzene. With mercuric chloride, diphenyltitanium forms phenylmercuric chloride in 42% yield. Oxygen inserts into the titanium–phenyl bonds, although not quantitatively. On oxidation with chloroform only 1 mole of benzene is produced from each mole of bis(phenyl)titanium. It forms grayish black monoadducts with THF and ammonia (L8, L9).

The end product of the reaction between Cp₂TiCl₂ and phenyllithium is also bis(phenyl)titanium (L10, L11). The phenyl radicals abstract hydrogen

$$Cp_2TiCl_2 + 2C_6H_5Li \xrightarrow{\text{ether}} Cp_2Ti(C_6H_5)_2 + 2LiCl$$

$$Cp_2Ti(C_6H_5)_2 + 2C_6H_5Li \xrightarrow{\text{ether}} Ti(C_6H_5)_2 + 2C_5H_5Li + 2C_6H_5\cdot$$

from the solvent to form benzene. Latyaeva et al. (L10, L11) also conclude that in the presence of still more phenyllithium, lithium metal is formed catalytically. Bis(phenyl)titanium has been used to introduce titanium into a porphyrin (T42).

Other compounds of this type have been formed by the same general

$$Cp_2TiR_2 + LiR' \text{ (excess)} \longrightarrow RTiR'$$
$$R = CH_3, R' = C_6H_5,$$
$$R = C_6H_5, R' = CH_2C_6H_5$$
$$R = CH_2C_6H_5, R' = CH_2C_6H_5 \text{ or } CH_3$$

---

* Tetraphenylzirconium was recently reported (R30a) to decompose under these conditions to give the dimer [(C₆H₅)₂Zr·O(C₂H₅)₂]₂.

reaction (R21a, R31). Benzylphenyltitanium has been isolated as the bis-(etherate). Because of its high thermal stability, thermolysis occurs only on heating at 250° for 240 hr (R21c, R31).

K. Arene Complexes

The initial report on arenetitanium or -zirconium compounds was made in the patent literature in 1960 by Fischer and Hafner (F16), who claimed the preparation in solution of dibenzenetitanium (brown) and dibenzenezirconium (wine red).

Fischer and Röhrschied (F18) prepared hexamethylbenzene complexes by reduction of the metal tetrachloride with aluminum powder in the presence of aluminum chloride in hexamethylbenzene melt at 120°–130°. From the conductivity, molecular weight, and magnetic data the complexes, formulated as $\{M_3[C_6(CH_3)_6]_3Cl_6\}X$ (where $M = Ti$, $X = Cl$, PF_6; or $M = Zr$, $X = Cl$) were considered to be cyclic trimers. These air-sensitive compounds are susceptible to thermal degradation on warming above 20° in the case of the zirconium compound or 30° for the others. The titanium compounds are soluble in methylene dichloride as well as in ethanol and water.

The structure advanced for these compounds is that of a cyclic trimer of metal atoms interlinked with metal–metal bonds and double chloride bridges. One arene group is coordinated to each metal atom. The infrared spectra have been recorded and some assignments have been made (F18).

Treatment of $TiCl_4$ with aluminum and aluminum chloride in benzene, toluene, or mesitylene (M18) gave the π-arene complexes, arene·$TiCl_2$·$2AlCl_3$, which are believed to be active in nitrogen fixation (V20, V21).

Chapter VIII

The Reactions of Dinitrogen with Titanium (and Zirconium) Compounds

The interaction of nitrogen gas (dinitrogen) with transition metal compounds is a significant aspect of organometallic chemistry which has engaged the attention of many laboratories over the last few years. A fundamental study of this interaction is important in understanding the processes of nitrogen fixation, both biological and chemical. Until recently, this work indicated some distinction between the lighter transition metals, which under highly reducing conditions absorbed nitrogen which was converted to ammonia on solvolysis, and the heavier transition elements which generally formed stable dinitrogen complexes, in which reduction to ammonia could not be effected. Recent work has tended to lessen this distinction; dinitrogen complexes of titanium have now been isolated, while iron, manganese, and other metals have been shown to form complexes similar to those of titanium and with similar chemical reactivities. In addition, dinitrogen attached to tungsten has been reduced to hydrazine (C18).

A review by Chatt and Leigh (C19 and references therein) covers effectively the work on the higher transition elements; while recent articles by Vol'pin and Shur (V15, V27) offer a comprehensive overview of the chemical fixation of molecular nitrogen with emphasis on titanium.

A. Absorption of Nitrogen by Reduced Titanium Species

The study of the reactions of molecular nitrogen with transition metals under mild conditions of temperature and pressure parallels the development of the Ziegler–Natta catalysts for olefin polymerization. In fact, nitrogen is a weak inhibitor of olefin polymerization (S13), and Ziegler–Natta catalysts formed under nitrogen were known in the 1950's to give some ammonia on hydrolysis (C19).

There is little doubt that the process involves the formation under mild conditions of a nitride of a transition metal in a highly reduced form. Several steps have been distinguished by Vol'pin (V27): (1) reduction of a transition metal compound to a low-valent form; (2) absorption of dinitrogen by this compound to give either terminal or bridging complexes of N_2; (3) reduction to nitride (conventionally written as $[N^{3-}]$); and (4) solvolysis to give ammonia. At an intermediate stage hydrolysis to hydrazine is possible.

$$L_nM \xrightarrow[\text{(reduction)}]{e^-} L_mM \xrightarrow{N_2} [L_mMN_2] \text{ or } [L_mMN_2ML_m] \xrightarrow[\text{(reduction)}]{e^-}$$

$$[^{2-}N{-\!-}N^{2-}] \xrightarrow[\text{(reduction)}]{e^-} [N^{3-}]$$

$$\Big\downarrow H_2O \qquad\qquad\qquad \Big\downarrow H_2O$$

$$H_2NNH_2 \qquad\qquad\qquad NH_3$$

The lighter transition elements, titanium, vanadium, chromium, and molybdenum, are the most reactive, and the rest of this discussion will be confined to titanium which has received the most attention. As with the Ziegler–Natta systems, those compounds of titanium which have been reduced are halides, alkoxides, or cyclopentadienyl derivatives, while the reducing agents are Grignard reagents, lithium alkyls and aryls, aluminum alkyls, lithium or sodium/naphthalene, or metallic sodium, magnesium, or aluminum. The molar ratio of reducing agents to titanium is critical and must be 4:1 or even higher for best results. In addition, the solvent is all important, particularly for those systems in which hydrogen abstraction from solvent is said to occur.

1. The Vol'pin-Shur system

The Vol'pin–Shur system utilizes bis(cyclopentadienyl)titanium dichloride in ether, reduced with ethylmagnesium bromide in excess (5- to 9-fold). Undoubtedly, the first equivalent of Grignard reagent reduces the titanium compound to $(Cp_2TiCl)_2$ by alkylation, followed by elimination of ethyl groups as ethane and ethylene (B90). Addition of excess ethylmagnesium halide produces further alkyltitanium species, the nature of which has been the subject of many papers. Alkylated anionic complexes have been detected in the dark brown solution by ESR spectroscopy (B91) e.g., $[Cp_2Ti^{III}(C_2H_5)_2]^-$, the spectrum showing five lines due to interaction of the unpaired electron on titanium(III) with the α hydrogens of the ethyl group. Recently, the same author has isolated from this solution an ethyltitanium(III) complex, $[Cp_2Ti(C_2H_5)]_2 \cdot 6MgCl_2 \cdot 7(C_2H_5)_2O$, as a yellow crystalline solid (B43).

The nature of the product which absorbs nitrogen is still the subject of some discussion; it appears to be somewhat variable in composition depending on the conditions of the reaction, particularly temperature, solvent, amount of excess organometallic reducing agent, and the presence or absence of a reactant such as nitrogen. Evidence for a hydride as one of the initially formed species (formed by loss of ethylene) comes from ESR (B91, H14) and the formation of alkyl and borohydride derivatives in the presence of olefins (B43, K58) and diborane (B43), respectively.

Nevertheless, there is no compelling evidence which indicates that hydride is either the major compound present or the active species. Present opinion is against the possibility (C19), particularly since the ESR triplet signal indicative of hydride is still present after reaction with nitrogen has ceased (H14). In the presence of certain reactants, for example, nitrogen or carbon monoxide, evolution of ethane and ethylene from $(Cp_2TiC_2H_5)_2 \cdot 6MgCl_2 \cdot 7(C_2H_5)_2O$ occurs (B43) suggesting that the active species is $[(C_5H_5)_2Ti]_n$ (in which n is probably 2). Shilov and co-workers (B56) have recently isolated the complex, $(Cp_2Ti)_2N_2$, from the reaction between Cp_2TiCl, CH_3MgI, and nitrogen at $-70°$. Here again, dealkylation has occurred before, or concurrently with, nitrogen absorption. This complex gave nitrogen and hydrazine (50% of each) on treatment with methanolic HCl, and could be reduced further to a nitride with phenyllithium.

On the other hand, both Russian (S36) and Dutch (T20) workers have observed absorption of nitrogen at low temperatures ($-80°$ to $-100°$) by alkyl- and aryltitanium(III) compounds, of type Cp_2TiR, without loss of R (see later). Although subsequent reduction and hydrolysis gave hydrazine and ammonia, the nature of the intermediate is still not clear.*

Vol'pin and Shur suggested (V27) that the nitrogen-fixing activity of the system $Cp_2TiCl_2 + C_2H_5MgBr$ is due to a zerovalent titanium cyclopentadiene complex, $[(C_5H_6)_2Ti^0]$. This possibility was also considered by Maskill and Pratt (M24) on the basis of kinetic measurements, but was later discarded (M25) in favor of the monomeric hydrido compound, $(Cp_2TiH_2)^-$. In addition, no evidence for the existence of $[(C_5H_6)_2Ti^0]$ was found by Doisneau and Marchon (D27) in a polarographic study of Cp_2TiCl_2 reduction.

The subsequent reaction of the reduced titanium species with nitrogen has been studied under a variety of reaction conditions. The solution will absorb up to 1 gram atom of nitrogen per mole of titanium compound. Kinetic data obtained by Maskill and Pratt show that there is a change from a rate-

* Shilov now regards the complexes $(Cp_2TiR)_2N_2$ and $(Cp_2Ti)_2N_2$ as two successive intermediates in the reduction of dinitrogen to a nitride (B55a) and has isolated the paramagnetic complex $(Cp_2Ti)_2N_2MgCl$.

determining step which involves nitrogen at low partial pressures to one which does not, at pressures around 1 atm. The rate appears to be independent of nitrogen pressure up to 150 atm. Under 1 atm of nitrogen and in the presence of an excess of Grignard reagent, the reaction is second-order and the amount of nitrogen fixed increases approximately with the square of the titanium concentration. The rate-determining step under these conditions shows the low activation energy of 4 kcal/mole.

For maximum rate of fixation the titanium concentration should be as high as possible and the amount of Grignard reagent greater than a four- to fivefold excess. The use of pressures above 1 atm does not increase the yield, and variation in temperature also has little effect (M25).

Complexation of the absorbed nitrogen with the reduced titanium must occur to activate the nitrogen sufficiently to bring about reduction by excess organometallic reducing agent. In the absence of excess RMgX or Li/naphthalene, absorption of dinitrogen by low-valent titanium complexes is reversible in many cases (see later), the gaseous N_2 being liberated on destruction of the complex (B43, C19). In the presence of excess Grignard reagent, however, reduction of dinitrogen to nitride occurs.

Shilov et al. (N26) have shown that ammonia obtained from hydrolysis with H_2O of the system $Cp_2TiCl_2-C_2D_5MgBr-(C_2D_5)_2O$ contained no deuterium. This result confirmed the proposal that the nitrogen is firmly bound as metal nitride with no hydrogen attached to nitrogen until hydrolysis. In addition, no hydrogen transfer either from solvent or from alkyl group of the reducing agent occurs in this system.

Ammonia need not necessarily be the final product after hydrolysis. Recent reports from Shilov and co-workers (S36) and from the Gröningen group (T18) describe hydrazine as the major hydrolysis product after reduction of Cp_2TiCl_2 or Cp_2TiCl with isopropylmagnesium chloride in ether at low temperatures ($-60°$) in the presence of nitrogen. Ammonia appears to predominate at high temperatures ($>0°$).

Apart from Grignard reagents as reductants in the above system, other electropositive metals and their derivatives have been employed (see Table VIII-1). Using metallic sodium sand with Cp_2TiCl_2, van Tamelen et al. have detected several reduced species by visible spectroscopy (V8). Under argon at room temperature, these reduction products were formulated as Cp_2TiCl, $(Cp_2Ti)_n$ ($n = 1-2$), $[Cp(C_5H_4)TiH]_x$, and $[Cp(C_5H_4)TiH]_2$, the last being referred to as "stable titanocene," probably analogous to that of Brintzinger (B43) (see Chapter VII, Section A). Only the second product, $(Cp_2Ti)_n$ ($n = 1-2$), which could be prepared by treatment of Cp_2TiCl_2 in toluene with exactly two equivalents of sodium sand under argon over 6–10 days [after Salzmann and Mosimann (S4)], was found to react with nitrogen. The reaction below room temperature was rapid and reversible, giving a dark blue complex, $(Cp_2Ti)_2N_2$.

Under nitrogen $(Cp_2Ti)_n$ ($n = 1–2$) was the last product which could be detected spectroscopically; a black precipitate (nitride?) appeared at this stage and subsequent hydrolysis gave 0.6–0.7 mole NH_3 per mole Ti (V8).

The identity of the absorbing species as $(Cp_2Ti)_n$ was supported by Ungurenasu and Streba (U4), using the system $Cp_2TiCl_2 + Li(Na)Hg$. It was shown that 1 mole of N_2 was bonded to every 2 moles of titanium. Activation and reduction of coordinated nitrogen involved the consumption of extra reducing agent corresponding with the equation

$$Cp_2TiCl_2 + 4Li + \tfrac{1}{2}N_2 \longrightarrow [Cp_2TiN]Li_2 + 2LiCl$$

2. The van Tamelen System

Besides the cyclopentadienyltitanium compounds, titanium alkoxides have been found to be active in reactions with nitrogen. Vol'pin showed in 1965 that the tetraethoxide gave small yields of ammonia when reduced with an organometallic reducing agent (V22), and van Tamelin has developed more efficient systems based on titanium(IV) alkoxides reduced with metallic reducing agents (V9).

Initially, bis(*tert*-butoxo)titanium dichloride reduced by potassium metal in diglyme was shown to produce ammonia slowly over several weeks (V7). The active titanium species was shown to be titanium(II), since ammonia was not produced until 2 moles of potassium had been consumed. Experiments with perdeuterated tetrahydrofuran showed that this solvent was the source of the hydrogen in the ammonia (V10). Altering the alkoxo group had no appreciable effect on the ammonia yield, but changing to a nonethereal solvent effectively stopped the reaction. Systems based on titanium tetraisopropoxide reduced with alkali metal/naphthalene in tetrahydrofuran or diglyme retain their activity for nitrogen reduction through several cycles. Isopropanol is added as the proton source for controlled solvolysis when nitrogen uptake is complete after 30–60 min, the process becoming a cyclic one provided ammonia is removed and additional metallic sodium is added to regenerate the naphthalene complex. Over five such cycles a yield of ammonia corresponding to 170% was obtained, while addition of aluminum to isopropoxide raised this yield to 275%, presumably by transfer of nitride to aluminum (V7). The process may be represented as

$$N_2 + 6e^- + 6ROH \longrightarrow 2NH_3 + 6RO^-$$

In addition a significant uptake of nitrogen from air has been observed by van Tamelen using such a system (V7). Electrolytic reduction of molecular nitrogen has been demonstrated (V12) using an electrolysis medium of titanium tetraisopropoxide, naphthalene, tetrabutylammonium chloride, and aluminum isopropoxide. After electrolysis at 40 volts for 11 days, followed by hydrolysis with alkali, ammonia was obtained in 150–300% yield. The yield is less in the absence of naphthalene which is believed to function as an electron carrier, being reduced to naphthalide by the cathode and oxidized back to naphthalene by titanium species.

The system $Ti(O\text{-}iso\text{-}C_3H_7)_4$ + sodium/naphthalene in tetrahydrofuran also produces hydrazine on hydrolysis after 15–90 min, the amount varying with the Ti:Na ratio. The yield of hydrazine rose to a maximum of 15–19% with a ratio of 1:5–6, while the ratio $2NH_3:N_2H_4$ varied from 3.3 to 5.0 (V7). Substitution of other transition metals for titanium gave no hydrazine.

Vol'pin and co-workers (V20, V21) developed a system for reduction of dinitrogen using metallic aluminum as the reducing agent, aluminum halide as the Lewis acid, and titanium compounds (e.g., $TiCl_4$) as catalysts, with or without benzene as solvent. When the ratio of the three reagents was of the order $Al:AlX_3:TiCl_4 = 600:1000:1$, at a temperature of 130° and a nitrogen pressure of 100 atm, the amount of ammonia obtained on hydrolysis corresponded to 286 moles per mole of $TiCl_4$. Although the reaction is catalytic with regard to titanium, and does not occur at all in the absence of titanium compound, the yield of ammonia is proportional to and never exceeds the amount of aluminum halide present. Vol'pin sees the process as initial formation of Ti–N-bonded complexes, which react with aluminum compounds regenerating the low-valent titanium catalyst and forming Al–N bonds which hydrolyze to ammonia. A possible intermediate is the complex, $C_6H_6 \cdot TiCl_2 \cdot 2AlCl_3$, which is known to form on heating $TiCl_4$, Al, and $AlBr_3$ in benzene (M18, N19). Vol'pin found that this complex alone would react with nitrogen at 130° with formation of a compound of composition close to $C_6H_6(TiCl_2 \cdot 2AlCl_3)_3N$, which gave a stoichiometric yield of ammonia on hydrolysis (V20).

Other systems employing titanium compounds as "catalysts" and various reducing agents are shown in Table VIII-1.

In summary, the drawbacks of the present systems are the requirements for an excess of strong reducing agents and the resultant formation of very strong titanium nitride bonds which require solvolysis to liberate ammonia. In no case does the amount of ammonia obtained after solvolysis exceed the total metal concentration present, so that although the claim for "catalytic fixation" may be valid from the point of view of the transition metal, it is not valid from the point of view of total metals (e.g., Ti + Al).

For the catalytic fixation of nitrogen it is necessary to introduce the appropriate ligands for stabilizing the lower valent titanium compound and then to add a reagent to supply hydrogen to the nitrogen and to cleave the metal–nitrogen bond, liberating the active titanium site for further reaction.

Present work is aimed at complexing the dinitrogen in such a way as to

Table VIII-1

Nitriding Systems Which Give Ammonia (or Hydrazine) on Hydrolysis

System	Reagent ratio	Moles NH_3/mole Ti	Remarks	Refs.
$Cp_2TiCl_2 + C_2H_5MgBr$	1:>5	1	—	M25, N26, V4, V27
$Cp_2TiCl_2 + iso\text{-}C_3H_7MgBr$	1:>5	1	N_2H_4 formed at $-60°$, NH_3 at $>0°$	S36, S55a
$Cp_2TiCl + CH_3MgI$, then LiC_6H_5	1:2	1	$(Cp_2Ti)_2N_2$ is isolable intermediate	B56
$Cp_2TiCl_2 + $ Li/naphthalene	1:>4	1	—	H14
$Cp_2TiCl_2 + Mg + MgI_2$	1:9:5	1	—	V16
$Cp_2TiCl_2 + RLi$	1:>5	0.5	Amines also formed	V25, V26, V28
$Cp_2TiCl_2 + Na$	1:2	1	Visible; IR	V8
$Cp_2TiCl_2 + $ Li, Na, K, Rb, Cs, Mg, Ca, La, Ce	1:>2	0.3–0.85	ESR	B20
$CpTiCl_3 + $ Na, Mg, Ce	1:>2	0.35–0.7	ESR	B20
$Cp_2TiR_2 + H_2$ R = CH_3 or C_6H_5	—	0.32–0.36	—	V29
$Cp_2ZrCl_2 + $ Na or Li	1:4	~0.2	—	B20
$Ti(OC_2H_5)_4 + (iso\text{-}C_4H_9)_3Al$	1:6	0.33 (up to 11 with H_2)	—	V15, V22
$Ti(OC_2H_5)_4 + n\text{-}C_4H_9Li$ or C_2H_5MgBr	1:9	0.1 (less with H_2)	—	V22
$Ti(O\text{-}iso\text{-}C_3H_7)_4 + $ Na/naphthalene	1:5–6	1 [up to 5 with $Al(iso\text{-}C_3H_7)_3$]	N_2H_4 also formed	V7, V9, V15
$TiCl_4 + LiAlH_4 + AlBr_3$	—	125	—	V15
$TiCl_4 + Mg + MgI_2$	1:9:5	1.25	—	V16
$TiCl_4 + (iso\text{-}C_4H_9)_3Al$	1:9	0.25	—	V24
$TiCl_4 + Al + AlX_3$ (X = Cl, Br)	1:600:1000	286	—	V20, V21
$TiCl_4 + $ Li/naphthalene	1:10–15	2	—	H14
$TiCl_3 + Mg$ (in THF)	1:>2.5	1	Also absorbs H_2	Y2

255

weaken the multiple N–N bond. If the bond order (indicated by the infrared frequency) can be reduced sufficiently, the difficult cleavage of the first of the three N–N bonds should be possible under mild conditions.

B. Reactions of Individual Organotitanium Compounds with Dinitrogen

The work just discussed demonstrates that dinitrogen complexed to a low-valent transition metal can be reduced in the presence of an excess of organometallic reducing agent. It has been shown recently in several research schools that complexes of dinitrogen with organotitanium compounds, presumably similar to the intermediate complexes in fixation experiments, are capable of isolation and characterization.

Vol'pin, Shur, and collaborators (V29) showed that nitrogen would react with the thermolysis products of $Cp_2Ti(C_6H_5)_2$ and $Cp_2Ti(CH_3)_2$ at $0°–100°$ under 80 atm pressure of nitrogen to give compounds which hydrolyzed to give ammonia (approx. 0.2 mole NH_3 per mole Ti). At this temperature the N_2 is apparently reduced by the titanium compound.

The reversible absorption of dinitrogen by lower valent titanium compounds has now been demonstrated in several cases. As mentioned previously, compounds of the type $(Cp_2Ti)_n$, obtained by reduction of Cp_2TiCl_2 with alkali metals (V8) or from the gray-green hydride, $(Cp_2TiH)_x$ (B43), form the blue complex, $(Cp_2Ti)_2N_2$, below room temperature. No infrared absorption attributable to coordinated nitrogen has been observed (B43, M21), but in the visible spectrum a band around 17 kK is present. Both "edge-on" (B43) and "end-on" (B56) bonding of nitrogen have been proposed.

In contrast, van Tamelen et al. found that "titanocene" [from reduction of Cp_2TiCl_2 with sodium/naphthalene (W16)] in benzene at $20°–25°$ would absorb over 0.9 mole of nitrogen during a period of 3 weeks (V7, V10), giving a complex formulated as $[(C_5H_5)_2TiN_2]_2$, which showed in the infrared spectrum a band at 1960 cm^{-1} attributed to N–N stretch. When argon was flushed through the system after 30% uptake of N_2, the nitrogen was liberated with a corresponding loss of the 1960 cm^{-1} absorption. Addition of sodium/naphthalene to the dinitrogen complex caused reduction to nitride (V10).

At low temperature ($-100°$) in pentane, nitrogen was readily absorbed by bis(pentamethylcyclopentadienyl)titanium(II), $[\pi\text{-}C_5(CH_3)_5]_2Ti$, and was lost again at low pressures or higher temperatures. Although the initially formed unstable olive green product was thought to contain 2 moles of N_2 per mole of titanium, only the more stable purple-blue complex, $\{[\pi\text{-}C_5(CH_3)_5]_2Ti\}_2N_2$, was characterized (B43).

The triphenylphosphine complex, $[Cp_2TiP(C_6H_5)_3]_2$, prepared from the phosphine and either $(Cp_2Ti)_2$ or the hydride, $(Cp_2TiH)_2$, absorbed nitrogen at room temperature giving a green-brown solution. On heating this solution to 110° for 1 hr, only part of the nitrogen was released as such (0.07 mole per mole Ti); the remainder was reduced and appeared as ammonia (0.13 mole per mole Ti) on hydrolysis (B43).

Reduction occurred also with the ethyltitanium(III) complex, $(Cp_2TiC_2H_5)_2 \cdot 6MgCl_2 \cdot 7(C_2H_5)_2O$, isolated by Brintzinger (B43) as mentioned earlier. This complex lost the ethyl group (as ethane and ethylene) on exposure to nitrogen under pressure. Because of this, Brintzinger believes that the nitrogen-absorbing species was actually $(Cp_2Ti)_2$. The product was a black solid, which although not completely characterized, was probably a nitride since hydrolysis gave ammonia in the ratio $2NH_3 : 3Ti$.

Teuben and de Liefde Meijer (T18, T20) have recently observed reversible absorption of dinitrogen by a series of arylbis(cyclopentadienyl)titanium(III) compounds, Cp_2TiR (where $R = C_6H_5$, $2\text{-}CH_3C_6H_4$, $4\text{-}CH_3C_6H_4$, C_6F_5, or $C_6H_5CH_2$). The absorption in toluene, pentane, or ether was characterized by the appearance of a deep blue color due to an absorption band in the visible region around 17 kK. The intensities of the absorption at various temperatures in the case of $Cp_2TiC_6H_5$, are shown in Fig. VIII-1 [from ref. (T20)]. There was no observable reaction when $R = 2,6\text{-}(CH_3)_2C_6H_3$ or $2,4,6\text{-}(CH_3)_3C_6H_2$, possibly for steric reasons.

The spectra of the complexes from the other aryl derivatives are quite similar. On cooling a toluene solution of the phenyl complex to $-78°$, purple crystals were obtained (T20) and characterized as $[Cp_2Ti(C_6H_5)N]_n$. On the

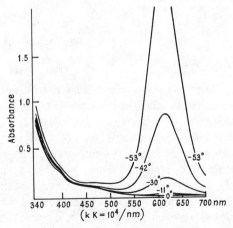

Fig. VIII-1. Absorption spectrum of $Cp_2TiC_6H_5$ in toluene (2×10^{-3} M) in a nitrogen atmosphere (1 atm) at various temperatures (T20).

Table VIII-2
Dinitrogen Complexes of Cp_2TiR^a

R in $(Cp_2TiR)_2N_2$	Heat of formation (kcal/mole)	Decomposition temp. (°C)	
		N_2 complex	Cp_2TiR
C_6H_5	−18	70	29
$2\text{-}CH_3C_6H_4$	−9	< −20	75
$3\text{-}CH_3C_6H_4$	−17	75	<20
$4\text{-}CH_3C_6H_4$	−20	65	<20
$2,6\text{-}(CH_3)_2C_6H_3$	—	Not formed	>180
C_6F_5	−17	20	106
$C_6H_5CH_2$	−14	60	43
$iso\text{-}C_3H_7{}^b$	−5	−70	—

^{a} Data from ref. (T18, T20).
^{b} From ref. (S36).

basis of the ready loss of nitrogen on solvolysis or warming, it is probable that the nitrogen molecule is intact in the complexes, which are, therefore, probably binuclear, i.e., $(Cp_2TiR)_2N_2$. The equilibrium is

$$2Cp_2TiR + N_2 \rightleftharpoons (Cp_2TiR)_2N_2$$

The heats of formation of the complexes have been calculated (T20) and lie in the range −9 to −20 kcal/mole. All are diamagnetic, contrary to the N_2-free compounds, and show no absorption in the infrared around 2000 cm^{-1} due to N≡N stretch. The dinitrogen is, therefore, symmetrically bound in one of two possible ways.

$$
\begin{array}{ccccc}
\text{R} & \text{Cp} & & \text{R} & \text{Cp} \\
| & \text{N} \quad | & & | & | \\
\text{Cp—Ti—|||—Ti—Cp} & \text{or} & \text{Cp—Ti—N≡N—Ti—Cp} \\
| & \text{N} \quad | & & | & | \\
\text{Cp} & \text{R} & & \text{Cp} & \text{R}
\end{array}
$$

The thermal stabilities of the complexes vary inversely with the stabilities of the nitrogen-free aryls, so that the least stable Cp_2TiR compounds form the most stable dinitrogen complexes (see Table VIII-2).

In many chemical reactions the complexes, $(Cp_2TiR)_2N_2$, readily lose N_2 and behave like the nitrogen-free aryl compounds; e.g., with HCl or with halogens at −78°, Cp_2TiX_2 is formed where X is a halogen.

Reduction of the complexed nitrogen occurs only when an excess of a strong reducing agent is added. Sodium/naphthalene in tetrahydrofuran reduces the complexed nitrogen giving, after acid hydrolysis, hydrazine and ammonia. The total amount of reduced nitrogen was found to increase with increasing Na:Ti ratio, reaching a maximum of one nitrogen atom per

Table VIII-3

Individual Organotitanium Compounds Which React with Nitrogen

Compound	Conditions	Proposed product	Other data	Refs.
$(Cp_2Ti)_2$	$-80°$ in toluene	$(Cp_2Ti)_2N_2{}^a$	Decomposition $> -20°$; visible 16.8 kK (ε 10^4); reduced by Li/naphthalene	B43, B56, U4
$[\pi\text{-}(CH_3)_5C_5]_2Ti$	$-100°$ in pentane	$\{[\pi\text{-}(CH_3)_5C_5]_2Ti\}_2N_2$	Decomposition $> 10°$; visible 17.6 kK; also absorbs H_2	B43
$[Cp_2TiP(C_6H_5)_3]_2$	Room temp. in toluene or solid	—	Green-brown solution	B43
$(Cp_2TiC_2H_5)_2 \cdot 6MgCl_2 \cdot 7(C_6H_5)_2O$	Room temp. in DME,^{b} 230 atm N_2	"Black nitride," C_2H_4 and C_2H_6 evolved during N_2 absorption	Hydrolysis with H_2SO_4 gave 0.6 mole NH_3 per mole Ti; "acts as source of $(Cp_2Ti)_2$"	B43
$Cp_2Ti(iso\text{-}C_3H_7)$	$-80°$ in ether	$(Cp_2TiC_3H_7)_2N_2$	Blue complex, liberates N_2 with HCl unless first treated with C_3H_7MgCl when N_2H_4 results	S36
Cp_2TiR R = C_6H_5, C_6F_5, and substituted phenyls in Table VIII-2	$-80°$ in toluene	$(Cp_2TiR)_2N_2$	Blue crystal complexes; visible $\sim$16.5 kK	T18, T20

^{a} A similar complex has been obtained by Shilov and co-workers (B56) from $Cp_2TiCl + CH_3MgI + N_2$ at $-70°$ (see text).
^{b} DME, 1,2-Dimethoxyethane.

titanium at a Na:Ti ratio of 2. One-half of the reduced nitrogen was obtained as hydrazine, the rest as ammonia.

Chivers and Ibrahim (C29) observed the development of a blue color in tetrahydrofuran solutions of the green, one-electron reduction products of $Cp_2Ti(Cl)R$ (where $R = CH_3$ or C_6F_5). These reduction products were presumed to be the organotitanium(III) compounds, Cp_2TiR, and since the blue color developed under an argon as well as under a nitrogen atmosphere it is not due to a nitrogen complex. It is believed that interaction with solvent or electrolyte, or alternatively, that isomerization induced by the solvent, had occurred. The complexes absorbed in the visible spectrum at 17.2 kK.

C. Nitrogen Incorporation into Organic Molecules

In all the above reactions dinitrogen has been reduced to ammonia or hydrazine via a transition metal intermediate. The possibility of introducing nitrogen into organic compounds by reaction of dinitrogen with organometallic compounds continues to be explored. For example, aromatic organolithium compounds react with nitrogen at room temperature in the presence of halides of titanium (and other transition metals). Hydrolysis of the reaction mixtures gives amines, the amino groups of which are in the ring positions previously occupied by lithium unless the presence of ortho substituents causes a shift to the meta position (V15, V28).

$$ArLi + N_2 \xrightarrow[\text{2. H}_2\text{O}]{\text{1. MX}_n} ArNH_2 + NH_3$$

$$MX_n = TiX_4, CpTiCl_3, Cp_2TiCl_2, Ti(OC_4H_9)_4, \text{etc.}$$

In like manner, ferrocenyllithium gives, in addition to biferrocenyl, ferrocenylamine in the presence of $Cp_2TiCl_2 + N_2$ (D28). In addition, aromatic hydrocarbons such as naphthalene and biphenyl have been converted to amino derivatives by treatment with nitrogen in the presence of metallic lithium or sodium and titanium compounds (e.g., $TiCl_4$) (V15).

The formation of aniline (1–2%) as well as ammonia in the reaction of $Cp_2Ti(C_6H_5)_2$ with dinitrogen has been observed by Shur, Berkovich, and Vol'pin (S37). Since the reaction was carried out at 80°–130° in benzene and ether, the reactant was probably the thermolysis product of $Cp_2Ti(C_6H_5)_2$, namely, $Cp_2Ti(C_6H_4)$, a phenylene, or dehydrobenzene derivative (D40). Formation of aniline is believed to occur by insertion of N_2 and subsequent reduction (S37, V15).

$$\left[Cp_2Ti \bigcirc \right] + N_2 \longrightarrow Cp_2Ti \underset{}{\overset{N=N}{\diagup}} \bigcirc \xrightarrow[H_2O]{e^-} C_6H_5NH_2 + NH_3$$

The production of amines under conditions of titanium–carbon bond formation in the presence of nitrogen would seem to indicate that the latter can participate in insertion reactions, just as carbon monoxide and ethylene do. For example, the systems Cp_2TiCl_2 + C_6H_5Li (excess) or $Cp_2Ti(C_6H_5)_2$ + C_6H_5Li react with nitrogen at room temperature, giving after hydrolysis approximately 0.15 and 0.65 mole of aniline, respectively, per mole of titanium compound (L10, L11, V27). Cp_2TiCl_2 with p-tolyllithium (excess) and m-tolyllithium similarly produced p-toluidine and m-toluidine on hydrolysis (together with ammonia). However, the production of all three isomeric toluidines from o-tolyllithium under the same conditions forced Vol'pin and collaborators to consider the possibility of a dehydrobenzene intermediate, e.g.,

$$-Ti\underset{}{\overset{CH_3}{\diagdown}}\bigcirc \rightleftharpoons -Ti-H + \bigcirc\overset{CH_3}{} \longrightarrow -Ti\underset{}{\overset{CH_3}{\diagup}}\bigcirc , \text{ etc.}$$

The possibilities have been discussed at length by Vol'pin (V15) without reaching any definite conclusions regarding mechanism. Aliphatic amines appear to be formed from alkyllithium compounds and Cp_2TiCl_2, either in very low yield or not at all.

Another system which allows incorporation of nitrogen into organic molecules is Cp_2TiCl_2 reduced under nitrogen with magnesium powder in tetrahydrofuran (V7, V11). On addition of carbonyl compounds, a reductive deoxygenation occurs giving, on hydrolysis, primary and secondary amines from ketones and aldehydes and benzonitrile from benzoyl chloride. The intermediate is probably a nitride.

$$[N^{3-}] + R\overset{|}{C}O \xrightarrow{H_2O} R-\overset{|}{C}HNH_2$$

The system has been reinvestigated by French workers (D28), using Cp_2TiCl_2 reduced by C_2H_5MgBr, Na/naphthalene, Mg/MgI_2, or Li/diglyme in the presence of nitrogen. Low yields ($<4\%$) of aromatic amines were obtained.

Attempted conversion of α-keto esters to α-amino esters gave only traces (0.4%) with phenylglyoxylic ester and none in the case of ethyl pyruvate.

Bibliography

A1.   Abel, E. W., and Jenkins, C. R., *J. Organometal. Chem.* **14**, 285 (1968).
A2.   Adema, E. H., *J. Polym. Sci., Part C* **16**, 3643 (1968).
A3.   Alcock, N. W., *J. Chem. Soc., A* p. 2001 (1967).
A4.   Aleksandrov, G. G., and Struchkov, Yu. T., *Zh. Strukt. Khim.* **12**, 667 (1971).
A5.   Alekseev, N. V., and Ronova, I. A., *Zh. Strukt. Khim.* **7**, 103 (1966).
A6.   Alexander, I. J., Allen, P. E. M., and Brown, J. K., *Eur. Polym. J.* **1**, 111 (1965).
A7.   Allegra, G., and Ganis, P., *Atti Accad. Naz. Lincei, Cl. Sci. Fis., Mat. Natur., Rend.* [8] **33**, 438 (1962).
A8.   Allegra, G., Ganis, P., Porri, L., and Corradini, P., *Atti Accad. Naz. Lincei, Cl. Sci. Fis., Mat. Natur., Rend.* [8] **30**, 44 (1961).
A9.   Allen, P. E. M., Brown, J. K., and Obaid, R. M. S., *Trans. Faraday Soc.* **59**, 1808 (1963).
A10. Allen, P. E. M., and Obiad, R. M. S., *Makromol. Chem.* **80**, 54 (1964).
A11. Alyea, E. C., Bradley, D. C., Lappert, M. F., and Sanger, A. R., *Chem. Cummun.* p. 1064 (1969).
A12. Anagnostopoulos, A., and Nicholls, D., *J. Inorg. Nucl. Chem.* **27**, 339 (1965).
A13. Anagnostopoulos, A., and Nicholls, D., *J. Inorg. Nucl. Chem.* **28**, 3045 (1966).
A14. Andrä, K., *Z. Chem.* **7**, 318 (1967).
A15. Andrä, K., *J. Organometal. Chem.* **11**, 567 (1968).
A16. Andrä, K., and Hille, E., *Z. Chem.* **8**, 65 (1968).
A17. Andrä, K., and Hille, E., *Z. Naturforsch. B* **24**, 169 (1969).
A18. Anton, E., Cleve, W., and Weise, W., Ger. (East) Patent Appl. 43,301 (1964).
A19. Arlman, E. J., and Cossee, P., *J. Catal.* **3**, 99 (1964).
A20. Armstrong, D. R., Perkins, P. G., and Stewart, J. J. P., *J. Chem. Soc., Dalton Trans.* p. 1972 (1972).
A20a. Attridge, C. J., Jackson, R., Maddock, S. J., and Thompson, D. T., *J. Chem. Soc. Chem. Commun.* p. 132, 1973.
A21. Aver'yanov, A. A., Fil'bert, D. V., Chirkov, N. M., Brikenshtein, A. A., Belov, G. P., and Gerasina, N. P., *Sin. Volokna* p. 44 (1969).
A22. Azuma, K., Shikata, K., Oba, S., Nishino, K., and Matsumura, T., *Kogyo Kagaku Zasshi* **68**, 347 (1965).

A23. Azuma, K., Shikata, K., and Yokokawa, K., Jap. Patent 68/13,048 (1968).
A24. Azuma, K., Shikata, K., Yokogawa, K., and Nakao, S., *Kogyo Kagaku Zasshi* 68, 1245 (1965).
B1.   Babkina, O. N., Grigoryan, E. A., D'yachkovskii, F. S., Shilov, A. E., and Shuvalova, N. I., *Zh. Fiz. Khim.* 43, 1759 (1969).
B2.   Badin, E. J., *J. Phys. Chem.* 63, 1791 (1959).
B3.   Badische Anilin-und Soda-Fabrik, A.-G., Fr. Patent 1,584,912 (1970).
B4.   Bainziger, N. C., Haight, H. W., and Doyle, J. R., *Inorg. Chem.* 3, 1535 (1964).
B5.   Ballard, D. G. H., Janes, W. H., and Medinger, T., *J. Chem, Soc., B* p. 1168 (1968).
B6.   Ballard, D. G. H., Janes, W. H., and Seddon, J. D., Brit. Patent 1,099,116 (1968).
B7.   Ballard, D. G. H., and van Lienden, P. W., *Chem. Commun.* p. 564 (1971).
B8.   Ballhausen, C. J., and Dahl, J. P., *Acta Chem. Scand.* 15, 1333 (1961).
B9.   Bamford, C. H., *J. Polym. Sci., Part C*, 4, 1571 (1964).
B10. Bamford, C. H., and Finch, C. A., *Z. Naturforsch. B* 17, 500 (1962).
B11. Barber, W. A., *J. Inorg. Nucl. Chem.* 4, 373 (1957).
B12. Barkdoll, A. E., and Lorenz, J. C., U.S. Patent 3,038,915 (1962).
B13. Bartelink, H. J. M., Bos, H., Smidt, J., Vrinssen, C. H., and Adema, E. H., *Rec. Trav. Chim. Pays-Bas* 81, 225 (1962).
B14. Bartelink, H. J. M., Bos, H., van Raayen, W., and Smidt, J., *Arch. Sci.* 14, 158 (1961),
B15. Bartlett, P. D., and Seidel, B., *J. Amer. Chem. Soc.* 83, 581 (1961).
B16. Bassi, I. W., Allegra, G., Scordamaglia, R., and Chioccola, G., *J. Amer. Chem. Soc.* 93, 3787 (1971).
B17. Batsanov, S. S., *Izv. Sib. Otd. Akad. Nauk SSSR, Ser. Khim. Nauk* p. 110 (1962).
B18. Bawn, C. E. H., and Gladstone, J., *Proc. Chem. Soc., London* p. 227 (1959).
B19. Baye, L. J., *Diss. Abstr.* 24, 3968 (1964).
B20. Bayer, E., and Schurig, V., *Chem. Ber.* 102, 3378 (1969).
B21. Beachell, H. C., and Butter, S. A., *Inorg. Chem.* 4, 1133 (1965).
B22. Becconsall, J. K., Job, B. E., and O'Brien, S., *J. Chem. Soc., A* p. 423 (1967).
B23. Becconsall, J. K., and O'Brien, S., *Chem. Commun.* p. 302 (1966).
B24. Beermann, C., *Angew. Chem.* 71, 195 (1959).
B25. Beermann, C., Ger. Patent 1,089,382 (1960).
B26. Beermann, C., Ger. Patent 1,091,331 (1960).
B27. Beermann, C., Ger. Patent 1,100,022 (1961).
B28. Beermann, C., and Bestian, H., Ger. Patent 1,026,964 (1958).
B29. Beermann, C., and Bestian, H., *Angew. Chem.* 71, 618 (1959).
B30. Beermann, C., and Bestian, H., U.S. Patent 3,037, 971 (1962).
B31. Beermann, C., Graf., R., and Bestian, H., Ger. Patent 1,023,766 (1958).
B32. Behar, D., and Feilchenfeld, H., *J. Organometal. Chem.* 4, 278 (1965).
B33. Behrens, H., and Brandl, H., *Z. Naturforsch. B* 22, 1216 (1967).
B34. Belov, G. P., Bogomolova, N. B., Tsvetkova, V. I., and Chirkov, N. M., *Kinet. Katal.* 8, 265 (1967).
B35. Belov, G. P., Kaltochikhina, S. K., Atanasova, L. I., Tsvetkova, V. I., and Chirkov, N. M., *Izv. Akad. Nauk SSSR, Ser. Khim.* p. 1275 (1966).
B36. Belov, G. P., Kuznetsov, V. I., Solov'eva, T. I., Chirkov, N. M., and Ivanchev, S. S., *Makromol. Chem.* 140, 213 (1970).
B37. Belov, G. P., Lisitskaya, A. P., and Chirkov, N. M., *Izv. Akad. Nauk SSSR, Ser. Khim.* p. 2650 (1968).
B38. Belov, G. P., Lisitskaya, A. P., Chirkov, N. M., and Tsvetkova, V. I., *Vysokomol. Soedin., Ser. A* 9, 1269 (1967).

B39. Belov, G. P., Raspopov, L. N., Lisitskaya, A. P., Tsvetkova, V. I., and Chirkov, N. M., *Vysokomol. Soedin.* **8**, 1568 (1966).
B40. Benedek, I., Asandei, N., and Simionescu, Cr. I., *Eur. Polym. J.* **7**, 995 (1971).
B41. Bercaw, J. E., and Brintzinger, H. H., *J. Amer. Chem. Soc.* **91**, 7301 (1969).
B42. Bercaw, J. E., and Brintzinger, H. H., *J. Amer. Chem. Soc.* **93**, 2045 (1971).
B43. Bercaw, J. E., Marvich, R. H., Bell, L. G., and Brintzinger, H. H., *J. Amer. Chem. Soc.* **94**, 1219 (1972).
B44. Berlin, A. A., Cherkashin, M. I., Kisilitsa, P. P., and Pirogov, O. N., *Vysokomol. Soedin., Ser A* **9**, 1835 (1967).
B45. Berthold, H. J., and Groh, G., *Z. Anorg. Allg. Chem.* **319**, 230 (1962).
B46. Berthold, H. J., and Groh, G., *Angew. Chem., Int. Ed. Engl.* **5**, 516 (1966).
B47. Berthold, H. J., and Groh, G., *Z. Anorg. Allg. Chem.* **372**, 292 (1970).
B48. Bestian, H., and Clauss, K., Ger. Patent 1,037,446 (1958).
B49. Bestian, H., and Clauss, K., *Angew. Chem., Int. Ed. Engl.* **2**, 704 (1963).
B50. Birmingham, J. M., Fischer, A. K., and Wilkinson, G., *Naturwissenschaften* **42**, 96 (1955).
B51. Blackmore, T., Bruce, M. I., Davidson, P. J., Iqbal, M. Z., and Stone, F. G. A., *J. Chem. Soc., A* p. 3153 (1970).
B52. Blumenthal, W., *Ind. Eng. Chem.* **55**, 51 (1963).
B53. Bochvar, D. A., and Chistyakov, A. L., *Zh. Strukt. Khim.* **9**, 267 (1968).
B54. Boor, J., *Macromol. Rev.* **2**, 115 (1967).
B55. Boor, J., *Ind. Eng. Chem., Prod. Res. Develop.* **9**, 437 (1970).
B55a. Borodko, Yu. G., Ivleva, I. N., Kachapina, L. M., Kvashina, E. F., Shilova, A. K., and Shilov, A. E., *J. Chem. Soc. Chem. Commun.* p. 169 (1973).
B56. Borodko, Yu.G., Ivleva, I. N., Kachapina, L. M., Salienko, S. I., Shilova, A. K., and Shilov, A. E., *J. Chem. Soc., Chem. Commun.* p. 1178 (1972).
B57. Bourn, A. J. R., Gilles, D. G., and Randall, E. W., *Proc. Chem. Soc., London* p. 200 (1963).
B58. Boustany, K. S., Bernauer, K., and Jacot-Guillarmod, A., *Helv. Chim. Acta* **50**, 1080 (1967).
B59. Boustany, K. S., Bernauer, K., and Jacot-Guillarmod, A., *Helv. Chim. Acta* **50**, 1305 (1967).
B60. Bower, B. K., and Tennent, H. G., *J. Amer. Chem. Soc.* **94**, 2512 (1972).
B61. Bradley, D. C., and Chivers, K. J., *J. Chem. Soc., A* p. 1967 (1968).
B62. Bradley, D. C., and Kasenally, A. S., *Chem. Commun.* p. 1430 (1968).
B63. Bradley, H. B., and Dowell, L. G., *Anal. Chem.* **30**, 548 (1958).
B64. Brainina, E. M., and Dvoryantseva, G. G., *Izv. Akad. Nauk SSSR, Ser. Khim.* p. 442 (1967).
B65. Brainina, E. M., Dvoryantseva, G. G., and Freidlina, R. Kh., *Dokl. Akad. Nauk SSSR* **156**, 1375 (1964).
B66. Brainina, E. M., Fedorov, L. A., and Minacheva, M. Kh., *Dokl Akad. Nauk SSSR* **196**, 1085 (1971).
B67. Brainina, E. M., and Freidlina, R. Kh., *Izv. Akad. Nauk SSSR, Otd. Khim. Nauk* p. 835 (1963).
B68. Brainina, E. M., and Freidlina, R. Kh., *Izv. Akad. Nauk SSSR, Ser Khim.* p. 1421 (1964).
B69. Brainina, E. M., Freidlina, R. Kh., and Nesmeyanov, A. N., *Dokl. Akad. Nauk SSSR* **154**, 1113 (1964).
B70. Brainina, E. M., Gambaryan, N. P., Lokshin, B. V., Petrovskii, P. V., Struchkov, Yu. T., and Kharlamova, E. N., *Izv. Akad. Nauk SSSR, Ser. Khim.* p. 87 (1972).

B70a. Brainina, E. M., Minacheva, M. Kh., and Fedorov, L. A., *Izv. Akad. Nauk SSSR, Ser. Khim.* p. 2356 (1971).

B71. Brainina, E. M., Minacheva, M. Kh., and Freidlina, R. Kh., *Izv. Akad. Nauk SSSR, Otd. Khim. Nauk.* p. 1716 (1961).

B72. Brainina, E. M., Minacheva, M. Kh., and Freidlina, R. Kh., *Izv. Akad. Nauk SSSR, Ser. Khim.* p. 1877 (1965).

B73. Brainina, E. M., Minacheva, M. Kh., and Lokshin, B. V., *Izv. Akad. Nauk SSSR, Ser. Khim.* p. 817 (1968).

B74. Brainina, E. M., Minacheva, M. Kh., Lokshin, B. V., Fedin, E. I., and Petrovskii, P. V., *Izv. Akad. Nauk SSSR, Ser. Khim.* p. 2492 (1969).

B75. Brainina, E. M., Mortikova, E. I., Petrashkevich, L. A., and Freidlina, R. Kh., *Dokl. Akad. Nauk SSSR* **169**, 335 (1966).

B76. Brantley, J. C., U.S. Patent 2,983,741 (1961).

B77. Brantley, J. C., Morehouse, E. L., and Parts, L., U.S. Patent Appl. 3,306,919 (1958).

B78. Braterman, P. S., and Cross, R. J., *J. Chem. Soc., Dalton Trans.* p. 657 (1972).

B79. Braterman, P. S., and Wilson, V. A., *J. Organometal. Chem.* **31**, 131 (1971).

B80. Braterman, P. S., Wilson, V. A., and Joshi, K. K., *J. Chem. Soc., A* p. 191 (1971).

B81. Braterman, P. S., Wilson, V. A., and Joshi, K. K., *J. Organometal. Chem.* **31**, 123 (1971).

B82. Brauer, D. J., and Krüger, C., *J. Organometal. Chem.* **42**, 129 (1972).

B83. Braye, E. H., Hübel, W., and Caplier, I., *J. Amer. Chem. Soc.* **83**, 4406 (1961).

B84. Breederveld, H., and Waterman, H. I., U.S. Patent 3,089,886 (1963).

B85. Breil, H., and Wilke, G., *Angew. Chem., Int. Ed. Engl.* **5**, 898 (1966).

B86. Breslow, D. S., U.S. Patent 2,827,446 (1958).

B87. Breslow, D. S., U.S. Patent 2,924,593 (1960).

B88. Breslow, D. S., and Newburg, N. R., *J. Amer. Chem. Soc.* **79**, 5072 (1957).

B89. Breslow, D. S., and Newburg, N. R., *J. Amer. Chem. Soc.* **81**, 81 (1959).

B90. Brintzinger, H., *J. Amer. Chem. Soc.* **88**, 4305 (1966).

B91. Brintzinger, H. H., *J. Amer. Chem. Soc.* **89**, 6871 (1967).

B92. Brintzinger, H. H., and Bartell, L. S., *J. Amer. Chem. Soc.* **92**, 1105 (1970).

B93. Brintzinger, H. H., and Bercaw, J. E., *J. Amer. Chem. Soc.* **92**, 6182 (1970).

B94. Bruce, M. I., Harbourne, D. A., Waugh, F., and Stone, F. G. A., *J. Chem. Soc. A* p. 356 (1968).

B95. Bruce, M. I., and Thomas, M. A., *Org. Mass. Spectrom.* **1**, 835 (1968).

B96. Brüser, W., Thiele, K.-H., Zdunneck, P., and Brune, F., *J. Organometal. Chem.* **32**, 335 (1971).

B97. Bürger, H., and Neese, H.-J., *J. Organometal. Chem.* **20**, 129 (1969).

B98. Bürger, H., and Neese, H.-J., *J. Organometal. Chem.* **21**, 381 (1970).

B99. Bürger, H., and Neese, H.-J., *J. Organometal. Chem.* **36**, 101 (1972).

B100. Burmeister, J. L., Deardorff, E. A., Jensen, A., and Christiansen, V. H., *Inorg. Chem.* **9**, 58 (1970).

B101. Burmeister, J. L., Deardorff, E. A., and van Dyke, C. E., *Inorg. Chem.* **8**, 170 (1969).

B102. Bush, M. A., and Sim, G. A., *J. Chem. Soc., A* p. 2225 (1971).

C1. Cahours, M. A., *Ann. Chim. (Paris)* [3] **62**, 280 (1861).

C2. Calderazzo, F., Salzmann, J. J., and Mosimann, P., *Inorg. Chim. Acta.* **1**, 65 (1967).

C3. Calderon, J. L., Cotton, F. A., DeBoer, B. G., and Takats, J., *J. Amer. Chem. Soc.* **92**, 3801 (1970).

C4.   Calderon, J. L., Cotton, F. A., DeBoer, B. G., and Takats, J., *J. Amer. Chem. Soc. Soc.* **93**, 3592 (1971).

C5.   Calderon, J. L., Cotton, F. A., and Takats, J., *J. Amer. Chem. Soc.* **93**, 3587 (1971).

C6.   Cameron, T. S., Prout, C. K., Rees, G. V., Green, M. L. H., Joshi, K. K., Davies, G. R., Kilbourn, B. T., Braterman, P. S., and Wilson, V. A., *Chem. Commun.* p. 14 (1971).

C7.   Cannell, L. G., *J. Amer. Chem. Soc.* **94**, 6867 (1972).

C8.   Canty, A. J., Coutts, R. S. P., and Wailes, P. C., *Aust. J. Chem.* **21**, 807 (1968).

C9.   Cardin, D. J., Keppie, S. A., Kingston, B. M., and Lappert, M. F., *Chem. Commun.* p. 1035 (1967).

C9a.   Carraher, Jr., C. E., and Bajah, S. T., *Polymer* **14**, 42 (1973).

C9b.   Carraher, Jr., C. E., and Nordin, R. J., *Makromol. Chem.* **164**, 87 (1973).

C10. Carrick, W. L., Chasar, A. G., and Smith, J. J., *J. Amer. Chem. Soc.* **82**, 5319 (1960).

C11. Carrick, W. L., Karol, F. J., and Karapinka, G. L., U.S. Patent Appl. 3,173,902 (1959).

C12. Carrick, W. L., Karol, F. J., and Karapinka, G. L., Brit. Patent 890,955 (1962).

C13. Carter, J. C., *Proc. Int. Conf. Coord. Chem., 8th, 1964* p. 178, (1964).

C13a. Causse, J., Tabacchi, R., and Jacot-Guillarmod, A., *Helv. Chim. Acta* **55**, 1560 (1972).

C14. Cetinkaya, B., Lappert, M. F., and McMeeking, J., *Chem. Commun.* p. 215 (1971).

C15. Chandra, G., and Lappert, M. F., *Inorg. Nucl. Chem. Lett.* **1**, 83 (1965).

C16. Chandra, G., and Lappert, M. F., *J. Chem. Soc., A* p. 1940 (1968).

C17. Charalambous, J., Frazer, M. J., and Newton, W. E., *J. Chem. Soc., A* p. 2487 (1971).

C18. Chatt, J., Heath, G. A., and Richards, R. L., *J. Chem. Soc., Chem. Commun.* p. 1010 (1972).

C19. Chatt, J., and Leigh, G. J., *Chem. Soc. Rev.* **1**, 121 (1972).

C20. Chatt, J., and Shaw, B. L., *J. Chem. Soc., London* p. 705 (1959).

C21. Chatt, J., and Shaw, B. L., *J. Chem. Soc., London* p. 285 (1961).

C22. Chaudhari, M. A., and Stone, F. G. A., *J. Chem. Soc., A* p. 838 (1966).

C23. Chaudhari, M. A., Treichel, P. M., and Stone, F. G. A., *J. Organometal. Chem.* **2**, 206 (1964).

C24. Chien, J. C. W., *J. Amer. Chem. Soc.* **81**, 86 (1959).

C25. Chien, J. C. W., *J. Phys. Chem.* **67**, 2477 (1963).

C26. Chien, J. C. W., *J. Catal.* **23**, 71 (1971).

C27. Chien, J. C. W., and Boss, C. R., *J. Amer. Chem. Soc.* **83**, 3767 (1961).

C28. Ch'iu, T.-W., Chu, Y.-F., Tang, H.-M., Hsiung, F.-C., and Kung, H.-Y., *Wu Li Hsueh Pao* **17**, 600 (1961); *Chem. Abstr.* **59**, 12328f (1963).

C29. Chivers, T., and Ibrahim, E. D., *Can. J. Chem.* **51**, 815 (1973).

C30. Clark, R. J. H., "The Chemistry of Titanium and Vanadium." Elsevier, Amsterdam, 1968.

C31. Clark, R. J. H., and Coles, M., *Chem. Commun.* p. 1587 (1971).

C31a. Clark, R. J. H., and Coles, M. A., *J. Chem. Soc., Dalton Trans.* p. 2454 (1972).

C32. Clark, R. J. H., and McAlees, A. J., *J. Chem. Soc., A* p. 2026 (1970).

C33. Clark, R. J. H., and McAlees, A. J., *Inorg. Chem.* **11**, 342 (1972).

C34. Clark, R. J. H., and McAlees, A. J., *J. Chem. Soc., Dalton Trans.* p. 640 (1972).

C35. Clauss, K., *Justus Liebigs Ann. Chem.* **711**, 19 (1968).

C36. Clauss, K., and Beermann, C., Ger. Patent 1,046,048 (1958).

C37. Clauss, K., and Beermann, C., *Angew. Chem.* **71**, 627 (1959).

C38. Clauss, K., and Bestian, H., *Justus Liebigs. Ann. Chem.* **654**, 8 (1962).

C39. Cohen, S. C., and Massey, A. G., *J. Organometal. Chem.* **10**, 471 (1967).

C40. Collier, M. R., Lappert, M. F., and McMeeking, J., *Abstr. Int. Conf. Coord. Chem.*, *13th, 1970* p. 321 (1970).

C41. Collier, M. R., Lappert, M. F., and McMeeking, J., *Inorg. Nucl. Chem. Lett.* **7**, 689 (1971).

C41a. Collier, M. R., Lappert, M. F., and Pearce, R., *J. Chem. Soc. Dalton Trans.* p. 445 (1973).

C42. Collier, M. R., Lappert, M. F., and Truelock, M. M., *J. Organometal. Chem.* **25**, C36 (1970).

C43. Corradini, P., and Allegra, G., *J. Amer. Chem. Soc.* **81**, 5510 (1959).

C44. Corradini, P., and Bassi, I. W., *Atti. Accad. Naz. Lincei, Cl. Sci. Fis., Mat. Natur., Rend.* [8] **24**, 43 (1958).

C45. Corradini, P., and Sirigu, A., *Inorg. Chem.* **6**, 601 (1967).

C46. Cossee, P., *J. Catal.* **3**, 80 (1964); *Rec. Trav. Chim. Pays-Bas* **85**, 1151 (1966).

C47. Coutts, R. S. P., Ph.D. Thesis, University of Melbourne (1972).

C48. Coutts, R. S. P., Kautzner, B., and Wailes, P. C., *Aust. J. Chem.* **22**, 1137 (1969).

C49. Coutts, R. S. P., Martin, R. L., and Wailes, P. C., *Aust. J. Chem.* **24**, 1079 (1971).

C50. Coutts, R. S. P., Martin, R. L., and Wailes, P. C., *Aust. J. Chem.* **24**, 2533 (1971).

C51. Coutts, R. S. P., Martin, R. L., and Wailes, P. C., *Aust. J. Chem.* **25**, 1401 (1972).

C52. Coutts, R. S. P., Martin, R. L., and Wailes, P. C., *Aust. J. Chem.* **26**, 47 (1973).

C53. Coutts, R. S. P., Martin, R. L., and Wailes, P. C., *J. Organometal. Chem.* **50**, 145 (1973).

C54. Coutts, R. S. P., Martin, R. L., and Wailes, P. C., *Aust. J. Chem.* **26**, 941 (1973).

C55. Coutts, R. S. P., Martin, R. L., and Wailes, P. C., *Inorg. Nucl. Chem. Lett.* **9**, 49 (1973).

C56. Coutts, R. S. P., Martin, R. L., and Wailes, P. C., *Inorg. Nucl. Chem. Lett.* **9**, 981 (1973).

C57. Coutts, R. S. P., Martin, R. L., and Wailes, P. C., *J. Organometal. Chem.* **47** 375 (1973).

C58. Coutts, R. S. P., Martin, R. L., and Wailes, P. C., *Aust. J. Chem.* **26**, 2101 (1973).

C59. Coutts, R. S. P., Martin, R. L., and Wailes, P. C., *Aust. J. Chem.* (1973) (in preparation).

C60. Coutts, R. S. P., and Surtees, J. R., *Aust. J. Chem.* **19**, 387 (1966).

C61. Coutts, R. S. P., Surtees, J. R., Swan, J. M., and Wailes, P. C., *Aust. J. Chem.* **19**, 1377 (1966).

C62. Coutts, R. S. P., and Wailes, P. C., *Aust. J. Chem.* **19**, 2069 (1966).

C63. Coutts, R. S. P., and Wailes, P. C., *Inorg. Nucl. Chem. Lett.* **3**, 1 (1967).

C64. Coutts, R. S. P., and Wailes, P. C., *Aust. J. Chem.* **20**, 1579 (1967).

C65. Coutts, R. S. P., and Wailes, P. C., *Chem. Commun.* p. 260 (1968).

C66. Coutts, R. S. P., and Wailes, P. C., *Aust. J. Chem.* **21**, 373 (1968).

C67. Coutts, R. S. P., and Wailes, P. C., *Aust. J. Chem.* **21**, 1181 (1968).

C68. Coutts, R. S. P., and Wailes, P. C., *Aust. J. Chem.* **21**, 2199 (1968).

C69. Coutts, R. S. P., and Wailes, P. C., *Aust. J. Chem.* **22**, 1547 (1969).

C70. Coutts, R. S. P., and Wailes, P. C., *Advan. Organometal. Chem.* **9**, 135 (1970).

C71. Coutts, R. S. P., and Wailes, P. C., *Aust. J. Chem.* **24**, 1075 (1971).

C72. Coutts, R. S. P., and Wailes, P. C., unpublished results.

C73. Coutts, R. S. P., Wailes, P. C., and Kingston, J. V., *Chem. Commun.* p. 1170 (1968).

C74. Coutts, R. S. P., Wailes, P. C., and Kingston, J. V., *Aust. J. Chem.* **23**, 463 (1970).

C75. Coutts, R. S. P., Wailes, P. C., and Kingston, J. V., *Aust. J. Chem.* **23**, 469 (1970).

C76. Cucinella, S., Mazzei, A., Marconi, W., and Busetto, C., *J. Macromol., Sci., Chem.* **4**, 1549 (1970).
D1.   Dahl, L. F., de Gil, E. R., and Feltham, R. D., *J. Amer. Chem. Soc.* **91**, 1653 (1969).
D2.   D'Alelio, G. F., U.S. Patent 3,060,161 (1962).
D3.   D'Alelio, G. F., U.S. Patent 3,075,957. (1963).
D4.   D'Alelio, G. F., U.S. Patent 3,090,776 (1963).
D5.   Davies, G. R., Jarvis, J. A. J., Kilbourn, B. T., and Pioli, A. J. P., *Chem. Commun.* p. 677 (1971).
D6.   Davies, G. R., Jarvis, J. A. J., and Kilbourn, B. T., *Chem. Commun.* p. 1511 (1971).
D7.   Davies, G. R., and Kilbourn, B. T., *J. Chem. Soc., A* p. 87 (1971).
D8.   Davies, N., James, B. D., and Wallbridge, M. G. H., *J. Chem. Soc., A* p. 2601 (1969).
D9.   Davis, B. R., and Bernal, I., *J. Organometal. Chem.* **30**, 75 (1971).
D10. Davison, A., and Rakita, P. E., *Inorg. Chem.* **9**, 289 (1970).
D10a. Debaerdemaeker, T., Hellner, E., Kutoglu, A., Schmidt, M., and Wilhelm, E., *Naturwissenschaften* **60**, 300 (1973).
D11. de Butts, E. H., U.S. Patent 2,992,212 (1961).
D12. de Butts, E. H., U.S. Patent 3,021,349 (1962).
D13. de Butts, E. H., and Matlack, A. S., U.S. Patent 3,021,319 (1962).
D14. de Liefde Meijer, H. J., and Jellinek, F., *Inorg. Chim. Acta* **4**, 651 (1970).
D15. Deluzarche, A., *Ann. Chim. (Paris)* [13] **6**, 661 (1961).
D16. Demarcay, E., *C. R. Acad. Sci.* **80**, 51 (1875).
D17. Dessy, R. E., King, R. B., and Waldrop, M., *J. Amer. Chem. Soc.* **88**, 5112 (1966).
D18. de Vries, H., *Rec. Trav. Chim. Pays-Bas* **80**, 866 (1961).
D19. Dicarlo, E., and Swift, H. E., *J. Phys. Chem.* **68**, 551 (1964).
D20. Dickson, R. S., and West, B. O., *Aust. J. Chem.* **14**, 555 (1961).
D21. Diedrich, B., Ger. Patent 1,924,049 (1970).
D22. Dierks, H., and Dietrich, H., *Acta Crystallogr., Sect. B* **24**, 58 (1968).
D23. Dietrich, H., and Soltwisch, M., *Angew. Chem., Int. Ed. Engl.* **8**, 765 (1969).
D24. Dijkgraaf, C., and Rousseau, J. P. G., *Spectrochim. Acta, Part A* **25**, 1455 (1969).
D25. Dillard, J. G., *Inorg. Chem.* **8**, 2148 (1969).
D26. Dillard, J. G., and Kiser, R. W., *J. Organometal. Chem.* **16**, 265 (1969).
D27. Doisneau, R. G., and Marchon, J.-C., *J. Electroanal. Chem. Interfacial Electrochem.* **30**, 487 (1971).
D28. Dormond, A., Leblanc, J.-C., Le Moigne, F., and Tirouflet, J., *C. R. Acad. Sci., Ser. C* **274**, 1707 (1972).
D29. Douglas, W. E., and Green, M. L. H., *J. Chem. Soc., Dalton Trans.* p. 1796 (1972).
D30. Doyle, G., and Tobias, R. S., *Inorg. Chem.* **6**, 1111 (1967).
D31. Doyle, G., and Tobias, R. S., *Inorg. Chem.* **7**, 2484 (1968).
D32. Drozdov, G. V., Bartashev, V. A., Maksimova, T. P., and Kozlova, N. V., *Zh. Obshch. Khim.* **37**, 2558 (1967).
D33. Drozdov, G. V., Klebanskii, A. L., and Bartashev, V. A., *Zh. Obshch. Khim.* **32**, 2390 (1962).
D34. Drozdov, G. V., Klebanskii, A. L., and Bartashev, V. A., *Zh. Obshch. Khim.* **33**, 2422 (1963).
D35. Druce, P. M., Kingston, B. M., Lappert, M. F., Spalding, T. R., and Srivastava, R. C., *J. Chem. Soc., A* p. 2106 (1969).
D36. Druce, P. M., Kingston, B. M., Lappert, M. F., Srivastava, R. C., Frazer, M. J., and Newton, W. E., *J. Chem. Soc., A* p. 2814 (1969).

D37. Drucker, A., and Daniel, J. H., Jr., *J. Polym. Sci.* 37, 553 (1959).
D38. Drummond, F. O., Jr., Ph.D. Thesis, University of Texas, Austin (1968).
D39. Dubsky, G. J., Boustany, K. S., and Jacot-Guillarmod, A., *Chimia* 24, 17 (1970).
D40. Dvorak, J., O'Brien, R. J., and Santo, W., *Chem. Commun.* p. 411 (1970).
D41. Dvoryantseva, G. G., Sheinker, Yu. N., Nesmeyanov, A. N., Nogina, O. V., Lazareva, N. A., and Dubovitskii, V. A., *Dokl. Akad. Nauk SSSR* 161, 603 (1965).
D42. D'Yachkovskii, F. S., *Vysokomol. Soedin.* 7, 114 (1965).
D43. D'Yachkovskii, F. S., and Krushch, N. E., *Zh. Ohshch. Khim.* 41, 1779 (1971).
D44. D'Yachkovskii, F. S., Khrushch, N. E., and Shilov, A. E., *Kinet. Katal.* 9, 1006 (1968).
D45. D'Yachkovskii, F. S., Shilova, A. K., and Shilov, A. E., *Vysokomol. Soedin.* 8, 308 (1966).
D46. D'Yachkovskii, F. S., Shilova, A. K., and Shilov, A. E., *J. Polym. Sci., Part C* 16, 2333 (1967).
D47. Dzhabiev, T. S., Sabirova, R. D., and Shilov, A. E., *Kinet Katal.* 5, 441 (1964).
D48. Dzhabiev, T. S., and Shilov, A. E., *Zh. Strukt. Khim.* 6, 302 (1965).
D49. Dzhabiev, T. S., and Shilov, A. E., *Zh. Strukt. Khim.* 10, 549 (1969).
E1. Edgecombe, F. H. C., *Nature (London)* 198, 1085 (1963).
E2. Elder, M., *Inorg. Chem.* 8, 2103 (1969).
E3. Elder, M., Evans, J. G., and Graham, W. A. G., *J. Amer. Chem. Soc.* 91, 1245 (1969).
E4. Ellermann, J., and Poersch, F., *Angew. Chem., Int. Ed. Engl.* 6, 355 (1967).
E5. Epstein, E. F., and Bernal, I., *Chem. Commun.* p. 410 (1970).
E5a. Epstein, E. F., and Bernal, I., *Inorg. Chim. Acta* (in press).
E6. Epstein, E. F., Bernal, I., and Köpf, H., *J. Organometal. Chem.* 26, 229 (1971).
E7. Eysel, H. H., Siebert, H., Groh., G., and Berthold, H. J., *Spectrochim. Acta, Part A* 26, 1595 (1970).
F1. Fachinetti, G., and Floriani, C., *Chem. Commun.* p. 654 (1972).
F2. Farbwerke Hoechst, A.-G., Brit. Patent 858,540 (1961).
F3. Farbwerke Hoechst, A.-G., Brit. Patent 858,541 (1961).
F4. Farbwerke Hoechst, A.-G., Belg. Patent 661,389 (1965).
F5. Farbwerke Hoechst, A.-G., Neth. Patent 6,507,858 (1965).
F6. Farbwerke Hoechst, A.-G., Neth. Patent Appl. 6,602,863 (1966).
F7. Feay, D. C., Belg. Patent 635,987 (1964).
F8. Feay, D. C., Belg. Patent 635,988 (1964).
F9. Feay, D. C., and Fujioka, G. S., *Abstr. Int. Conf. Coord. Chem., 10th, 1967* p. 1 (1967).
F10. Feld, R., and Cowe, P. L., "The Organic Chemistry of Titanium." Butterworth, London, 1965.
F11. Fel'dblyum, V. Sh., and Obeshchalova, N. V., *Dokl. Akad. Nauk SSSR* 172, 368 (1967).
F12. Felten, J. J., and Anderson, W. P., *J. Organometal. Chem.* 36, 87 (1972).
F13. Fischer, A. K., and Wilkinson, G., *J. Inorg. Nucl. Chem.* 2, 149 (1956).
F14. Fischer, E. O., and Amtmann, R., *J. Organometal. Chem.* 9, P15 (1967).
F15. Fischer, E. O., and Fontana, S., *J. Organometal. Chem.* 40, 159 (1972).
F16. Fischer, E. O., and Hafner, W., Brit. Patent 829,574 (1960).
F17. Fischer, E. O., and Löchner, A., *Z. Naturforsch. B* 15, 266 (1960).
F18. Fischer, E. O., and Röhrscheid, F., *J. Organometal. Chem.* 6, 53 (1966).
F19. Fischer, R. F., U.S. Patent 2,952,670 (1960).
F20. Floria, V. D., and Rokosz, L. F., U.S. Patent 3,362,945 (1968).

F21. Floriani, C., and Fachinetti, G., *J. Chem. Soc., Chem. Commun.* p. 790 (1972).
F22. Fowles, G. W. A., Rice, D. A., and Wilkins, J. D., *J. Chem. Soc., A* p. 1920 (1971).
F23. Frazer, M. J., and Newton, W. E., *Inorg. Chem.* 10, 2137 (1971).
F24. Freidlina, R. Kh., Brainina, E. M., Minacheva, M. Kh., and Nesmeyanov, A. N., *Izv. Akad. Nauk SSSR, Ser. Khim.* p. 1417 (1964).
F25. Freidlina, R. Kh., Brainina, E. M., and Nesmeyanov, A. N., *Dokl. Akad. Nauk SSSR* 138, 1369 (1961).
F26. Freidlina, R. Kh., Brainina, E. M., Petrashkevich, L. A., and Minacheva, M. Kh., *Izv. Akad. Nauk SSSR, Ser. Khim.* p. 1396 (1966).
F27. Fritz, H. P., *Advan. Organometal. Chem.* 1, 262 (1964).
F28. Fritz, H. P., and Paulus, E. F., *Z. Naturforsch. B* 18, 435 (1963).
F29. Fushman, E. A., Gerasina, M. P., Utkin, S. P., Raspopov, L. N., Chirkov, N. M., Brikenshtein, Kh. M. A., and Tsvetkova, V. I., *Plast. Massy* p. 3 (1966).
F30. Fushman, E. A., Potyagailo, E. D., and Chirkov, N. M., *Izv. Akad. Nauk SSSR, Ser. Khim.* 4, 715 (1971).
F31. Fushman, E. A., Tsvetkova, V. I., and Chirkov, N. M., *Dokl. Akad. Nauk SSSR* 164, 1085 (1965).
F32. Fushman, E. A., Tsvetkova, V. I., and Chirkov, N. M., *Izv. Akad. Nauk SSSR, Ser. Khim.* p. 2075 (1965).
G1.   Ganis, P., and Allegra, G., *Atti Accad. Naz. Lincei, Cl. Sci. Fis., Mat. Natur., Rend.* [8] 33, 304 (1962).
G2.   Geetha, S., and Joseph, P. T., *Indian J. Chem.* 7, 518 (1969).
G3.   Gevaert-Agfa, N. V., Neth. Patent 6,603,202 (1966).
G4.   Giannini, U., and Cesca, S., *Tetrahedron Lett.* p. 19 (1960).
G5.   Giannini, U., and Zucchini, U., *Chem. Commun.* p. 940 (1968).
G6.   Giannini, U., Zucchini, U., and Albizzati, E., *J. Polym. Sci., Part B* 8, 405 (1970).
G7.   Giddings, S. A., U.S. Patent 3,030,394 (1962).
G8.   Giddings, S. A., U.S. Patent 3,030,395 (1962).
G9.   Giddings, S. A., U.S. Patent Appl. 3,226,363 (1962).
G10. Giddings, S. A., U.S. Patent Appl. 3,226,369 (1962).
G11. Giddings, S. A., *Inorg. Chem.* 3, 684 (1964).
G12. Giddings, S. A., *Inorg. Chem.* 6, 849 (1967).
G13. Giddings, S. A., U.S. Patent 3,347,887 (1967).
G14. Giddings, S. A., and Best, R. J., *J. Amer. Chem. Soc.* 83, 2393 (1961).
G15. Giggenbach, W., and Brubaker, C. H., Jr., *Inorg. Chem.* 7, 129 (1968).
G16. Gilman, H., and Jones, R. G., *J. Org. Chem.* 10, 505 (1945).
G17. Gilman, H., Jones, R. G., and Woods, L. A., *J. Amer. Chem. Soc.* 76, 3615 (1954).
G18. Gordon, H. B., *Diss. Abstr. Int. B* 30, 4561 (1970).
G19. Gorsich, R. D., *J. Amer. Chem. Soc.* 80, 4744 (1958).
G20. Gorsich, R. D., *J. Amer. Chem. Soc.* 82, 4211 (1960).
G21. Gorsich, R. D., U.S. Patent 2,952,697 (1960).
G22. Gorsich, R. D., U.S. Patent 3,072,691 (1963).
G23. Gorsich, R. D., U.S. Patent 3,080,305 (1963).
G24. Gorsich, R. D., U.S. Patent 3,161,629 (1964).
G25. Gray, A. P., *Can. J. Chem.* 41, 1511 (1963).
G26. Gray, A. P., Callear, A. B., and Edgecombe, F. H. C., *Can. J. Chem.* 41, 1502 (1963).
G27. Gray, D. R., and Brubaker, C. H., Jr., *Abstr. 161st Nat. Meet., Amer. Chem. Soc., Los Angeles* p. INOR 66 (1971).

G28. Gray, D. R., and Brubaker, C. H., Jr., *Inorg. Chem.* **10**, 2143 (1971).
G29. Green, J. C., Green, M. L. H., and Prout, C. K., *Chem. Commun.* p. 421 (1972).
G30. Green, M. L. H., and Lucas, C. R., *J. Chem. Soc., Dalton Trans.* p. 1000 (1972).
G31. Grigoryan, E. A., D'yachkovskii, F. S., Khvostik, G. M., and Shilov, A. E., *Vysokomol. Soedin. Ser. A* **9**, 1233 (1967).
G32. Grigoryan, E. A., D'yachkovskii, F. S., Semenova, N. M., and Shilov, A. E., *Kinet. Mech. Polyreactions, Int. Symp. Macromol. Chem., Prepr.* **2**, 267 (1969).
G33. Grigoryan, E. A., D'yachkovskii, F. S., and Shilov, A. E., *Vysokomol. Soedin.* **7**, 145 (1965).
G33a. Grigoryan, E. A., Semenova, N. M., and D'yachkovskii, F. S., *Izv. Akad. Nauk SSSR, Ser. Khim.* p. 2719 (1972).
G34. Groenewege, M. P., *Z. Phys. Chem. (Frankfurt am Main)* [N.S.] **18**, 147 (1958).
G35. Gubin, S. P., and Smirnova, S. A., *J. Organometal. Chem.* **20**, 229 (1969).
G36. Gubin, S. P., and Smirnova, S. A., *J. Organometal. Chem.* **20**, 241 (1969).
G36a. Guggenberger, L. J., and Tebbe, F. N., *J. Amer. Chem. Soc.* (in press).
G37. Gumboldt, A., and Jastrow, H., Ger. Patent Appl. 1,203, 775 (1964).
G38. Gutmann, V., and Meller, A., *Monatsh. Chem.* **91**, 519 (1960).
G39, Gutmann, V., and Nedbalek, E., *Monatsh. Chem.* **88**, 320 (1957).
G40. Gutmann, V., and Nowotny, H., *Monatsh. Chem.* **89**, 331 (1958).
G41. Gutmann, V. and Schöber, G., *Monatsh. Chem.* **93**, 1353 (1962).
G42. Guzman, I. Sh., Sharaev, O. K., Tinyakova, E. I., and Dolgoplosk, B. A., *Izv. Akad. Nauk SSSR, Ser. Khim.* p. 661 (1971).
G43. Guzman, I. Sh., Sharaev, O. K., Tinyakova, E. I., and Dolgoplosk, B. A., *Dokl. Akad. Nauk SSSR*, **202**, 1329 (1972).
H1. Hamprecht, G., Grubert, H., and Schwarzmann, M., Ger. Patent 1,153,750 (1963).
H2. Hamprecht, G., and Taglinger, L., Ger. Patent 1,153,749 (1963).
H3. Hanlan, J. F., and McCowan, J. D., *Can. J. Chem.* **50**, 747 (1972).
H4. Heck, R. F., *J. Org. Chem.* **30**, 2205 (1965).
H5. Heins, E., Hinck, H., Kaminsky, W., Oppermann, G., Raulinat, P., and Sinn, H., *Makromol. Chem.* **134**, 1 (1970).
H6. Helmholdt, R. B., Jellinek, F., Martin, H. A., and Vos, A., *Rec. Trav. Chim. Pays-Bas* **86**, 1263 (1967).
H6a. Hencken, G., and Weiss, E., *Chem. Ber.* **106**, 1747 (1973).
H7. Hengge, E., and Zimmermann, H., *Angew. Chem., Int. Ed. Engl.* **7**, 142 (1968).
H8. Henrici-Olivé, G., and Olivé, S., *Angew. Chem., Int. Ed. Engl.* **6**, 790 (1967).
H9. Henrici-Olivé, G., and Olivé, S., *Kolloid-Z. Z. Polym.* **228**, 43 (1968).
H10. Henrici-Olivé, G., and Olivé, S., *Angew. Chem., Int. Ed. Engl.* **7**, 386 (1968).
H11. Henrici-Olivé, G., and Olivé, S., *Angew. Chem., Int. Ed. Engl.* **7**, 821 (1968).
H12. Henrici-Olivé, G., and Olivé, S., *Chem. Commun.* p. 113 (1969).
H13. Henrici-Olivé, G., and Olivé, S., *Makromol. Chem.* **121**, 70 (1969).
H14. Henrici-Olivé, G., and Olivé, S., *Angew. Chem., Int. Ed. Engl.* **8**, 650 (1969).
H15. Henrici-Olivé, G., and Olivé, S., *J. Organometal. Chem.* **16**, 339 (1969).
H16. Henrici-Olivé, G., and Olivé, S., *J. Organometal. Chem.* **17**, 83 (1969).
H17. Henrici-Olivé, G., and Olivé, S., *J. Organometal. Chem.* **19**, 309 (1969).
H18. Henrici-Olivé, G., and Olivé, S., *J. Polym. Sci., Part C* **22**, 965 (1969).
H18a. Henrici-Olivé, G., and Olivé, S., *Advan. Polym. Sci.* **6**, 421 (1969).
H19. Henrici-Olivé, G., and Olivé, S., *J. Organometal. Chem.* **21**, 377 (1970).
H20. Henrici-Olivé, G., and Olivé, S., *J. Organometal. Chem.* **23**, 155 (1970).
H21. Herman, D. F., Ger. Patent 1,032,924 (1958).
H22. Herman, D. F., U.S. Patent 2,886,579 (1959).

H23. Herman, D. F., U.S. Patent 2,963,471 (1960).
H24. Herman, D. F., U.S. Patent 3,027,392 (1962).
H25. Herman, D. F., and Nelson, W. K., *J. Amer. Chem. Soc.* **74**, 2693 (1952).
H26. Herman, D. F., and Nelson, W. K., *J. Amer. Chem. Soc.* **75**, 3877 (1953).
H27. Herman, D. F., and Nelson, W. K., *J. Amer. Chem. Soc.* **75**, 3882 (1953).
H28. Hethington, D., and Aplington, J. P., unpublished results appearing in ref. L2.
H28a. Hewitt, B. J., Holliday, A. K., and Puddephatt, R. J., *J. Chem. Soc. Dalton Trans.* p. 801 (1973).
H29. Higashi, K., and Takahashi, K., Jap. Patent 62/12,141 (1962).
H30. Higashi, K., and Takahashi, K., Jap. Patent 63/5,991 (1963).
H31. Hillman, M., and Weiss, A. J., *J. Organometal. Chem.* **42**, 123 (1972).
H32. Hirai, H., Hiraki, K., Noguchi, I., and Makishima, S., *J. Polym. Sci., Part A* **8**, 147 (1970).
H33. Hoard, J. L., and Silverton, J. V., *Inorg. Chem.* **2**, 235 (1963).
H34. Hofstee, H. K., van Oven, H. O., and de Liefde Meijer, H. J., *J. Organometal. Chem.* **42**, 405 (1972).
H35. Holloway, H., *Chem. Ind. (London)* p. 214 (1962).
H36. Hong, P., Sonogashira, K., and Hagihara, N., *Bull. Chem. Soc. Jap.* **39**, 1821 (1966).
H37. Howe, J. J., and Pinnavaia, T. J., *J. Amer. Chem. Soc.* **92**, 7342 (1970).
H38. Hsiung, H. -S., *Diss. Abstr.* **21**, 3629 (1961).
H39. Hsiung, H. -S., and Brown, G. H., *J. Electrochem. Soc.* **110**, 1085 (1963).
H40. Hubel, K. W., Braye, E. H., and Caplier, I. H., U.S. Patent Appl. 3,151,140 (1960).
H41. Hunt, C. C., and Doyle, J. R., *Inorg. Nucl. Chem. Lett.* **2**, 283 (1966).
H42. Hunt, D. F., Russell, J. W., and Torian, R. L., *J. Organometal. Chem.* **43**, 175 (1972).
I1.   Idemitsu Kosan Co. Ltd., Fr. Patent 1,540,898 (1968).
I2.   Israeli, Y. J., *Bull. Soc. Chim. Fr.* p. 837 (1966).
I3.   Issleib, K., and Bätz, G., *Z. Anorg. Allg. Chem.* **369**, 83 (1969).
I4.   Issleib, K., and Häckert, H., *Z. Naturforsch. B* **21**, 519 (1966).
I5.   Issleib, K., Wille, G., and Krech, F., *Angew. Chem., Int. Ed. Engl.* **11**, 527 (1972).
J1.   Jacot-Guillarmod, A., Tabacchi, R., and Porret, J., *Helv. Chim. Acta* **53**, 1491 (1970).
J2.   James, B. D., Nanda, R. K., and Wallbridge, M. G. H., *J. Chem. Soc., A* p. 182 (1966).
J3.   James, B. D., Nanda, R. K., and Wallbridge, M. G. H., *Chem. Commun.* p. 849 (1966).
J4.   James, B. D., Nanda, R. K., and Wallbridge, M. G. H., *Inorg. Chem.* **6**, 1979 (1967).
J5.   James, B. D., Nanda, R. K., and Wallbridge, M. G. H., *Abstr. 155th Meet., Amer. Chem. Soc., San Francisco*, p. M83 (1968).
J6.   James, T. A., and McCleverty, J. A., *J. Chem. Soc., A* p. 3318 (1970).
J7.   Jenkins, A. D., Lappert, M. F., and Srivastava, R. C., *J. Organometal. Chem.* **23**, 165 (1970).
J8.   Jenkins, A. D., Lappert, M. F., and Srivastava, R. C., *Eur. Polym. J.* **7**, 289 (1971).
J9.   Jensen, A., *Proc. Int. Conf. Coord. Chem., 7th, 1962* p. 254 (1962).
J10. Jensen, A., Christiansen, V. H., Hansen, J. F., Likowski, T., and Burmeister, J. L., *Acta Chem. Scand.* **26**, 2898 (1972).

J11. Jensen, A., and Basolo, F., *J. Amer. Chem. Soc.* **81**, 3813 (1959).
J12. Jensen, A., and Jørgensen, E., *J. Organometal. Chem.* **7**, 528 (1967).
J13. Job, B. E., and Pioli, A. J. P., Ger. Patent 1,954,479 (1970).
J14. Job, B. E., Pioli, A. J. P., and Medinger, T., Ger. Patent, 1,963,072 (1970).
J15. Joseph, P. T., and Mathew, P. M., *Indian J. Chem.* **9**, 175 (1971).
J16. Joshi, K. K., Ger. Patent 1,930,674 (1970) (Brit. Patent Appl. 1968).
J17. Joshi, K. K., Wardle, R., and Wilson, V. A., *Inorg. Nucl. Chem. Lett.* **6**, 49 (1970).
K1.   Kaar, H., Kirret, O., and Schwindlerman, G., *Eesti NSV Tead. Akad. Toim. Fuus.-Mat. Tehnikatead. Seer.* **12**, 295 (1963); *Chem. Abstr.* **60**, 10792h (1964).
K2.   Kaar, H., Kirret, O., and Schwindlerman, G., *Eesti NSV Tead. Akad. Toim. Fuus.-Mat.Tehnikatead. Seer.* **12**, 414 (1963); *Chem. Abstr.* **60**, 15985h (1964).
K3.   Kaar, H., and Schwindlerman, G., *Eesti NSV Tead. Akad. Toim. Fuus.-Mat. Tehnikatead. Seer.* **13**, 154 (1964); *Chem. Abstr.* **62**, 13241g (1965).
K4.   Kablitz, H. -J., Kallweit, R., and Wilke, G., *J. Organometal. Chem.* **44**, C49 (1972).
K4a. Kablitz, H.-J., and Wilke, G., *J. Organometal. Chem.* **51**, 241 (1973).
K5.   Kalechits, I. V., Lipovich, V. G., and Shmidt, F. K., *Kinet. Katal.* **9**, 24 (1968).
K6.   Kalechits, I. V., and Shmidt, F. K., *Kinet. Katal.* **7**, 614 (1966).
K7.   Kalinnikov, V. T., Larin, G. M., Ubozhenko, O. D., Zharkikh, A. A., Latyaeva, V. N., and Lineva, A. N., *Dokl. Akad. Nauk SSSR* **199**, 95 (1971).
K8.   Kambara, H., and Sudo, Y., Jap. Patent 67/13,091 (1967).
K9.   Karapinka, G. L., Smith, J. J., and Carrick, W. L., *J. Polym. Sci.* **50**, 143 (1961).
K10. Karol, F. J., and Carrick, W. L., *J. Amer. Chem. Soc.* **83**, 2654 (1961).
K11. Katz, T. J., and Acton, N., *Tetrahedron Lett.* **28**, 2497 (1970).
K12. Kaufman, D., Ger. Patent 1,032,738 (1958).
K13. Kaufman, D., U.S. Patent 2,911,424 (1959).
K14. Kaufman, D., U.S. Patent 2,922,802 (1960).
K15. Kautzner, B., Wailes, P. C., and Weigold, H., *Chem. Commun.* p. 1105 (1969).
K16. Kealy, T. J., and Pauson, P. L., *Nature (London)* **168**, 1039 (1951).
K17. Kenworthy, J. G., and Myatt, J., *Chem. Commun.* p. 447 (1970).
K18. Kenworthy, J. G., Myatt, J., and Symons, M. C. R., *J. Chem. Soc., A* p. 1020 (1971).
K19. Kenworthy, J. G., Myatt, J., and Symons, M. C. R., *J. Chem. Soc., A* p. 3428 (1971).
K20. Kenworthy, J. G., Myatt, J., and Todd, P. F., *Chem. Commun.* p. 263 (1969).
K21. Kenworthy, J. G., Myatt, J., and Todd, P. F., *J. Chem. Soc., B* p. 791 (1970).
K22. Kepert, D. L., "The Early Transition Metals." Academic Press, New York, 1972.
K23. Khachaturov, A. S., Bresler, L. S., and Poddubnyi, I. Ya., *J. Organometal. Chem.* **42**, C18 (1972).
K24. Kharlamova, E. N., Brainina, E. M., and Gur'yanova, E. N., *Izv. Akad. Nauk SSSR, Ser. Khim.* p. 2621 (1970).
K25. Khodzhemirov, V. A., Ostrovsky, V. E., Zabolotskaya, E. V., and Medvedev, S. S., *Vysokomol. Soedin., Ser. A* **13**, 1662 (1971).
K26. King, R. B., *J. Amer. Chem. Soc.* **91**, 7211 and 7217 (1969).
K27. King, R. B., and Bisnette, M. B., *J. Organometal. Chem.* **8**, 287 (1967).
K28. King, R. B., and Eggers, C. A., *Inorg. Chem.* **7**, 340 (1968).
K29. King, R. B., and Kapoor, R. N., *J. Organometal. Chem.* **15**, 457 (1968).
K30. Kingston, B. M., and Lappert, M. F., *Inorg. Nucl. Chem. Lett.* **4**, 371 (1968).
K31. Kingston, B. M., and Lappert, M. F., *J. Chem. Soc., Dalton Trans.* p. 69 (1972).

K32. Klanberg, F., Muetterties, E. L., and Guggenberger, L. J., *Inorg. Chem.* **7**, 2272 (1968).

K33. Knox, G. R., Munro, J. D., Pauson, P. L., Smith, G. H., and Watts, W. E., *J. Chem. Soc., London* p. 4619 (1961).

K33a. Kocman, V., Dvorak, J., Nyburg, S. C., and O'Brien, R. J., private communication.

K34. Kocman, V., Rucklidge, J. C., O'Brien, R. J., and Santo, W., *Chem. Commun.* p. 1340 (1971).

K35. Kogerman, A., and Martinson, K., *Eesti NSV Tead. Akad. Toim. Keem., Geol.* **18**, 50 (1969); *Chem. Abstr.* **70**, 106909p (1969).

K36. Kogerman, A., and Martinson, H., *Eesti NSV Tead. Akad. Toim., Keem, Geol.* **18**, 56 (1969); *Chem. Abstr.* **70**, 106910g (1969).

K37. Kogerman, A., Martinson, H., Evseev, T., and Kongas, A., *Eesti NSV Tead. Akad. Toim., Keem., Geol.* **18**, 232 (1969); *Chem. Abstr.* **71**, 124986m (1969).

K38. Kolomnikov, I. S., Lobeeva, T. S., Gorbachevskaya, V. V., Aleksandrov, G. G., Struckhov, Yu. T., and Vol'pin, M. E., *Chem. Commun.* p. 972 (1971).

K39. Köpf, H., *J. Organometal. Chem.* **14**, 353 (1968).

K40. Köpf, H., *Z. Naturforsch. B* **23**, 1531 (1968).

K41. Köpf, H., *Chem. Ber.* **102**, 1509 (1969).

K42. Köpf, H., *Angew. Chem., Int. Ed. Engl.* **10**, 134 (1971).

K43. Köpf, H., and Block, B., *Z. Naturforsch. B* **23**, 1534 (1968).

K44. Köpf, H., and Block, B., *Z. Naturforsch. B* **23**, 1536 (1968).

K45. Köpf, H., and Block, B., *Chem. Ber.* **102**, 1504 (1969).

K46. Köpf, H., Block, B., and Schmidt, M., *Z. Naturforsch. B* **22**, 1077 (1967).

K47. Köpf, H., Block, B., and Schmidt, M., *Chem. Ber.* **101**, 272 (1968).

K48. Köpf, H., and Räthlein, K. H., *Angew. Chem., Int. Ed. Engl.* **8**, 980 (1969).

K49. Köpf, H., and Schmidt, M., *Z. Anorg. Allg. Chem.* **340**, 139 (1965).

K50. Köpf, H., and Schmidt, M., *J. Organometal. Chem.* **4**, 426 (1965).

K51. Köpf, H., and Schmidt, M., *Angew. Chem., Int. Ed. Engl.* **4**, 953 (1965).

K52. Köpf, H., and Schmidt, M., *J. Organometal. Chem.* **10**, 383 (1967).

K53. Korshak, V. V., Sladkov, A. M., Luneva, L. K., and Girshovich, A. S., *Vysokomol. Soedin.* **5**, 1284 (1963).

K54. Korshunov, I. A., and Malyugina, N. I., *Zh. Obshch. Khim.* **34**, 734 (1964).

K55. Kozina, I. Z., and Subbotina, A. I., *Zavod. Lab.* **36**, 670 (1970); *Chem. Abstr.* **73**, 83677g (1970).

K56. Kraitizer, I., McTaggart, K., and Winter, G., *Aust. Dep. Munitions Paint Notes* **2**, 304 (1947).

K57. Kraitizer, I., McTaggart, K., and Winter, G., *J. Counc. Sci. Ind. Res.* **21**, 328 (1948).

K58. Krashina, E. V., Borod'ko, Yu. G., Plakhova, E. I., Rukhadze, Sh. M., and Shilov, A. E., *Izv. Akad. Nauk SSSR, Ser. Khim.* p. 936 (1970).

K58a. Krieger, J. K., Deutch, J. M., and Whitesides, G. M., *Inorg. Chem.* **12**, 1535 (1973).

K59. Kroon, P. A., and Helmholdt, R. B., *J. Organometal. Chem.* **25**, 451 (1970).

K60. Kühlein, K., and Clauss, K., *Angew. Chem., Int. Ed. Engl.* **8**, 387 (1969).

K61. Kulishov, V. I., Bokii, N. G., and Struchkov, Yu. T., *Zh. Strukt. Khim.* **11**, 700 (1970).

K61a. Kulishov, V. I., Bokii, N. G., and Struchkov, Yu. T., *Zh. Strukt. Khim.* **13**, 1110 (1972).

K62. Kulishov, V. I., Brainina, E. M., Bokiy, N. G., and Struchkov, Yu. T., *Chem. Commun.* p. 475 (1970).

K63. Kulishov, V. I., Brainina, E. M., Bokiy, N. G., and Struchkov, Yu. T., *J. Organometal. Chem.* **36**, 333 (1972).

K64. Kutoglu, A., *Z. Anorg. Allg. Chem.* **390**, 195 (1972).

L1. La Lau, C., *Rec. Trav. Chim. Pays-Bas* **84**, 429 (1965).

L2. Langford, C. H., and Aplington, J. P., *J. Organometal. Chem.* **4**, 271 (1965).

L3. Lanovskaya, L. M., Pravikova, N. A., Gantmakher, A. R., and Medvedev, S. S., *Vysokomol. Soedin., Ser. A* **11**, 1157 (1969).

L4. Lappert, M. F., and Poland, J. S., *Chem. Commun.* p. 1061 (1969).

L5. Lappert, M. F., and Sanger, A. R., *J. Chem. Soc., A* p. 874 (1971).

L6. Lappert, M. F., and Sanger, A. R., *J. Chem. Soc., A* p. 1314 (1971).

L7. Larsen, E. M., *Advan. Inorg. Chem. Radiochem.* **13**, 1 (1970).

L8. Latyaeva, V. N., Razuvaev, G. A., and Kilyakova, G. A., *Zh. Obshch. Khim.* **35**, 1498 (1965).

L9. Latyaeva, V. N., Razuvaev, G. A., Malisheva, A. V., and Kiljakova, G. A., *J. Organometal. Chem.* **2**, 388 (1964).

L10. Latyaeva, V. N., Vyshinskaya, L. I., Shur, V. B., Fedorov, L. A., and Vol'pin, M. E., *Dokl. Akad. Nauk SSSR* **179**, 875 (1968).

L11. Latyaeva, V. N., Vyshinskaya, L. I., Shur, V. B., Fyodorov, L. A., and Vol'pin, M. E., *J. Organometal. Chem.* **16**, 103 (1969).

L12. Latyaeva, V. N., Vyshinskaya, L. I., and Vyshinskii, N. N., *Sb. Dokl. Sib. Soveshch. Spektrosk., 3rd, 1964* p. 248 (1964); *Chem. Abstr.* **68**, 87367a (1968).

L13. Lehmkuhl, H., Leuchte, W., Janssen, E., Mehler, K., Kintopf, S., and Eisenbach, W., *Angew. Chem., Int. Ed. Engl.*, **10**, 843 (1971).

L14. Lehmkuhl, H., and Mehler, K., *J. Organometal. Chem.* **25**, C44 (1970).

L15. Leleu, J., and Riou, M., Fr. Patent 1,157,196 (1958).

L16. Lemarchand, D., M'baye, N., and Braun, J., *J. Organometal. Chem.* **39**, C69 (1972).

L17. Lindner, E., Lorenz, I.-P., and Vitzthum, G., *Chem. Ber.* **103**, 3182 (1970).

L18. Locke, J., and McCleverty, J. A., *Inorg. Chem.* **5**, 1157 (1966).

L19. Lokshin, B. V., and Brainina, E. M., *Zh. Strukt Khim.* **12**, 1001 (1971).

L20. Long, W. P., *J. Amer. Chem. Soc.* **81**, 5312 (1959).

L21. Long, W. P., *J. Polym. Sci.* **45**, 250 (1960).

L22. Long, W. P., Ger. Patent 2,049,477 (1971).

L23. Long, W. P., and Breslow, D. S., *J. Amer. Chem. Soc.*, **82**, 1953 (1960).

L23a. Lucas, C. R., Green, M., Forder, R. A., and Prout, K., *J. Chem. Soc., Chem. Commun.* p. 97 (1973).

L24. Lynch, M. A., and Brantley, J. C., Brit. Patent 785,760 (1957).

M1. McCleverty, J. A., James, T. A., and Wharton, E. J., *Inorg. Chem.* **8**, 1340 (1969).

M1a. McCowan, J. D., *Can. J. Chem.* **51**, 1083 (1973).

M2. McCowan, J. D., and Hanlan, J. F., *Can. J. Chem.* **50**, 755 (1972).

M3. Machida, K., and Adachi, N., Jap. Patent 68/15,823 (1968).

M4. McHugh, K. L., and Smith, J. O., U.S. Patent Appl. 3,242,081 (1963).

M5. Maki, A. H., and Randall, E. W., *J. Amer. Chem. Soc.* **82**, 4109 (1960).

M6. Malatesta, A., *Can. J. Chem.* **37**, 1176 (1959).

M7. Malatesta, A., *J. Polym. Sci.* **51**, S45 (1961).

M8. Marconi, W., Santostasi, M. L., and De Malde, M., *Chim. Ind. (Milan)* **44**, 229 (1962).

M9. Marconi, W., Santostasi, M. L., and De Malde, M., *Chim. Ind. (Milan)* **44**, 235 (1962).

M10. Marti, M. G., and Reichert, K. H., *Makromol. Chem.* **144**, 17 (1971).

M11. Martin, H. A., and de Jongh, R. O., *Chem. Commun.* p. 1366 (1969).
M12. Martin, H. A., and de Jongh, R. O., *Rec. Trav. Chim. Pays-Bas* **90**, 713 (1971).
M13. Martin, H. A., and Jellinek, F., *Angew. Chem., Int. Ed. Engl.* **3**, 311 (1964).
M14. Martin, H. A., and Jellinek, F., *J. Organometal. Chem.* **8**, 115 (1967).
M15. Martin, H. A., and Jellinek, F., *J. Organometal. Chem.* **12**, 149 (1968).
M16. Martin, H. A., Lemaire, P. J., and Jellinek, F., *J. Organometal. Chem.* **14**, 149 (1968).
M17. Martin, H. A., van Gorkom, M., and de Jongh, R. O., *J. Organometal. Chem.* **36**, 93 (1972).
M18. Martin, H., and Vohwinkel, F., *Chem. Ber.* **94**, 2416 (1961).
M19. Martin, R. L., and Winter, G., *J. Chem. Soc., London* p. 4709 (1965).
M20. Martinson, H., and Kolk, A., *Chem. Abstr.* **68**, 50113p (1968); *Eesti NSV Tead. Akad. Toim. Keem., Geol.* **16**, 221 (1967).
M21. Marvich, R. H., and Brintzinger, H. H., *J. Amer. Chem. Soc.* **93**, 2046 (1971).
M22. Masai, H., Sonohashira, K., and Hagihara, N., *Chem. Abstr.* **69**, 66948w (1968); *Mem. Inst. Sci. Ind. Res., Osaka Univ.* **25**, 117 (1968).
M23. Masai, H., Sonogashira, K., and Hagihara, N., *Bull. Chem. Soc. Jap.* **41**, 750 (1968).
M24. Maskill, R., and Pratt, J. M., *Chem. Commun.* p. 950 (1967).
M25. Maskill, R., and Pratt, J. M., *J. Chem. Soc., A* p. 1914 (1968).
M26. Maslowsky, E., Jr., and Nakamoto, K., *Appl. Spectrosc.* **25**, 187 (1971).
M27. Matkovskii, P. E., Belov, G. P., Lisitskaya, A. P., Russiyan, L. N., Brikenstein, Kh.-M. A., Gerasina, M. P., and Chirkov, N. M., *Vysokomol. Soedin., Ser. A* **12**, 1662 (1970).
M28. Matkovskii, P. E., Belov, G. P., Russiyan, L. N., Lisitskaya, A. P., Kissin, V. V., Solov'eva, T. I., Brikenshtein, A. A., and Chirkov, N. M., *Vysokomol. Soedin., Ser A* **12**, 2286 (1970).
M29. Matsuzaki, K., and Yasukawa, T., *J. Organometal. Chem.* **10**, P9 (1967).
M30. Matthews, J. D., Singer, N., and Swallow, A. G., *J. Chem. Soc., A* p 2545, (1970).
M31. Matthews, J. D., and Swallow, A. G., *Chem. Commun.* p. 882 (1969).
M32. Medvedeva, A. V., Ryabenko, D. M., Zayarnaya, R. F., and Fridenberg, A. E., U.S.S.R. Patent 166,689 (1964).
M32a. Melmed, K. M., Coucouvanis, D., and Lippard, S. J., *Inorg. Chem.* **12**, 232 (1973).
M32b. Merijanian, A., Mayer, T., Helling, J. F., and Klemick, F., *J. Org. Chem.* **37**, 3945 (1972).
M33. Meshkova, I. N., Tsvetkova, V. I., and Chirkov, N. M., *Izv. Akad. Nauk SSSR, Ser. Khim.* p. 77 (1966).
M34. Minacheva, M. Kh., and Brainina, E. M., *Izv. Akad. Nauk. SSSR, Ser. Khim.* p. 139 (1972).
M35. Minacheva, M. Kh., Brainina, E. M., and Fedorov, L. A., *Izv. Akad. Nauk. SSSR, Ser. Khim.* p. 1104 (1969).
M36. Minacheva, M. Kh., Brainina, E. M., and Freidlina, R. Kh., *Dokl. Akad. Nauk SSSR* **173**, 581 (1967).
M37. Minacheva, M. Kh., Fedorov, L. A., Brainina, E. M., and Freidlina, R. Kh., *Dokl. Akad. Nauk SSSR* **200**, 598 (1971).
M38. Mingos, D. M. P., *J. Chem. Soc. Chem. Commun.* p. 165 (1972).
M39. Minsker, K. S., Sangalov, Yu. A., and Razuwayev, G. A., *J. Polym. Sci., Part C* p. 1489 (1967).

M40. Miyaki, A., Kondo, H., and Aoyama, M., *Angew. Chem., Int. Ed. Engl.* **8**, 520 (1969).

M41. Moffitt, W., *J. Amer. Chem. Soc.* **76**, 3386 (1954).

M42. Montecatini, Ital. Patent 588,211 (1957).

M43. Montecatini, Brit. Patent 875,078 (1958).

M44. Montecatini, Ital. Patent 602,210 (1960).

M45. Montecatini, Brit. Patent 900,615 (1962).

M46. Montecatini, Brit. Patent 926,994 (1963).

M47. Morehouse, E. L., Brit. Patent 797,151 (1958).

M48. Morehouse, E. L., U.S. Patent 3,071,605 (1963).

M49. Mowat, W., Shortland, A., Yagupsky, G., Hill, N. J., Yagupsky, M., and Wilkinson, G., *J. Chem. Soc., Dalton Trans.* p. 533 (1972).

M50. Mowat, W., and Wilkinson, G., *J. Organometal. Chem.* **38**, C35 (1972).

M50a. Mowat, W., and Wilkinson, G., *J. Chem. Soc. Dalton Trans.* p. 1120 (1973).

M50b. Mrowca, J. J., *U.S. Patent* 3,728,365 (1973).

M51. Muir, K. W., *J. Chem. Soc., A* p. 2663 (1971).

M52. Müller, J., and Mertschenk, B., *Chem. Ber.* **105**, 3346 (1972).

M53. Müller, J., and Thiele, K.-H., *Z. Anorg. Allg. Chem.* **362**, 120 (1968).

M54. Murahashi, S., Kamachi, M., and Wakabayashi, N., *J. Polym. Sci., Part B* **7**, 135 (1969).

M55. Murray, J. G., *J. Amer. Chem. Soc.* **81**, 752 (1959).

M56. Murray, J. G., *J. Amer. Chem. Soc.* **83**, 1287 (1961).

N1. Nanda, R. K., and Wallbridge, M. G. H., *Inorg. Chem.* **3**, 1798 (1964).

N2. National Lead Co., Brit. Patent 779,490 (1957).

N3. National Lead Co., Brit. Patent 793,354 (1958).

N4. National Lead Co., Brit. Patent 793,355 (1958).

N5. National Lead Co., Brit. Patent 798,001 (1958).

N6. National Lead Co., Brit. Patent 800,528 (1958).

N7. National Lead Co., Brit. Patent 833,805 (1960).

N8. National Lead Co., Brit. Patent 855,202 (1960).

N9. National Lead Co., Brit. Patent 856,434 (1960).

N10. National Lead Co., Brit. Patent 858,930 (1961).

N11. Natta, G., and Corradini, P., *Angew. Chem.* **72**, 39 (1960).

N12. Natta, G., Corradini, P., and Bassi, I. W., *J. Amer. Chem. Soc.* **80**, 755 (1958).

N13. Natta, G., Dall'asta, G., Mazzanti, G., Giannini, U., and Cesca, S., *Angew. Chem.* **71**, 205 (1959).

N14. Natta, G., and Mazzanti, G., *Tetrahedron* **8**, 86 (1960).

N15. Natta, G., Mazzanti, G., Corradini, P., Giannini, U., and Cesca, S., *Atti Acad. Naz. Lincei, Cl. Sci. Fis., Mat. Natur., Rend* [8] **26**, 150 (1959).

N16. Natta, G., Mazzanti, G., and Giannini, U., Ger. Patent 1,084,918 (1960).

N17. Natta, G., Mazzanti, G., Giannini, U., and Cesca, S., *Angew. Chem.* **72**, 39 (1960).

N18. Natta, G., Mazzanti, G., Giannini, U., and Cesca, S., Ger. Patent Appl. 1,105,416 (1960).

N19. Natta, G., Mazzanti, G., and Pregaglia, G., *Gazz. Chim. Ital.* **89**, 2065 (1959).

N20. Natta, G., Pino, P., and Mazzanti, G., Ital. Patent 579,112 (1958).

N21. Natta, G., Pino, P., Mazzanti, G., and Giannini, U., *J. Amer. Chem. Soc.* **79** 2975 (1957).

N22. Natta, G., Pino, P., Mazzanti, G., and Giannini, U., *J. Inorg. Nucl. Chem.* **8**, 612 (1958).

N23. Natta, G., Pino, P., Mazzanti, G., Giannini, U., Mantica, E., and Peraldo, M., *Chim. Ind. (Milan)* **39**, 19 (1957).
N24. Natta, G., Pino, P., Mazzanti, G., Giannini, U., Mantica, E., and Peraldo, M., *J. Polym. Sci.* **26**, 120 (1957).
N25. Natta, G., Pino, P., Mazzanti, G., and Lanzo., R., *Chim. Ind. (Milan)* **39**, 1032 (1957).
N26. Nechiporenko, G. N., Tabrina, G. M., Shilova, A. K., and Shilov, A. E., *Dokl. Akad. Nauk SSSR* **164**, 1062 (1965).
N27. Neese, H.-J., and Bürger, H., *J. Organometal. Chem.* **32**, 213 (1971).
N27a. Nesmeyanov, A. N , Bryukhova, E. V., Alymov, I. M., Nogina, O. V., and Dubovitskii, V. A., *Izv. Akad. Nauk SSSR, Ser. Khim.* p. 1671 (1972).
N28. Nesmeyanov, A. N., Dubovitskii, V. A., Nogina, O. V., and Bochkarev, V. N., *Dokl. Akad. Nauk SSSR* **165**, 125 (1965).
N28a. Nesmeyanov, A. N., Fedin, E. I., Fedorov, L. A., and Petrovskii, P. V., *Zh. Strukt. Khim.* **13**, 1033 (1972).
N29. Nesmeyanov, A. N., Fedin, E. I., Nogina, O. V., Kochetkova, N. S., Dubovitsky, V. A., and Petrovsky, P. V., *Tetrahedron. Suppl.* **8, pt 2**, 389 (1966).
N30. Nesmeyanov, A. N., Fedin, E. I., Petrovskii, P. V., Dubovitskii, V. A., Nogina, O. V., and Lazareva, N. A., *Dokl. Akad. Nauk SSSR* **163**, 659 (1965).
N31. Nesmeyanov, A. N., Materikova, R. B., Brainina, E. M., and Kochetkova, N. S., *Izv. Akad. Nauk SSSR, Ser. Khim.* p. 1323 (1969).
N32. Nesmeyanov, A. N., and Nogina, O. V., *Izv. Akad. Nauk SSSR, Otd. Khim. Nauk.* p. 831 (1963).
N33. Nesmeyanov, A. N., Nogina, O. V., and Berlin, A. M., *Dokl. Akad. Nauk SSSR* **134**, 607 (1960).
N34. Nesmeyanov, A. N., Nogina, O. V., and Berlin, A. M., *Izv. Akad. Nauk SSSR, Otd. Khim. Nauk.* p. 804 (1961).
N35. Nesmeyanov, A. N., Nogina, O. V., Berlin, A. M., Girshovich, A. S., and Shatalov, G. V., *Izv. Akad. Nauk SSSR, Otd. Khim. Nauk* p. 2146 (1961).
N36. Nesmeyanov, A. N., Nogina, O. V., and Dubovitskii, V. A., *Izv. Akad. Nauk SSSR, Otd. Khim. Nauk* p. 1481 (1962).
N37. Nesmeyanov, A. N., Nogina, O. V., and Dubovitskii, V. A., *Izv. Akad. Nauk SSSR, Ser. Khim.* p. 527 (1968).
N38. Nesmeyanov, A. N., Nogina, O. V., Dubovitskii, V. A., Kvasov, B. A., Petrovskii, P. V., and Lazareva, N. A., *Izv. Akad. Nauk. SSSR, Ser. Khim.* p. 2729 (1971).
N38a. Nesmeyanov, A. N., Nogina, O. V., Fedin, E. I., Dubovitskii, V. A., Kvasov, B. A., and Petrovskii, P. V., *Dokl. Akad. Nauk SSSR* **205**, 857 (1972).
N39. Nesmeyanov, A. N., Nogina, O. V., Lazareva, N. A., and Dubovitskii, V. A., *Izv. Akad. Nauk SSSR, Ser. Khim.* p. 808 (1967).
N40. Nesmeyanov, A. N., Nogina, O. V., Lazareva, N. A., Dubovitskii, V. A., and Lokshin, B. V., *Izv. Akad. Nauk SSSR, Ser. Khim.* p. 2482 (1971).
N41. Nesmeyanov, A. N., Nogina, O. V., Lokshin, B. V., and Dubovitskii, V. A., *Dokl. Akad. Nauk SSSR* **182**, 844 (1968).
N42. Noltes, J. G., and van der Kerk, G. J. M., *Rec. Trav. Chim. Pays-Bas* **81**, 39 (1962).
N43. North, A. M., *Proc. Roy. Soc., Ser. A* **254**, 408 (1960).
N44. Nöth, H., and Hartwimmer, R., *Chem. Ber.* **93**, 2238 (1960).
N45. Nöth, H., and Hartwimmer, R., *Chem. Ber.* **93**, 2246 (1960).
N46. Nöth, H., Voitländer, J., and Nussbaum, M., *Naturwissenschaften* **47**, 57 (1960).
N47. Nozawa, Y., and Takeda, M., *Kogyo Kagaku Zasshi* **71**, 189 (1968).

N48. Nozawa, Y., and Takeda, M., *Kogyo Kagaku Zasshi* **72**, 2527 (1969).

O1. O'Brien, S., Janes, W. H., Taylor, K. A., and Todd, P. F., Brit. Patent 1,091,296 (1967).

O2. Oreshkin, I. A., Chernenko, G. M., Tinyakova, E. I., and Dolgoplosk, B. A., *Dokl. Akad. Nauk SSSR* **169**, 1102 (1966).

O3. Overberger, C. G., Diachkovsky, F. S., and Jarovitzky, P. A., *J. Polym. Sci., Part A* **2**, 4113 (1964).

O4. Overberger, C. G., and Jarovitzky, P. A., *J. Polym. Sci., Part A* **3**, 1483 (1965).

O5. Overberger, C. G., Jarovitzky, P. A., and Mukamal, H., *J. Polym. Sci., Part A* **5**, 2487 (1967).

P1. Paul, R. C., Kaushal, R., and Pahil, S. S., *J. Indian Chem. Soc.* **44**, 995 (1967).

P2. Pauson, P. L., and Knox, G. R., U.S. Patent 3,278,514 (1962).

P3. Pechiney-Compagnie, Fr. Patent 1,157,195 (1958).

P4. Pellizer, G., and De Alti, G., *J. Inorg. Nucl. Chem.* **29**, 1565 (1967).

P5. Pinnavaia, T. J., Howe, J. J., and Butler, E. D., *J. Amer. Chem. Soc.* **90**, 5288 (1968).

P6. Pinnavaia, T. J., and Lott, A. L., *Inorg. Chem.* **10**, 1388 (1971).

P7. Pino, P., and Mazzanti, G., U.S. Patent Appl. 3,000,870 (1958).

P8. Pino, P., and Mazzanti, G., Ital. Patent 583,219 (1958).

P9. Pioli, A. J. P., Hollyhead, W. B., and Todd, P. F., Ger. Patent 2,026,032 (1971).

P10. Piper, T. S., and Wilkinson, G., *J. Inorg. Nucl. Chem.* **3**, 104 (1956).

P11. Plyusnin, A. N., Uvarov, B. A., Tsvetkova, V. I., and Chirkov, N. M., *Izv. Akad. Nauk SSSR, Ser. Khim.* p. 2324 (1969).

P12. Porri, L., Corradini, P., Morero, D., and Allegra, G., *Chim. Ind.* (*Milan*) **42**, 487 (1960).

R1. Ralea, R., Ungurenasu, C., and Cihodaru, S., *Rev. Roum. Chim.* **12**, 861 (1967).

R2. Ralea, R., Ungurenasu, C., and Maxim, I., *Rev. Roum. Chim.* **12**, 523 (1967).

R3. Rausch, M. D., *Inorg. Chem.* **3**, 300 (1964).

R4. Rausch, M. D., *Pure Appl. Chem.* **30**, 523 (1972).

R5. Rausch, M. D., and Ciappenelli, D. J., *J. Organometal. Chem.* **10**, 127 (1967).

R6. Rausch, M. D., and Klemann, L. P., *Chem. Commun.* p. 354 (1971).

R7. Razuvaev, G. A., and Bobinova, L. M., *Tr. Khim. Khim. Tekhnol.* **1**, 654 (1958).

R8. Razuvaev, G. A., and Bobinova, L. M., *Dokl. Akad. Nauk SSSR* **150**, 325 (1963).

R9. Razuvaev, G. A., and Bobinova, L. M., *Dokl. Akad. Nauk SSSR* **152**, 1363 (1963).

R10. Razuvaev, G. A., Bobinova, L. M., and Etlis, V. S., *Dokl. Akad. Nauk SSSR* **127**, 581 (1959).

R11. Razuvaev, G. A., Domrachev, G. A., Suvotova, O. N., and Abakumova, L. G., *J. Organometal. Chem.* **32**, 113 (1971).

R12. Razuvaev, G. A., and Latyaeva, V. N., *Usp. Khim.* **34**, 585 (1965).

R13. Razuvaev, G. A., and Latyaeva, V. N., *Organometal. Chem. Rev.* **2**, 349 (1967).

R14. Razuvaev, G. A., Latyaeva, V. N., and Kilyakova, G. A., *Dokl. Akad. Nauk SSSR* **203**, 126 (1972).

R15. Razuvaev, G. A., Latyaeva, V. N., Kilyakova, G. A., and Batalov, A. P., *Dokl. Akad. Nauk SSSR* **185**, 369 (1969).

R16. Razuvaev, G. A., Latyaeva, V. N., Kilyakova, G. A., and Mal'kova, G. Ya., *Dokl. Akad. Nauk SSSR* **191**, 620 (1970).

R17. Razuvaev, G. A., Latyaeva, V. N., and Lineva, A. N., *Dokl. Akad. Nauk SSSR* **187**, 340 (1969).

R18. Razuvaev, G. A., Latyaeva, V. N., and Lineva, A. N., *Zh. Obshch. Khim.* **40**, 1804 (1970).

R19. Razuvaev, G. A., Latyaeva, V. N., and Lineva, A. N., *Zh. Obshch. Khim.* **41**, 1556 (1971).

R20. Razuvaev, G. A., Latyaeva, V. N., and Malysheva, A. V., *Dokl. Akad. Nauk SSSR* **173**, 1353 (1967).

R21. Razuvaev, G. A., Latyaeva, V. N., and Malysheva, A. V., *Zh. Obshch. Khim.* **37**, 2339 (1967).

R21a. Razuvaev, G. A., Latyaeva, V. N., Vasil'eva, G. A., and Vyshinskaya, L. I., *Zh. Obshch. Khim.* **42**, 1306 (1972).

R21b. Razuvaev, G. A., Latyaeva, V. N., Vasil'eva, G. A., and Vyshinskaya, L. I., *Izv. Akad. Nauk SSSR, Ser. Khim.* p. 1658 (1972).

R21c. Razuvaev, G. A., Latyaeva, V. N., Vasil'eva, G. A., and Vyshinskaya, L. I., *Synth. Inorg. Metal-Org. Chem.* **2**, 33 (1972).

R21d. Razuvaev, G. A., Latyaeva, V. N., Vasil'eva, G. A., Vyshinskaya, L. I., and Mal'kova, G. Ya., *Dokl. Akad. Nauk SSSR* **206**, 1127 (1972).

R22. Razuvaev, G. A., Latyaeva, V. N., and Vyshinskaya, L. I., *Dokl. Akad. Nauk SSSR* **134**, 612 (1960).

R23. Razuvaev, G. A., Latyaeva, V. N., and Vyshinskaya, L. I., *Dokl. Akad. Nauk SSSR* **138**, 1126 (1961).

R24. Razuvaev, G. A., Latyaeva, V. N., and Vishinskaya, L. I., *Zh. Obshch. Khim.* **31**, 2667 (1961).

R25. Razuvaev, G. A., Latyaeva, V. N., and Vyshinskaya, L. I., *Zh. Obshch. Khim.* **32**, 1354 (1962).

R26. Razuvaev, G. A., Latyaeva, V. N., and Vyshinskaya, L. I., *Tr. Khim. Khim. Tekhnol.* **4**, 616 (1962); *Chem. Abstr.* **58**, 2463h (1963).

R27. Razuvaev, G. A., Latyaeva, V. N., and Vyshinskaya, L. I., *Dokl. Akad. Nauk SSSR* **159**, 383 (1964).

R28. Razuvaev, G. A., Latyaeva, V. N., and Vyshinskaya, L. I., *Zh. Obshch. Khim.* **35**, 169 (1965).

R29. Razuvaev, G. A., Latyaeva, V. N., Vyshinskaya, L. I., Denisova, T. P., and Gorelow, Yu. P., *Dokl. Akad. Nauk SSSR* **202**, 1090 (1972).

R30. Razuvaev, G. A., Latyaeva, V. N., Vyshinskaya, L. I., and Kilyakova, G. A., *Zh. Obshch. Khim.* **36**, 1491 (1966).

R30a. Razuvaev, G. A., Latyaeva, V. N., Vishinskaya, L. I., and Rabinovitch, A. M., *J. Organometal. Chem.* **49**, 441 (1973).

R31. Razuvaev, G. A., Latyaeva, V. N., Vyshinkaya, L. I., and Vasil'eva, G. A., *Zh. Obshch. Khim.* **40**, 2033 (1970).

R32. Razuvaev, G. A., Latyaeva, V. N., Vyshinskaya, L. I., and Vyshinskii, N. N., *Dokl. Akad. Nauk SSSR* **156**, 1121 (1964).

R33. Razuvaev, G. A., Minsker, K. S., Latyaeva, V. N., and Sangalov, Yu. A., *Dokl. Akad. Nauk SSSR* **163**, 906 (1965).

R34. Reagan, W. J., and Burg, A. B., *Inorg. Nucl. Chem. Lett.* **7**, 741 (1971).

R35. Reid, A. F., Scaife, D. E., and Wailes, P. C., *Spectrochim. Acta* **20**, 1257 (1964).

R36. Reid, A. F., Shannon, J. S., Swan, J. M., and Wailes, P. C., *Aust. J. Chem.* **18**, 173 (1965).

R37. Reid, A. F., and Wailes, P. C., *J. Organometal. Chem.* **2**, 329 (1964).

R38. Reid, A. F., and Wailes, P. C., *Aust. J. Chem.* **18**, 9 (1965).

R39. Reid, A. F., and Wailes, P. C., *Aust. J. Chem.* **19**, 309 (1966).

R40. Reid, A. F., and Wailes, P. C., *Inorg. Chem.* **5**, 1213 (1966).

R41. Reynolds, L. T., and Wilkinson, G., *J. Inorg. Nucl. Chem.* **9**, 86 (1959).
R42. Robertson, R. E., and McConnell, H. M., *J. Amer. Chem. Soc.* **64**, 70 (1960).
R43. Rodriguez, L. A. M., and van Looy, H. M., *J. Polym. Sci., Part A-1* **4**, 1951 and 1971 (1966).
R44. Röhl, H., and Eversmann, W., Ger. Patent 1,153,365 (1963).
R45. Röhl, H., Lange, E., and Eversmann, W., Ger. Patent 1,153,366 (1963).
R46. Röhl, H., Lange, E., Gössl, T., and Roth, G., *Angew. Chem., Int. Ed. Engl.* **1**, 117 (1962).
R47. Ronova, I. A., and Alekseev, N. V., *Dokl. Akad Nauk SSSR* **174**, 614 (1967).
R48. Ronova, I. A., and Alekseev, N. V., *Dokl. Akad. Nauk SSSR* **185**, 1303 (1969).
R49. Ronova, I. A., Alekseev, N. V., Gapotchenko, N. I., and Struchkov, Yu. T., *J. Organometal. Chem.* **25**, 149 (1971).
R50. Rosenblum, M., "Chemistry of the Iron Group Metallocenes," Part 1. Wiley (Interscience), New York, 1965.
R51. Roshchupkina, O. S., Dubovitskii, V. A., and Borod'ko, Yu. G., *Zh. Strukt. Khim.* **12**, 1007 (1971).
R52. Roshchupkina, O. S., Khrushch, N. E., D'yachkovskii, F. S., and Borod'ko, Yu. G., *Zh. Fiz. Khim.* **45**, 1329 (1971).
R53. Ruskin, S. L., U.S. Patent 2,951,796 (1960).
R54. Ryabov, A. V., Latyaeva, V. N., Tikhonova, Z. A., Ivanova, Yu. A., and Odnosevtsev, A. I., *Vysokomol. Soedin., Ser. B* **11**, 49 (1969).
S1. Saito, T., *Chem. Commun.* p. 1422 (1971).
S2. Salzmann, J.-J., *Helv. Chim. Acta* **51**, 526 (1968).
S3. Salzmann, J.-J., *Helv. Chim. Acta* **51**, 903 (1968).
S4. Salzmann, J.-J., and Mosimann, P., *Helv. Chim. Acta* **50**, 1831 (1967).
S5. Samuel, E., *Bull. Soc. Chim. Fr.* p. 3548 (1966).
S6. Samuel, E., *J. Organometal. Chem.* **19**, 87 (1969).
S6a. Samuel, E., Ferner, R., and Bigorgne, M., *Inorg. Chem.* **12**, 881 (1973).
S7. Samuel, E., and Setton, R., *C. R. Acad. Sci., Ser. C* **254**, 308 (1962).
S8. Samuel, E., and Setton, R., *C. R. Acad. Sci., Ser. C* **256**, 443 (1963).
S9. Samuel, E., and Setton, R., *J. Organometal. Chem.* **4**, 156 (1965).
S10. Sastri, V. S., Chakrabarti, C. L., and Willis, D. E., *Can. J. Chem.* **47**, 587 (1969).
S11. Saunders, L., and Spirer, L., *Polymer* **6**, 635 (1965).
S12. Schäfer, W., and Thiele, K.-H., *Z. Anorg. Allg. Chem.* **381**, 205 (1971).
S13. Schindler, A., *J. Polym. Sci., Part C* **4**, 81 (1963).
S14. Schmid, G., Petz, W., Arloth, W., and Nöth, H., *Angew. Chem., Int. Ed. Engl.* **6**, 696 (1967).
S15. Schmidt, M., Block, B., Block, H. D., Köpf, H., and Wilhelm, E., *Angew. Chem., Int. Ed. Engl.* **7**, 632 (1968).
S16. Schmidt, M., and Wilhelm, E., *Chem. Commun.* p. 1111 (1970).
S17. Schramm, C. H., and Frühauf, E. J., Fr. Patent 1,397,533 (1965).
S18. Schwab, G. -M., Fischer, E. O., and Voitländer, J., *Naturwissenschaften* **41**, 228 (1954).
S19. Schwab, G. M., and Voitländer, J., *Z. Phys. Chem. (Frankfurt am Main)* [N.S.] **3**, 341 (1955).
S19a. Schwartz, J., and Sadler, J. E., *J. Chem. Soc. Chem. Commun.* p. 172 (1973).
S20. Schwarzmann, M., and Hamprecht, G., Ger. Patent 1,102,735 (1961).
S21. Scott, D. R., and Matsen, F. A., *J. Phys. Chem.* **72**, 16 (1968).
S22. Seleznev, K. A., and Shatalina, G. A., *Tr. Khim. Khim. Tekhnol.*, 402 (1962); *Chem. Abstr.* **59**, 8124c (1963).

S23. Semin, G. K., Nogina, O. V., Dubovitskii, V. A., Babushkina, T. A., Bryukhova, E. V., and Nesmeyanov, A. N., *Dokl. Akad. Nauk SSSR* **194**, 101 (1970).
S24. Sen, D. N., and Kantak, U. N., *J. Indian Chem. Soc.* **46**, 358 (1969).
S25. Sen, D. N., and Kantak, U. N., *Proc. Chem. Symp.*, *1st Sept. 1969* Vol. 1, p. 122 (1970); *Chem. Abstr.* **74**, 27640y (1971).
S26. Sen, D. N., and Kantak, U. N., *Indian. J. Chem.* **9**, 254 (1971).
S27. Seyferth, D., Hofmann, H. P., Burton, R., and Helling, J. F., *Inorg. Chem.* **1**, 227 (1962).
S28. Sharma, K. M., Anand, S. K., Multani, R. K., and Jain, B. D., *J. Organometal. Chem.* **23**, 173 (1970).
S29. Sharma, K. M., Anand, S. K., Multani, R. K., and Jain, B. D., *J. Organometal. Chem.* **25**, 447 (1970).
S30. Shikata, K., Nakao, S., and Azuma, K., *Kogyo Kagaku Zasshi* **68**, 1251 (1965).
S31. Shikata, K., Nishino, K., and Azuma, K., *Kogyo Kagaku Zasshi* **68**, 352, (1965).
S32. Shikata, K., Nishino, K., and Azuma, K., *Kogyo Kagaku Zasshi* **68**, 490 (1965).
S33. Shikata, K., Nishino, K., Azuma, K., and Takegami, Y., *Kogyo Kagaku Zasshi* **68**, 358 (1965).
S34. Shikata, K., Yokogawa, K., Nakao, S., and Azuma, K., *Kogyo Kagaku Zasshi* **68**, 1248 (1965).
S35. Shilov, A. E., Shilova, A. K., and Bobkov, B. N., *Vysokomol. Soedin.* **4**, 1688 (1962).
S36. Shilov, A. E., Shilova, A. K., Kvashina, E. F., and Vorontsova, T. A., *Chem. Commun.* p. 1590 (1971).
S37. Shur, V. B., Berkovich, E. G., and Vol'pin, M. E., *Izv. Akad. Nauk SSSR, Ser. Khim.* p. 2358 (1971).
S38. Shustorovich, E. M., and Dyatkina, M. E., *Russ. J. Inorg. Chem.* **4**, 251 (1959).
S39. Shuto, Y., Uchiyama, M., and Sugahara, H., *Kinet. Mech. Polyreactions, Int. Symp. Macromol. Chem., Prepr.* **2**, 401 (1969); *Chem. Abstr.* **75**, 64425c (1971).
S40. Siegert, F. W., and de Liefde Meijer, H. J., *J. Organometal. Chem.* **20**, 141 (1969).
S41. Siegert, F. W., and de Liefde Meijer, H. J., *J. Organometal. Chem.* **23**, 177 (1970).
S42. Siegert, F. W., and de Liefde Meijer, H. J., *Rec. Trav. Chim. Pays-Bas* **89**, 764 (1970).
S43. Simionescu, C., Asandei, N., and Benedek, I., *Eur. Polym. J.* **7**, 1561 (1971).
S44. Simionescu, C., Asandei, N., Benedek, I., and Ungurenasu, C., *Eur. Polym. J.* **5**, 449 (1969).
S45. Sinn, H., and Kolk, E., *J. Organometal. Chem.* **6**, 373 (1966).
S46. Sinn, H., and Oppermann, G., *Angew. Chem., Int. Ed. Engl.* **5**, 962 (1966).
S47. Skapski, A. C., and Troughton, P. G. H., *Acta Crystallogr., Sect. B* **26**, 716 (1970).
S48. Skapski, A. C., Troughton, P. G. H., and Sutherland, H. H., *Chem. Commun.* p. 1418 (1968).
S49. Skelcey, J. S., *Diss. Abstr.* **22**, 4177 (1962).
S50. Sloan, C. L., and Barber, W. A., *J. Amer. Chem. Soc.* **81**, 1364 (1959).
S51. Sloan, C. L., and Barber, W. A., Ger. Patent 1,072,246 (1959).
S52. Sloan, M. F., Matlack, A. S., and Breslow, D. S., *J. Amer. Chem. Soc.* **85**, 4014 (1963).
S53. Smirnova, S. A., and Gubin, S. P., *Izv. Akad. Nauk SSSR, Ser. Khim.* p. 1890 (1969).
S54. Smith, A. E., *Acta Crystallogr.* **18**, 331 (1965).

S55. Société belge du titane S.A., Belg. Patent 550,573 (1956).
S56. Société belge du titane S.A., Belg. Patent 559,046 (1957).
S57. Société belge du titane S.A., Belg. Patent 559,047 (1957).
S58. Société belge du titane S.A., Belg. Patent 559,048 (1957).
S59. Société belge du titane S.A., Belg. Patent 560,503 (1957).
S60. Sonogashira, K., and Hagihara, N., *Bull. Chem. Soc. Jap.* **39**, 1178 (1966).
S61. Stepovik, L. P., Shilova, A. K., and Shilov, A. E., *Dokl. Akad. Nauk SSSR* **148**, 122 (1963).
S62. Stern, R., Hillion, G., and Sajus, L., *Tetrahedron Lett.* p. 1561 (1969).
S63. Stezowski, J. J., and Eick, H. A., *J. Amer. Chem. Soc.* **91**, 2890 (1969).
S64. Strohmeier, W., Landsfeld, H., and Gernert, F., *Z. Elektrochem.* **66**, 823 (1962).
S65. Studiengesellschaft Kohle m.b.H., Neth. Patent 6,409,180 (1965).
S66. Studiengesellschaft Kohle m.b.H., Brit. Patent 1,128,128 (1968).
S67. Subbotin, A. I., and Kozina, I. Z., *Tr. Khim. Khim. Tekhnol.* p. 153 (1968).
S68. Sudo, Y., Takeshita, Y., Konuma, Y., Nishimura, T., Higashimori, N., Ogawa, M., and Tachibana, H., Jap. Patent 70/24,787 (1970).
S69. Sudo, Y., Uchiyama, M., Sugawara, H., Takeshita, Y., Tomori, S., Ogawa, M., Nishimura, T., Jap. Patent 70/21,716 (1970).
S70. Sugahara, H., and Shuto, Y., *J. Organometal. Chem.* **24**, 709 (1970).
S71. Sullivan, M. F., and Little, W. F., *J. Organometal. Chem.* **8**, 277 (1967).
S72. Summers, L., and Uloth, R. H., *J. Amer. Chem. Soc.* **76**, 2278 (1954).
S73. Summers, L., Uloth, R. H., and Holmes, A., *J. Amer. Chem. Soc.* **77**, 3604 (1955).
S74. Surtees, J. R., *Chem. Commun.* p. 567 (1965).
T1. Tabacchi, R., Boustany, K. S., and Jacot-Guillarmod, A., *Helv. Chim. Acta* **53**, 1971 (1970).
T2. Tabacchi, R., and Jacot-Guillarmod, A., *Chimia* **24**, 271 (1970).
T3. Tabacchi, R., and Jacot-Guillarmod, A., *Helv. Chim. Acta* **53**, 1977 (1970).
T4. Tabacchi, R., and Jacot-Guillarmod, A., *Chima* **25**, 326 (1971).
T5. Takeda, M., Iimura, K., Nozawa, Y., Hisatome, M., and Koide, N., *J. Polym. Sci., Part C* **23**, 741 (1968).
T6. Takegami, Y., Suzuki, T., and Aamada, K., *Kogyo Kagaku Zasshi* **72**, 1720 (1969).
T7. Takegami, Y., Suzuki, T., and Okazaki, T., *Bull. Chem. Soc. Jap.* **42**, 1060 (1969).
T8. Takegami, Y., Ueno, T., Suzuki, T., and Fuchizaki, Y., *Bull. Chem. Soc. Jap.* **41**, 2637 (1968).
T9. Takiguchi, T., and Suzuki, H., *Bull. Chem. Soc. Jap.* **41**, 2810 (1968).
T10. Tămas, V., and Bodea, C., *Rev. Roum. Chim.* **14**, 1591 (1969).
T11. Tamborski, C., Soloski, E. J., and Dec, S. M., *J. Organometal. Chem.* **4**, 446 (1965).
T12. Tatlow, J. C., *Endeavour* **22**, 89 (1963).
T13. Tazima, Y., and Yuguchi, S., *Bull. Chem. Soc. Jap.* **39**, 404 (1966).
T14. Tebbe, F. N., and Guggenberger, L. J., private communication to P. C. Wailes.
T15. Tebbe, F. N., and Guggenberger, L. J., *J. Chem. Soc. Chem. Commun.* p. 227 (1973).
T16. Tel'noi, V. I., Rabinovich, I. B., Latyaeva, V. N., and Lineva, A. N., *Dokl. Akad. Nauk SSSR* **197**, 1348 (1971).
T17. Tel'noi, V. I., Rabinovich, I. B., Tikhonov, V. D., Latyaeva, V. N., Vyshinskaya L. I., and Razuvaev, G. A., *Dokl. Akad. Nauk SSSR* **174**, 1374 (1967).
T18. Teuben, J. H., *J. Organometal. Chem.* **57**, 159 (1973).

T19. Teuben, J. H., and de Liefde Meijer, H. J., *J. Organometal. Chem.* **17**, 87 (1969).
T20. Teuben, J. H., and de Liefde Meijer, H. J., *Rec. Trav. Chim. Pays-Bas* **90**, 360 (1971).
T21. Teuben, J. H., and de Liefde Meijer, H. J., *Rec. Trav. Chim. Pays-Bas* (1973) (in press).
T22. Thayer, J. S., *Organometal. Chem. Rev.* **1**, 157 (1966).
T23. Thayer, J. S., private communication to R. S. P. Coutts (1967).
T24. Thiele, K.-H., *Pure Appl. Chem.* **30**, 575 (1972).
T25. Thiele, K.-H., and Jacob, K., *Z. Anorg. Allg. Chem.* **356**, 195 (1968).
T25a. Thiele, K.-H., Köhler, E., and Adler, B., *J. Organometal. Chem.* **50**, 153 (1973).
T26. Thiele, K.-H., and Krüger, J., *Z. Anorg. Allg. Chem.* **383**, 272 (1971).
T27. Thiele, K.-H., Milowski, K., Zdunneck, P., Müller, J., and Rau, H., *Z. Chem.* **12**, 187 (1972).
T28. Thiele, K.-H., and Müller, J., *Z. Chem.* **4**, 273 (1964).
T29. Thiele, K.-H., and Müller, J., *Z. Anorg. Allg. Chem.* **362**, 113 (1968).
T30. Thiele, K.-H., Müller, J., Jacob, K., Schroeder, S., Anton, E., and Lehmann, H. D., Ger. (East) Patent 76,673 (1970).
T31. Thiele, K.-H., and Schäfer, W., *Z. Anorg. Allg. Chem.* **379**, 63 (1970).
T32. Thiele, K.-H., and Zdunneck, P., *Z. Chem.* **10**, 152 (1970).
T33. Thiele, K.-H., Zdunneck, P., and Baumgart, D., *Z. Anorg. Allg. Chem.* **378**, 62 (1970).
T34. Thomas, J. L., and Hayes, R. G., *Inorg. Chem.* **11**, 348 (1972).
T35. Tille, D., *Z. Naturforsch. B* **25**, 1358 (1970).
T36. Titaan, N. V., Neth. Patent 95,973 (1960).
T37. Titaan, N. V., Neth. Patent 100,434 (1962).
T38. Tkachev, V. V., and Atovmyan, L. O., *Zh. Strukt. Khim.* **13**, 287 (1972).
T39. Tokuyama Soda Co. Ltd., Brit. Patent 1,009,117 (1965).
T40. Treichel, P. M., Chaudhari, M. A., and Stone, F. G. A., *J. Organometal. Chem.* **1**, 98 (1963).
T41. Tsutsui, M., and Ariyoshi, J., *J. Polym. Sci., Part A* **3**, 1729 (1965).
T42. Tsutsui, M., Velapoldi, R. A., Suzuki, K., Vohwinkel, F., Ichikawa, M., and Koyano, T., *J. Amer. Chem. Soc.* **91**, 6262 (1969).
U1. Uchida, Y., Furuhata, K., and Yoshida, S., *Bull. Chem. Soc. Jap.* **44**, 1966 (1971).
U2. Ungurenasu, C., and Cecal, A., *J. Inorg. Nucl. Chem.* **31**, 1735 (1969).
U3. Ungurenasu, C., and Haiduc, I., *Rev. Roum. Chim.* **13**, 957 (1968).
U4. Ungurenasu, C., and Streba, E., *J. Inorg. Nucl. Chem.* **34**, 3753 (1972).
V1. Valcher, S., and Mastragostino, M., *J. Electroanal. Chem. Interfacial Electrochem.* **14**, 219 (1967).
V2. van der Hende, J. H., and Baird, W. C., *J. Amer. Chem. Soc.* **85**, 1009 (1963).
V3. van Oven, H. O., *J. Organometal. Chem.* **55**, 309 (1973).
V4. van Oven, H. O., and de Liefde Meijer, H. J., *J. Organometal. Chem.* **19**, 373 (1969).
V5. van Oven, H. O., and de Liefde Meijer, H. J., *J. Organometal. Chem.* **23**, 159 (1970).
V6. van Oven, H. O., and de Liefde Meijer, H. J., *J. Organometal. Chem.* **31**, 71 (1971).
V7. van Tamelen, E. E., *Accounts Chem. Res.* **3**. 361 (1970).
V8. van Temelen, E. E., Cretney, W., Klaentschi, N., and Miller, J. S., *J. Chem. Soc. Chem. Commun.* p. 481 (1972).
V9. van Tamelen, E. E., Fechter, R. B., and Schneller, S. W., *J. Amer. Chem. Soc.* **91**, 7196 (1969).

V10. van Tamelen, E. E., Fechter, R. B., Schneller, S. W., Boche, G., Greeley, R. H., and Åkermark, B., *J. Amer. Chem. Soc.* **91**, 1551 (1969).
V11. van Tamelen, E. E., and Rudler, H., *J. Amer. Chem. Soc.* **92**, 5253 (1970).
V12. van Tamelen, E. E., and Seeley, D. A., *J. Amer. Chem. Soc.* **91**, 5194 (1969).
V13. Verkouw, H. T., and van Oven, H. O., *J. Organometal. Chem.* **59**, 259 (1973).
V14. Vohwinkel, F., *Trans. N. Y. Acad. Sci.* [2] **26**, 446 (1964).
V15. Vol'pin, M. E., *Pure Appl. Chem.* **30**, 607 (1972).
V16. Vol'pin, M. E., Belyi, A. A., and Shur, V. B., *Izv. Akad. Nauk SSSR, Ser. Khim.* p. 2225 (1965).
V17. Vol'pin, M. E., Belyi, A. A., Shur, V. B., Lyakhovetsky, Yu. I., Kudryavtsev, R. V., and Bubnov, N. N., *Dokl. Akad. Nauk SSSR,* **194**, 577 (1970).
V18. Vol'pin, M. E., Belyi, A. A., Shur, V. B., Lyakhovetsky, Yu. I., Kudryavtsev, R. V., and Bubnov, N. N., *J. Organometal. Chem.* **27**, C5 (1971).
V19. Vol'pin, M. E., Dubovitskii, V. A., Nogina, O. V., and Kursanov, D. N., *Dokl. Akad. Nauk SSSR* **151**, 1100 (1963).
V20. Vol'pin, M. E., Ilatovskaya, M. A., Kosyakova, L. V., and Shur, V. B., *Dokl. Akad. Nauk SSSR* **180**, 103 (1968).
V21. Vol'pin, M. E., Ilatovskaya, M. A., Kosyakova, L. V., and Shur, V. B., *Chem. Commun.* p. 1074 (1968).
V22. Vol'pin, M. E., Ilatovskaya, M. A., Larikov, E. I., Khidekel, M. L., Shvetsov, Yu. A., and Shur, V. B., *Dokl. Akad. Nauk SSSR* **164**, 331 (1965).
V23. Vol'pin, M. E., and Kursanov, D. N., *Angew. Chem.* **75**, 1034 (1963).
V24. Vol'pin, M. E., and Shur, V. B., *Dokl. Akad. Nauk SSSR* **156**, 1102 (1964).
V25. Vol'pin, M. E., and Shur, V. B., *Vestn. Akad. Nauk SSSR,* p. 51 (1965).
V26. Vol'pin, M. E., and Shur, V. B., *Izv. Akad. Nauk SSSR, Ser. Khim.* p. 1873 (1966).
V27. Vol'pin, M. E., and Shur, V. B., *in* "Organometallic Reactions," Vol. 1, p. 55. Wiley (Interscience), New York, 1970.
V28. Vol'pin, M. E., Shur, V. B., Kudryavtsev, R. V., and Prodayko, L. A., *Chem. Commun.* p. 1038 (1968).
V29. Vol'pin, M. E., Shur, V. B., Latyaeva, V. N., Vyshinskaya, L. I., and Shul'gaitser, L. A., *Izv. Akad. Nauk SSSR, Ser. Khim.* p. 385 (1966).
V30. von Ammon, R., Kanellakopulos, B., Fischer, R. D., and Laubereau, P., *Inorg. Nucl. Chem. Lett.* **5**, 219 (1969).
V31. von Ammon, R., Kanellakopulos, B., Schmid, G., and Fischer, R. D., *J. Organometal. Chem.* **25**, C1 (1970).
V32. Vyshinskii, N. N., Ermolaeva, T. I., Latyaeva, V. N., Lineva, A. N., and Lukhton. N. E., *Dokl. Akad. Nauk SSSR* **198**, 1081 (1971).
W1. Wailes, P. C., and Kautzner, B., unpublished results (1969).
W2. Wailes, P. C., and Weigold, H., *J. Organometal. Chem.* **24**, 405 (1970).
W3. Wailes, P. C., and Weigold, H., *J. Organometal. Chem.* **24**, 413 (1970).
W4. Wailes, P. C., and Weigold, H., *J. Organometal. Chem.* **24**, 713 (1970).
W5. Wailes, P. C., and Weigold, H., *J. Organometal. Chem.* **28**, 91 (1971).
W6. Wailes, P. C., Weigold, H., and Bell, A. P., *J. Organometal. Chem.* **27**, 373 (1971).
W7. Wailes, P. C., Weigold, H., and Bell, A. P., *J. Organometal Chem.* **33**, 181 (1971).
W8. Wailes, P. C., Weigold, H., and Bell, A. P., *J. Organometal. Chem.* **34**, 155 (1972).
W9. Wailes, P. C., Weigold, H., and Bell, A. P., *J. Organometal. Chem.* **43**, C29 (1972).

W10. Wailes, P. C., Weigold, H., and Bell, A. P., *J. Organometal. Chem.* **43**, C32 (1972).
W11. Wailes, P. C., Weigold, H., and Bell, A. P., unpublished results.
W12. Wasserman, E., Snyder, L. C., and Yager, W. A., *J. Chem. Phys.* **41**, 1763 (1964).
W13. Waters, J. A., and Mortimer, G. A., *J. Organometal. Chem.* **22**, 417 (1970).
W14. Waters, J. A., Vickroy, V. V., and Mortimer, G. A., *J. Organometal. Chem.* **33**, 41 (1971).
W15. Watt, G. W., and Baye, L. J., *J. Inorg. Nucl. Chem.* **26**, 2099 (1964).
W16. Watt, G. W., Baye, L. J., and Drummond, F. O., *J. Amer. Chem. Soc.* **88**, 1138 (1966).
W17. Watt, G. W., and Drummond, F. O., *J. Amer. Chem. Soc.* **88**, 5926 (1966).
W18. Watt, G. W., and Drummond, F. O., *J. Amer. Chem. Soc.* **92**, 826 (1970).
W19. Weber, H., Ring, W., Hochmuth, U., and Franke, W., *Justus Liebigs Ann. Chem.* **681**, 10 (1965).
W20. White, D. A., *J. Inorg. Nucl. Chem.* **33**, 691 (1971).
W21. Wilke, G., Bogdanovic, B., Hardt, P., Heimbach, P., Keim, W., Kröner, M., Oberkirch, W., Tanaka, K., Steinrücke, E., Walter, D., and Zimmermann, H., *Angew. Chem., Int. Ed. Engl.* **5**, 151 (1966).
W22. Wilke, G., Mueller, E. W., Kroner, M., Heimbach, P., and Breil, H., Fr. Patent 1,320,729 (1963).
W23. Wilkinson, G., *Pure Appl. Chem.* **30**, 627 (1972).
W24. Wilkinson, G., and Birmingham, J. M., *J. Amer. Chem. Soc.* **76**, 4281 (1954).
W25. Wilkinson, G., Pauson, P. L., Birmingham, J. M., and Cotton, F. A., *J. Amer. Chem. Soc.* **75**, 1011 (1953).
W26. Wiman, R. E., and Rubin, I. D., *Makromol. Chem.* **94**, 160 (1966).
W27. Winkhaus, G., and Uhrig, H., *Z. Anal. Chem.* **200**, 14 (1964).
W28. Wozniak, B., Ruddick, J. D., and Wilkinson, G., *J. Chem. Soc., A* p. 3116 (1971).
Y1. Yagupsky, G., Mowat, W., Shortland, A., and Wilkinson, G., *Chem. Commun.* p. 1369 (1970).
Y2. Yamamoto, A., Ookawa, M., and Ikeda, S., *Chem. Commun.* p. 841 (1969).
Y3. Yasufuku, K., and Yamazaku, H., *Bull. Chem. Soc. Jap.* **45**, 2664 (1972).
Y4. Yokokawa, K., and Azuma, K., *Bull. Chem. Soc. Jap.* **38**, 859 (1965).
Y5. Yoshino, A., Shuto, Y., and Iitaka, Y., *Acta Crystallogr., Sect. B* **26**, 744 (1970).
Z1. Zakharkin, L. I., Kalinin, V. N., and Podvisotskaya, L. S., *Izv. Akad. Nauk SSSR, Ser. Khim* p. 679 (1968).
Z2. Zdunneck, P., and Thiele, K.-H., *J. Organometal. Chem.* **22**, 659 (1970).
Z3. Zefirova, A. K., and Shilov, A. E., *Dokl. Akad. Nauk SSSR*, **136**, 599 (1961).
Z4. Zefirova, A. K., Tikhomirova, N. N., and Shilov, A. E., *Dokl. Akad. Nauk SSSR* **132**, 1082 (1960).
Z5. Zeinstra, J. D., and de Boer, J. L., *J. Organometal. Chem.* **54**, 207 (1973).
Z6. Ziegler, K., Holzkamp, E., Breil, H., and Martin, H., *Angew. Chem.* **67**, 541 (1955).
Z7. Zucchini, U., Albizzati, E., and Giannini, U., *J. Organometal. Chem.* **26**, 357 (1971).
Z8. Zucchini, U., Giannini, U., Albizzati, E., and D'Angelo, R., *Chem. Commun.* p. 1174 (1969).

Appendix

1. Crystal and molecular structures of $Cp_2TiS_2(CH)_2$ (1), $[\pi\text{-}1,3\text{-}(C_6H_5)_2C_5H_3]_2TiCl_2$ (2), $Cp[\pi\text{-}(CH_3)_5C_5]TiCl_2$ (3), $Cp_2TiCl_2ZnCl_2TiCp_2$ (4), and $(\text{fluorenyl})_2ZrCl_2$ (5) have been determined. In each case the metal atom is in a distorted tetrahedral environment. In bis(fluorenyl)-zirconium dichloride one of the fluorenyl ligands is pentahapto-bonded while the other shows an unusual trihapto configuration (5).

2. Bis(benzene)titanium: An authentic sample has now been made by electron beam evaporation of titanium metal followed by cocondensation with benzene at 77°K (6).

3. The synthesis of tetra(neopentyl)hafnium completes the series; the order of thermal stabilities is Hf > Zr > Ti, neopentane being the sole detectable thermolysis product (7). Two alkyltitanium(III) compounds of enhanced stability, $Cp_2TiCH_2Si(CH_3)_3$ and $Cp_2TiCH_2C(CH_3)_3$, have been reported (8).

4. The metal(IV) derivatives, $Cp_2M[N(CN)_2]_2$, $Cp_2M[C(CN)_3]_2$, and $Cp_2M[ONC(CN)_2]_2$, where M = Ti, Zr, Hf, have been prepared from Cp_2MCl_2 and the silver "pseudohalide" anion (9).

5. Details of the synthesis and PMR spectra of methyl, phenyl, and penta-fluorophenyl derivatives of bis(cyclopentadienyl) metals have been published (10). Some bis(indenyl)metal compounds were included.

6. Derivatives of ferrocene substituted in one or both rings with tris(dialkyl-amido)titano groups have been isolated and their PMR spectra assigned (11).

7. Mono-, di-, and trihaloacetic acids displace two cyclopentadienyl rings from $(C_5H_5)_4M$, where M = Zr, Hf, giving $Cp_2M(OCOR)_2$, in which the carboxylato groups are monodentate (12).

8. The metallocycles, bis(indenyl)titanafluorene (10) and bis(cyclopentadienyl)octafluorozirconafluorene (13) were isolated from the reaction between the 2,2'-dilithiobiphenyl and Cp_2MCl_2, while from $LiN[Si(CH_3)_3]_2$ and the corresponding titanium halide,

$$Cp_2Ti \underset{\underset{\underset{Si(CH_3)_3}{|}}{N}}{\overset{CH_2}{\diagup}} {\diagdown} Si(CH_3)_2$$

was obtained (14).

9. Loss of CO from $Cp_2Ti(CO)_2$ occurs readily during equilibration with Cp_2TiCl_2 or $TiCl_4$. In the presence of cyclohexylisocyanide, the complexes $Cp_2TiCl \cdot L$ and $CpTiCl_2 \cdot 2L$, where $L = C_6H_{11}NC$, result (15). With azobenzene the titanium carbonyl undergoes an oxidative addition reaction (16) giving

$$Cp_2Ti \underset{\underset{C_6H_5}{\diagdown}}{\overset{\overset{C_6H_5}{\diagup}}{N}}{\diagup}{\overset{|}{\underset{N}{}}}$$

10. Several papers describe physical measurements carried out on cyclopentadienyl compounds. The mass spectra of Cp_2TiX_2, where X = F, Cl, Br, I, and of $CpTiX_3$, π-$CH_3C_5H_4TiX_3$, and π-$(CH_3)_5C_5TiX_3$, where X = Cl, Br, or OC_2H_5, show that the molecular ions decompose by two routes leading to elimination of X and of a cyclopentadienyl ligand. In the case of π-$(CH_3)_5C_5TiX_3$, an intense peak, $(M–HX)^+$, indicates loss of a proton from one CH_3 leading to a fulvene-type structure (17).

   PMR spectra of a number of compounds of type Cp_2TiRCl and Cp_2TiR_2, where R = C_6H_5 or $CH_2C_6H_5$, also of $CpTi(CH_2C_6H_5)_3$, have been measured in different solvents. The results of Beachell and Butter are critically discussed (18).

11. An authoritative article by Ballard describes the use in polymerization of vinyl monomers and olefins of π- and σ-bonded compounds of titanium and zirconium (19). Emphasis is placed on allyl and benzyl derivatives and the active catalysts obtained by treatment of the benzyl derivatives with silanols.

REFERENCES

1. Kutoglu, A., *Acta Cryst.* **B29**, 2891 (1973).
2. van Soest, T. C., Rappard, J. C. C. W., and Royers, E. C., *Cryst. Struct. Commun.* **2**, 451 (1973).

3. Khotsyanova, T. L., and Kuznetsov, S. I., *J. Organometal. Chem.* **57**, 155 (1973).
4. Vonk, C. G., *J. Cryst. Mol. Struct.* **3**, 201 (1973).
5. Wunderlich, J. A., Kowala, C., Wailes, P. C., and Weigold, H., unpublished results.
6. Benfield, F. W. S., Green, M. L. H., Ogden, J. S., and Young, D., *J. Chem. Soc. Chem. Commun.*, p. 866 (1973).
7. Davidson, P. J., Lappert, M. F., and Pearce, R., *J. Organometal. Chem.* **57**, 269 (1973).
8. Chivers, T., and Ibrahim, E. D., *Abst. 6th Int. Conf. Coord. Chem.*, University of Massachusetts, August 1973, abstract 232.
9. Issleib, K., Köhler, H., and Wille, G., *Z. Chem*, **13**, 347 (1973).
10. Samuel, E. and Rausch, M. D., *J. Amer. Chem. Soc.*, **95**, 6263 (1973).
11. Bürger, H., and Kluess, C., *J. Organometal. Chem.* **56**, 269 (1973).
12. Brainina, E. M., Bryukhova, E. V., Lokshin, B. V., and Alimov, N. S., *Izv. Akad. Nauk SSSR Ser. Khim.*, p. 891 (1973).
13. Gardner, S. A., Gordon, H. B., and Rausch, M. D., *J. Organometal. Chem.* **60**, 179 (1973).
14. Bennett, C. R., and Bradley, D. C., *J. Chem. Soc. Chem. Commun.*, p. 29 (1974).
15. Floriani, C., and Fachinetti, G., *J. Chem. Soc., Dalton Trans.*, p. 1954 (1973).
16. Fachinetti, G., Fochi, G., and Floriani, C., *J. Organometal. Chem.* **57**, C51, (1973).
17. Nesmeyanov, A. N., Nekrasov, Yu. S., Sizoi, V. F., Nogina, O. V., Dubovitsky, V. A., and Sirotkina, Ye. I., *J. Organometal. Chem.* **61**, 225 (1973).
18. Glivicky, A., and McCowan, J. D., *Can. J. Chem.* **51**, 2609 (1973).
19. Ballard, D. G. H., *Advan. Catal.* **23**, 263 (1973).

Subject Index

For the convenience of the reader the more informative page numbers are shown in boldface type while tables are shown in italics.

Dienes
  formation of allyl derivatives from, 171,
    183, 215, 218
  polymerization of, 171, *190–193*
β-Diketonates, 39, 43, **66**, **67**, 70, 129, 132,
  134–142, 213, 217
Dinitrogen, *see* Nitrogen, di
Dioxane, *see* Complexes
Dipivalomethane, *see* β-Diketonates
Dipole moments, 54, 60, 120, 125, 137,
  166
Dipyridyl, *see* Bipyridyl
Dithiocarbamates, 47, 49, **213**, *216*, 222

E

Electrochemical syntheses, *see also* Polarography, 107, 178, 197, 202, 240,
  253, 260
Electronic spectroscopy, *see* Ultraviolet-
  visible
Electronic structure, 229
Electron spin resonance spectroscopy,
  174, 175, 176, 177, 202, 209, 212,
  215, 221, 222, 223, 224, 237, 244,
  250, 251
Electron diffraction, *see* Structure
Equilibrium constants, 113, 120
Ethyl derivatives, *see* Alkyl
Ethylene
  conversion to crotyl, 220
  elimination of, 173, 215, 250, 251
  polymerization of, 187, 189, *190–193*,
    194
Ethylenediamine, *see* Complexes

F

Far infrared, *see* Infrared
Fluorenyltitanium compounds, 59
Fluorenylzirconium compounds, 116, 119,
  123, 127, 287
Fluoro compounds
  alkoxides and phenoxides, 64, 129
  β-diketonates, 39, 122, 137, 140–142
  perfluoroaryl, 13, 28, **89**, *90*, 97, 156,
    221, 257, 258, 287
  perfluorocarboxylates, 43, 68, 142–144,
    205

Fluxional compounds, 106–108, 136, 137,
  141, 156, 166, 168–170, 171
Fulvenes, 56, 288

G

Gallium compounds, 178, 179, *180*, *181*
Germanium compounds, 57, 92, 93, 101,
  104, 105, 163, 164, 224
Grignard reagents
  reaction with cyclopentadienyltitanium
    compounds, 31, 52, 89, 92, 179, **215**,
    244, 245, 250
  use in nitrogen fixation, 250–252, 255,
    261
  use in polymerization, 193

H

Hafnocene, 243
Halides, organometal, *see* Alkyl, Aryl,
  Benzyl, Bis(cyclopentadienyl), Cy-
  clooctatetraene, Indenyl, Fluorenyl,
  Mono(cyclopentadienyl), Tris(cy-
  clopentadienyl)
Halogenation, 31, 34, 87, 91
Hapto, *see* Cyclopentadienyl
Heats of combustion, 123
Heats of formation, 123, 258
Hydrazine, formation from nitrogen, 250,
  252, 254, 258
Hydrazoic acid, on Cp₂TiCl, 211
Hydrides of titanium, **179**, **182–184**, *186*,
  218, 220, 228, 229, 231–235, 237–
  240, 251, 252
Hydrides of zirconium
  benzyl, 111, 112
  bis(cyclopentadienyl) derivatives, 128,
    146–153, 226, 228, 242
  cyclooctatetraene zirconium dihydride,
    148, 170, **171**
  reactions with alkenes and alkynes, 149–
    151
  reactions with ketones and acids, 127,
    130, 143, 149
  reaction with CO, 150, 242
  reaction with CH₂Cl₂, 150
Hydrogen
  reaction with titanium alkyls, 91, 182,
    183, 234, 235, 238, 255
  reaction with zirconium alkyls, 155, 227